Global Climate Change and Agricultural Production

Global Climate Change and Agricultural Production

Direct and Indirect Effects of Changing Hydrological, Pedological and Plant Physiological Processes

Edited by
FAKHRI BAZZAZ
Harvard University, Cambridge, MA, USA

WIM SOMBROEK
FAO, Rome, Italy

Published by the
FOOD AND AGRICULTURE ORGANIZATION
OF THE UNITED NATIONS
and
JOHN WILEY & SONS
Chichester • New York • Brisbane • Toronto • Singapore

Copyright © 1996 by FAO, Rome, Italy

Published 1996 by John Wiley & Sons Ltd,
Baffins Lane, Chichester,
West Sussex PO19 1UD, England

National 01243 779777
International (+44) 1243 779777
e-mail (for orders and customer service enquiries): cs-books@wiley.co.uk
Visit our Home Page on http://www.wiley.co.uk
or http://www.wiley.com

All rights reserved. No part of this book may be reproduced, stored in a retrieval system, or transmitted, in any form or by any means, electronic, mechanical, photocopying, recording, scanning or otherwise, except under the terms of the Copyright, Designs and Patents Act 1988 or under the terms of a licence issued by the Copyright Licensing Agency, 90 Tottenham Court Road, London W1P 9HE, UK, without the permission in writing of the Publisher and the copyright owner.

Other Wiley Editorial Offices

John Wiley & Sons, Inc., 605 Third Avenue,
New York, NY 10158-0012, USA

Jacaranda Wiley Ltd, 33 Park Road, Milton,
Queensland 4064, Australia

John Wiley & Sons (Canada) Ltd, 22 Worcester Road,
Rexdale, Ontario M9W 1L1, Canada

John Wiley & Sons (Asia) Pte Ltd, 2 Clementi Loop #02-01,
Jin Xing Distripark, Singapore 129809

Library of Congress Cataloging-in-Publication Data

Global climate change and agricultural production : direct and
 indirect effects of changing hydrological soil and plant
 physiological processes / edited by Fakhri Bazzaz, Wim Sombroek.
   p. cm.
 Includes bibliographical references and index.
  ISBN 0-471-95763-1
  1. Climatic changes. 2. Agriculture. 3. Crops and climate.
 4. Meteorology, Agricultural. I. Bazzaz, F. A. (Fakhri A.)
 II. Sombroek, Wim G. III. Food and Agriculture Organization of the
 United Nations.
 S600.7.C54G58 1996
 338.1'4—dc20   96–19503
                                                          CIP

British Library Cataloguing in Publication Data

A catalogue record for this book is available from the British Library

ISBN 0-471-95763-1 (cloth)
ISBN 0-471-96927-3 (paper)

Produced from camera-ready-copy supplied by the editors
Printed and bound in Great Britain by Bookcraft (Bath) Ltd
This book is printed on acid-free paper responsibly manufactured from sustainable forestation, for which at least two trees are planted for each one used for paper production.

CONTENTS

List of Contributors		vii
Preface and Acknowledgements		ix
1	The Climate Change – Agriculture Conundrum *W.G. Sombroek and R. Gommes*	1
2	The Effects of Changes in the World Hydrological Cycle on Availability of Water Resources *T.E. Evans*	15
3	The Effects of Global Change on Soil Conditions in Relation to Plant Growth and Food Production *R. Brinkman and W.G. Sombroek*	49
4	The CO_2 Fertilization Effect: Higher Carbohydrate Production and Retention as Biomass and Seed Yield *L.H. Allen Jr., J.T. Baker and K.J. Boote*	65
5	The Effects of Elevated CO_2 and Temperature Change on Transpiration and Crop Water Use *S.C. Van de Geijn and J. Goudriaan*	101
6	Effects of Higher Day and Night Temperatures on Growth and Yields of Some Crop Plants *Y.P. Abrol and K.T. Ingram*	123
7	Adverse Effects of Elevated Levels of Ultraviolet (UV)-B Radiation and Ozone (O_3) on Crop Growth and Productivity *S.V. Krupa and H. Jäger*	141
8	Combined Effects of Changing CO_2, Temperature, UV-B Radiation and O_3 on Crop Growth *M.H. Unsworth and W.E. Hogsett*	171

9	The Potential Effects of Climate Change on World Food Production and Security *G. Fischer, K. Frohberg, M.L. Parry and C. Rosenzweig*	199
10	Climate Change, Global Agriculture and Regional Vulnerability *J. Reilly*	237
11	Integrating Land-use Change and Evaluating Feedbacks in Global Change Models: the IMAGE 2 Approach *R. Leemans, G.J. Van den Born and L. Bouwman*	267
12	Global Change Impacts on Agriculture, Forestry and Soils: The Programme of the Global Change and Terrestrial Ecosystems Core Project of IGBP *B. Tinker, J. Goudriaan, P. Teng, M. Swift, S. Linder, J. Ingram and S. Van de Geijn*	295
13	Global Climatic Change and Agricultural Production: an Assessment of Current Knowledge and Critical Gaps *F.A. Bazzaz and W.G. Sombroek*	319

List of Contributors

Yash P. Abrol
Division of Plant Physiology, Indian Agricultural Research Institute, New Delhi 110 012, India

L. Hartwell Allen, Jr.
US Department of Agriculture, Agricultural Research Service, University of Florida, Gainesville, Florida 32611-0840, USA

Jeff T. Baker
Agronomy Department, University of Florida, Gainesville, Florida 32611-0840, USA

Fakhri A. Bazzaz
Department of Organismic and Evolutionary Biology, Harvard University, Cambridge, Massachusetts, USA

Ken J. Boote
Agronomy Department, University of Florida, Gainesville, Florida 32611-0840, USA

Lex Bouwman
Global Change Department, National Institute of Public Health and Environmental Protection, RIVM, PO Box 1, 3720 BA Bilthoven, The Netherlands

Robert Brinkman
Land and Water Development Division, FAO, Viale delle Terme di Caracalla, 00100 Rome, Italy

Terry E. Evans
Former Director, Mott MacDonald Co. now at 107 Ely Road, Littleport, Ely, CB 6 IHJ, UK

Günther Fischer
International Institute for Applied Systems Analysis, A-2361 Laxenburg, Austria

Klaus Frohberg
Institut für Agrarentwicklung in Mittel- und Osteuropa, Magdeburger Strasse 1, D-06112 Halle, Germany

René Gommes
Agrometeorology Group (Environmental Information Management Service), and Interdepartmental Working Group on Climate Change, FAO, Viale delle Terme di Caracalla, 00100 Rome, Italy

Jan Goudriaan
Wageningen Agricultural University, Dept. Theoretical Production Ecology, PO Box 9109, 6700 HB Wageningen, The Netherlands

William E. Hogsett
US Environmental Protection Agency, Environmental Research Laboratory, Corvallis, Oregon, 97331-6511, USA

Keith T. Ingram
University of Georgia, College of Agricultural and Environmental Sciences, Georgia Agricultural Experiment Station, Griffin, Georgia, USA

Hans-Jurg Jäger
Institut für Pflanzenökologie der Justus-Liebig-Universität, D-35392 Giessen, Germany

Sagar V. Krupa
Department of Plant Pathology, University of Minnesota, St. Paul, MN 55108, USA

Rik Leemans
Global Change Department, National Institute of Public Health and Environmental Protection, RIVM, PO Box 1, 3720 BA Bilthoven, The Netherlands

Martin L. Parry
Jackson Environment Institute, University College, 5 Gower Street, London, WC1

John Reilly
Natural Resources and Environment Division, Economic Research Service, USDA, 1301 New York Avenue, N.W., Room 524, Washington DC 20005-4788, USA

Cynthia Rosenzweig
Goddard Institute for Space Studies and Columbia University, 2880 Broadway, New York, NY 10025, USA

Wim G. Sombroek
Land and Water Development Division, and Interdepartmental Working Group on Climate Change, FAO, Viale delle Terme di Caracalla, 00100 Rome, Italy

Bernard Tinker
Department of Plant Sciences, University of Oxford, South Parks Road, Oxford OX1 3RB, UK

Michael H. Unsworth
Center for Analysis of Environmental Change, Oregon State University, Weniger Hall 283, Corvallis, Oregon 97331-6511, USA

Siebe C. van de Geijn
Research Institute for Agrobiology and Soil Fertility (AB-DLO), PO Box 14, 6700 AA Wageningen, The Netherlands

Gert Jan van den Born
Global Change Department, National Institute of Public Health and Environmental Protection, RIVM, PO Box 1, 3720 BA, Bilthoven, The Netherlands

Preface and Acknowledgements

In its second Assessment Report (December 1995) the Intergovernmental Panel on Climate Change (IPCC), established by the World Meteorological Organization and the United Nations Environment Programme, concluded that 'the balance of evidence suggests a discernible human influence on global climate'. A change in climatic conditions will affect agricultural production systems the world over. Until now, the projections of the Food and Agriculture Organization of the United Nations (FAO) on the state of agriculture in the forthcoming decades, such as *Agriculture: Towards 2010* (FAO and John Wiley, 1995) have not included the potential effects of any anthropogenic climate change at global and regional levels. Instead, they concentrated on the expected increase in human populations, their basic needs and aspirations for increased well-being, and the associated demands on natural resources, especially land and water resources, to provide them with the necessary food, fibre, animal feed, forest products and living space.

Although there can be no doubt that the increase in human populations and their capacity to influence land cover and land use will remain of predominant influence, projections beyond 2010 will have to reckon with changes in agroclimatic conditions as a result of the enhanced greenhouse effect (including temperature rise, CO_2 increase, and nitrogen deposition) envisaged to be larger than any natural climatic variation in the last few thousand years.

With the financial support of the United Nations Environment Programme (UNEP), an Expert Consultation at FAO headquarters in Rome was held to discuss direct and indirect effects on agriculture at the regional level. Agriculture was defined in the broad sense to include crop growing, animal husbandry, forestry and fisheries – all within the mandate of FAO. However, for a number of reasons the Consultation and the resulting texts concentrated on changing climatic conditions for annual and perennial crop growing, with emphasis on the effects of changing hydrological, pedological and plant physiological processes. In doing so, it attempted to strike a balance between the negative effects of the anticipated climate change on natural and managed ecosystems – so often emphasized in the debate on the enhanced greenhouse effect – and the potential positive effects on plant production of higher temperatures, an increased CO_2 fertilization and higher water-use efficiency – which might constitute a blessing in disguise for the future of humanity.

It should be mentioned that the projections on the pace and severity of climate change, as the basis for models on land productivity discussed in several chapters, are still those of the IPCC First Assessment of 1990 and its Supplement of 1992 (global temperature rise till 2100 between 2 and 5°C, and a sea-level rise between 30 and 100 cm); in some cases, such as the projections for Africa, the models have even taken the maximum rather than the middle values of these changes as a starting point.

The new 1995 estimates of IPCC are lower (between 1 and 3.5°C temperature rise and between 15 and 50 cm sea-level rise). Modelling on the basis of the latter values still has to start. A reassessment of the effects on agricultural production will therefore be necessary in a few years' time. This is also in view of the ever-increasing amount of free-air measurements and experimental data on the effects of increasing atmospheric CO_2 concentrations on plant growth and soil conditions.

The editors gratefully acknowledge the contributions to the conclusions and recommendations[1] by the participants of the 1993 Expert Consultation, including all members of FAO's standing Working Group on Climate Change. We appreciate the willingness of the authors of the various chapters to update their texts with information that became available only after the meeting.

Special thanks are due to Ms. Linda See for the preparations for the meeting, Ms. Margaret Farrell for the conscientious handling of the large amount of correspondence involved, Dr. David Norse for his initial editing work, and Ms. Chrissi Smith-Redfern for the careful screening of the final texts.

Fakhri Bazzaz and **Wim Sombroek**
Rome, January 1996

---

[1] See: Highlights from an Expert Consultation on Global Climate Change and Agricultural Production, FAO, Rome, 1994. Available from Dr. R. Gommes, Agrometeorology Group, FAO, Viale delle Terme di Caracalla, 00100 Rome.

1. The Climate Change – Agriculture Conundrum

WIM G. SOMBROEK
Land and Water Development Division, and Interdepartmental Working Group on Climate Change, FAO, Rome, Italy

RENÉ GOMMES
Agrometeorology Group (Environmental Information Management Service), and Interdepartmental Working Group on Climate Change, FAO, Rome, Italy

The risks associated with climate change lie in the interaction of several systems with many variables that must be collectively considered. Agriculture (including crop agriculture, animal husbandry, forestry and fisheries) can be defined as one of the systems, and climate the other. If these systems are treated independently, this would lead to an approach which is too fragmentary. The issue is more global. It is now held as likely that human activities can affect climate, one of the components of the environment. Climate in turn affects agriculture, the source of all food consumed by human beings and domestic animals. It must be further considered that not only climate may be changing, but that human societies and agriculture develop trends and constraints of their own which climate change impact studies must take into consideration.

An expert meeting held at FAO Headquarters in Rome from 7 to 10 December 1993 considered the direct effects of changing hydrological, pedological and plant physiological processes on agricultural production and concentrated on mechanisms. This introductory chapter of its Proceedings looks beyond the technical aspects of agriculture and stresses some of the major goals of FAO.

Firstly, sustainability of agricultural and rural development. How will the links between environmental resources and demography be affected in the coming 50 years? Will it be possible, at the same time, to increase food production without irremediably losing environmental resources like soils or biodiversity?

Secondly, improved food security and nutrition, two members of a spiral which also includes rural poverty and demographic pressure. What are the prospects of breaking the vicious poverty circle in many developing countries under changing climate conditions?

THE WORLD AGRICULTURAL CONTEXT

In general, global food production has been growing faster than human population (Table 1.1). There are, however, marked disparities between continents. For example, in Africa local production of cereals cannot keep pace with population increase, and production of root and tuber crops is growing faster than that of the nutritionally more valuable cereals.

In contrast, arable land growth lags behind population growth, which indicates some intensification of production. In Asia, the upper limit of available land has been reached in several countries, resulting in very high cropping intensities and a dominant role for irrigation. In Latin America, the increase in arable land is achieved only at a high ecological cost (especially deforestation) which may have direct relevance to climate change.

According to a recent FAO prospective study covering the years 1988/1990 to 2010 (*World Agriculture: Towards 2010* or *AT-2010*; FAO, 1995a), 'the rate of growth of agricultural land will be further reduced during the next two decades. The pressures on *fresh* water *resources*, however, will be considerable, as will be those on the environment arising from the intensification of land use'.

A significant reduction in world population growth rates is foreseen from 1.8% (1980/1990) to 1.4% (2000/2010), roughly equivalent to an increase of the population doubling time from 40 to 50 years. The projected continuation of the high population growth in Africa can be related to slow economic development during the coming two decades, since in general a reduction of poverty precedes reduction of population growth.

AT-2010 also lists the following significant trends for the near future: (i) world production of cereals will continue to grow, but not in per capita terms; (ii) export cereals will undergo a modest growth in demand; (iii) the livestock sector in developing countries will continue to grow; (iv) root crops, tubers and plantains will retain their importance; (v) oil crops will undergo rapid growth in developing countries; and, most importantly, (vi) many developing countries will become net agricultural importers.

As indicated above for population increase, large differences exist in the rate of agricultural development among the developing continents, while differences among the developed countries tend to level out.

It is also likely that hunger and under-nutrition, which currently affects 800 million people (about 20% of the 4 000 million inhabitants of developing countries), will be reduced, but large pockets of malnutrition will persist. Pressure on environmental resources will continue to build up.

Table 1.1. Some trends in worldwide population, food production, arable and irrigated land (data from FAO, 1990). Europe and Asia do not include the former USSR. Arable land is defined as all agricultural land, excluding perennial crops and permanent rangeland. Irrigated land is in % of the land under annual and permanent crops, average value from 1981 to 1990. Fertilizer use is the 1980-1990 average (after Gommes, 1993)

Continent	Population	1961-1990 exponential growth rate (%)					% Cereal yield increase (1961-1970 and 1986-1990)	Fertilizer use (kg fertilizer per ha arable land)	Irrigated land
		Production			Arable land			
		Cereals	Tubers	Pulses				
Africa	2.83	1.95	2.74	1.58	0.44	32	1.2	5.8
North and Central America	1.52	2.32	1.30	1.53	-0.59	49	10.7	9.6
South America	2.34	2.80	0.51	0.71	1.53	44	2.6	5.8
Asia	2.11	3.30	1.95	0.26	0.07	86	14.5	31.0
Europe	0.51	2.37	-1.27	1.57	-0.53	75	64.3	10.8
Oceania	1.72	3.14	1.62	13.76	1.23	34	2.0	4.1
World	**1.88**	**2.66**	**0.78**	**0.83**	**0.12**	**65**	**9.6**	**15.2**

THE CHANGING AGRICULTURAL ENVIRONMENT

THE CLIMATE 'COMPLEX'

Climate constitutes a complex of inter-related variables. On average, through a set of regulatory mechanisms, a smooth change in one variable triggers smooth changes in most others. With the exception of possible qualitative and abrupt variations, which will be mentioned below, such inter-relations are independent of atmospheric carbon dioxide (CO_2). The latter and other greenhouse gases play a part largely through their effect on the radiation balance of the atmosphere.

There is only a weak link between such factors as cloudiness and wind. Temperature, evaporation and rain are strongly correlated, which illustrates the likely intensification of the hydrological cycle (Figure 1.1). Combined with the projected pressure on land and water use, competition for land and water will certainly become a key social and political issue.

Climate variability is likely to increase under global warming (Katz and Brown, 1992), both in absolute and in relative terms. This is linked with thresholds which affect the occurrence of many meteorological phenomena. For instance, tropical cyclones are 'fed' by water vapour evaporating from oceans at a temperature above 26 or 27°C. Therefore, higher average sea surface temperatures are bound to result in a higher frequency of tropical cyclones.

The rate of change itself is extremely important. For example, recent work (Lehman, 1993; Paillard and Labeyrie, 1994; Rahmsdorf, 1994; Holmes, 1995) on the saw-tooth temperature changes in the past as observed in the Arctic, raise concern that changes may occur abruptly, with average temperatures changing by 10 or 12°C in just a matter of decades. The mechanism of such changes is not clear as yet, but seems to involve the mechanical stability of ice sheets, and sudden changes in the hydrological cycle and the Atlantic conveyor current.

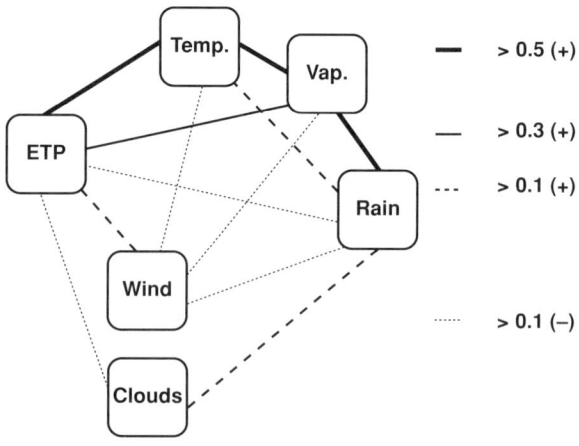

Figure 1.1. Some relationships between major climate variables (average temperature, *Temp.*; water vapour pressure, *Vap.*; rainfall, *Rain*; wind speed, *Wind*; cloudiness, *Clouds*; and evapo-transpiration potential, *ETP*). Solid and dotted lines indicate positive and negative correlations, respectively. The strength of the correlation decreases from double heavy lines to thin single lines. Computed from annual averages of 3 263 land stations, mostly from developing countries (FAO, 1995b)

THE CLIMATE CHANGE – AGRICULTURE CONUNDRUM

Such changes would, of course, be associated with dramatic changes in the distribution and quantities of ocean products, and cause havoc to established national fishery activities. They would also make adaptation to climate change, together with most agricultural planning, extremely difficult.

AGRICULTURAL GREENHOUSE GASES

In addition to water vapour, important greenhouse gases are carbon dioxide (CO_2), methane (CH_4), nitrous oxide (N_2O), tropospheric ozone (O_3) and chlorofluorocarbons (CFCs). The basic characteristics of the first three gases are given in Table 1.2. The degree to which these greenhouse gases stem from agricultural sources is also given. The exact ratio between these land-use related emissions and those from natural ecosystems (swamps, tundra), fossil fuels (coal, oil, gas) and geological sources (volcanoes) are a bone of contention among industrialized, oil producing and developing countries.

Moral rights and duties in relation to the international flow of development and conservation funds are involved. Therefore, better estimates based on exact measurements of the net greenhouse gas emissions from agricultural practices in developing countries are urgently required to assess responsibilities properly. The required reductions of emissions to achieve stabilization of atmospheric concentrations of current levels are believed to be >60% for CO_2, 15-20% for CH_4 and 70-80% for N_2O. This implies the need for up-to-date and complete information on land cover and land uses per country: not only the kind of crops grown but also their intensity, their rotation and the amount of inputs (energy, fertilizers).

Table 1.2. An overview of agricultural greenhouse gases with the trends as currently envisaged (adapted from IPCC, 1992; Houghton *et al.*, 1995; Keeling *et al.*, 1995). ppmv and ppbv stand for parts per 10^6 and parts per 10^9, respectively, by volume.

	CO_2	CH_4	N_2O
Atmospheric lifetime (yr)	120	14.5	120
Direct GWP [1]	1	24.5 [2]	320
Pre-industrial concentration [3]	280 ppmv	0.8 ppmv	288 ppbv
Present-day levels	360 ppmv	1.72 ppmv	310 ppbv
Current annual increase (%)	0.5	0.9	0.25
Major agricultural sources [4]	deforestation	-wetland rice -ruminants -biomass burning	-synthetic N fertilizers -animal excreta -biological N fixation
Percentage of global source stemming from agriculture	30	40	25
Predicted change 1990-2020	–	+	+

[1] GWP, Global Warming Potential, is the direct warming effect in relation to CO_2 at a time horizon of 100 years.
[2] Includes indirect effects through chemistry.
[3] About year 1750-1800.
[4] Activities responsible for emissions are projected to increase by: rice (+ 10%); ruminant population (+30%); synthetic fertilizer use (+20%); animal excreta and biological N fixation increasing but rate not specified.

The current trends of some of the agricultural sectors directly associated with greenhouse gas emission sources are listed in Table 1.3.

The loss of 'natural land', including tropical forests to agriculture, grazing, logging and urbanization, may continue, though at a slower pace (*AT-2010*). Many earlier estimates are now regarded with suspicion, as the borderline between crop agriculture, forest and cattle agriculture appears to be a fuzzy one.

Finally, both recent trends of fertilizer use and *AT-2010* projections indicate major increases in consumption. Even with appropriate measures to optimize fertilizer use, it is likely that N_2O losses from fertilizers will continue to increase.

Ecological and indirect climate effects

In qualitative terms, many indirect effects of climate change on agriculture can be conjectured. Most of them are estimated to be negative and they catch most of the attention of the media. These effects include:
- the overall predictability of weather and climate would decrease, making the day-to-day and medium-term planning of farm operations more difficult;
- loss of biodiversity from some of the most fragile environments, such as tropical forests and mangroves;
- sea-level rise (40 cm in the coming 100 years) would submerge some valuable coastal agricultural land;
- the incidence of diseases and pests, especially alien ones, could increase;
- present (agro) ecological zones could shift in some cases over hundreds of kilometres horizontally, and hundreds of metres altitudinally, with the hazard that some plants, especially trees, and animal species cannot follow in time, and that farming systems cannot adjust themselves in time;
- higher temperatures would allow seasonally longer plant growth and crop growing in cool and mountainous areas, allowing in some cases increased cropping and

Table 1.3. Growth rates between 1961 and 1990 in agricultural sectors responsible for greenhouse gas emissions (from FAO, 1990). Europe and Asia do not include the former USSR. Domestic ruminant numbers were computed as the sum of cattle, sheep, goats, camels and buffaloes

Continent	\multicolumn{4}{c}{1961–1990 exponential growth rate (%)}
	Ruminant numbers	Forested area	Rice area	Fertilizer consumption
Africa	1.29	−0.43	2.23	6.21
N and C America	−0.07	−0.02	1.50	3.29
S. America	1.29	−0.49	1.65	9.15
Asia	1.18	−0.59	0.62	9.54
Europe	0.33	0.25	0.78	2.75
Oceania	0.07	−1.15	5.88	1.25
World	**0.90**	**−0.26**	**0.74**	**5.35**

production. In contrast, in already warm areas climate change can cause reduced productivity;
- the current imbalance of food production between cool and temperate regions and tropical and subtropical regions could worsen.

PLANT PHYSIOLOGICAL DIRECT EFFECTS

The greenhouse gases CH_4, N_2O and chlorofluorocarbons (CFCs) have no known direct effects on plant physiological processes. They only change global temperature and are therefore not discussed further. Instead, concentration should be on the effects of increased CO_2, tropospheric O_3, increased UV-B through depleted stratospheric ozone, increased temperatures and the associated intensification of the hydrological cycle.

CARBON DIOXIDE

The CO_2 fertilization effect

CO_2 is an essential plant 'nutrient', in addition to light, suitable temperature, water and chemical elements such as N, P and K, and it is currently in short supply.

Higher concentrations of atmospheric CO_2, due to increased use of fossil fuels, deforestation and biomass burning, can have a positive influence on photosynthesis (Figure 1.2); under optimal growing conditions of light, temperature, nutrient and moisture supply, biomass production can increase, especially of plants with C_3 photosynthetic metabolism (Box 1.1), above and even more below ground (for details see chapter 4 by Allen *et al.*).

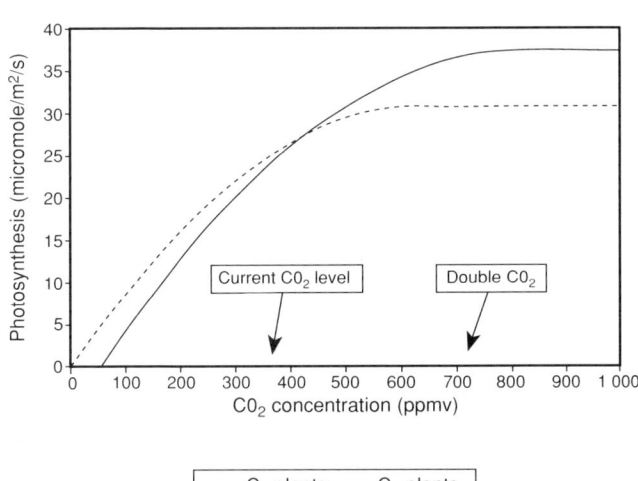

Figure 1.2. Schematic effect of CO_2 concentrations on C_3 and C_4 plants (after Wolfe and Erickson, 1993). The main mechanism of CO_2 fertilization is that it depresses photo-respiration, more so in C_3 than in C_4 plants

> **Box 1.1. The major agricultural crops and the three photosynthetic pathways**
>
> Plants are classified as C_3, C_4 or CAM according to the products formed in the initial phases of photosynthesis.
>
> C_3 species respond more to increased CO_2; C_4 species respond better than C_3 plants to higher temperature and their water-use efficiency increases more than for C_3 plants. There are some indications that enhancements can decline over time ('down-regulation') (see chapter by Allen [chapter 4 in this book]).
>
> C_3 plants: cotton, rice, wheat, barley, soybeans, sunflower, potatoes, most leguminous and woody plants, most horticultural crops and many weeds
>
> C_4 plants: maize, sorghum, sugar cane, millets, halophytes (i.e., salt-tolerant plants) and many tall tropical grasses, pasture, forage and weed species
>
> CAM plants (Crassulacean Acid Metabolism, an optional C_3 or C_4 pathway of photosynthesis, depending on conditions): cassava, pineapple, opuntia, onions, castor

A total of 10 to 20% of the approximate doubling of crop productivity over the past 100 years could be due to this effect (Tans et al., 1990) and forest growth or regrowth may have been stimulated as well. Further productivity increases may occur in the coming century, in the order of 30% or more where plant nutrients and moisture are adequate.

Higher CO_2 values would also mitigate the plant growth damage caused by pollutants such as NO_x and SO_2 because of smaller stomatal openings (see below). Higher percentages of starch in grasses improves their feeding quality, implying less need for feed mixes when silaging.

The CO_2 anti-transpirant effect (improved water-use efficiency WUE)

With increased atmospheric CO_2 the consumptive use of water becomes more efficient because of reduced transpiration. This is induced by a contraction of plant stomata and/or a decrease in the number of stomata per unit leaf area. This restricts the escape of water vapour from the leaf more than it restricts photosynthesis (Wolfe and Erickson, 1993; see details in chapter by Van de Geijn and Goudriaan).

With the same amount of available water, there could be more leaf area and biomass production by crops and natural vegetation. Plants could survive in areas hitherto too dry for their growth.

ULTRAVIOLET RADIATION

Increased ultraviolet radiation (UV-B, between 280 and 320 nanometres), due to depletion of the stratospheric ozone layer, mainly in the Antarctic region, may negatively affect terrestrial and aquatic photosynthesis and animal health. Over the last decade, a decrease of stratospheric ozone was observed at all latitudes (about 10% in winter, 0% during summer and intermediate values during spring and autumn). However, the 'Biological Action Factor' of UV-B can vary over several orders of magnitude with even slight changes in the amount and wavelength of UV-B.

The subject is treated in detail elsewhere in this book (see chapter by Unsworth and Hogsett, or Runeckles and Krupa, 1994). In particular it can be noted that:
- there are damaging effects of increasing UV-B on crops, animals and plankton growth. It has been reported that UV-B affects the ability of plankton organisms to control their vertical movements and to adjust to light levels;
- reductions in yield of up to 10% have been measured at experimentally very high UV-B values, and would be particularly effective in plants where the CO_2 fertilization effect is strongest. On the other hand, UV-B increase could increase the amount of plant internal compounds that act against pests.

TROPOSPHERIC OZONE

Tropospheric ozone originates about half from photochemical reactions involving nitrogen oxides (NO_x), methane or carbon monoxide, and half by downward movement of stratospheric ozone.

High ozone concentrations have toxic effects on both plant and animal life (German Bundestag, 1991; MacKenzie and El-Ashry, 1988; UNEP, 1993). It is likely that ozone, in conjunction with other photo-oxidants, is contributing towards the 'new type of forest damage' observed in Europe and the United States.

In the tropics, tropospheric ozone concentrations are generally lower than at northern mid-latitudes. However, this does not apply to periods when biomass burning releases precursor substances for the photochemical formation of ozone.

TEMPERATURE

In general, higher temperatures are associated with higher radiation and higher water use. It is relatively difficult to separate the physiological effects (at the level of plants and plant organs) of temperatures from the ecological ones (at the level of the field or of the region). There are both positive and negative impacts at the two levels, and only crop- and site-specific simulation can assess the global 'net' effect of temperature increases (for details see chapter by Abrol and Ingram). It is generally agreed that:
- rising temperatures – now estimated to be 0.2°C per decade, or 1°C by 2040 (Mitchell et al., 1995) with smallest increases in the tropics (IPCC, 1992) – would diminish the yields of some crops, especially if night temperatures are increased (the temperature increase since the mid-1940s is mainly due to increasing night-time temperatures, while CO_2-induced warming would result in an almost equally large rise in minimum and maximum temperatures (Kukla and Karl, 1993);
- higher temperatures could have a positive effect on growth of plants of the CAM

type. They would also strengthen the CO_2 fertilization effect and the CO_2 antitranspirant effect of C_3 and C_4 plants (see Box 1.1) unless plants get overheated;
- higher night temperature may increase dark respiration of plants, diminishing net biomass production;
- higher cold-season temperatures may lead to earlier ripening of annual crops, diminishing yield per crop, but would allow locally for the growth of more crops per year due to lengthening of the growing season. Winter kill of pests is likely to be reduced at high latitudes, resulting in greater crop losses and higher need for pest control;
- higher temperatures will allow for more plant growth at high latitudes and altitudes.

COMBINED EFFECTS AND SOME UNCERTAINTIES

The changes in CO_2, tropospheric ozone and increased UV-B do not necessarily occur simultaneously: CO_2 increase is worldwide, but with a strong seasonality in middle and higher latitudes; significant increase of UV-B is largely limited to subpolar regions (and mainly during the northern hemisphere winter months); high near-surface O_3 levels are restricted to the neighbourhood of major cities, airports, etc. (Seitz, 1994)

Box 1.2 illustrates some of the potential mechanisms which could account for either increased or decreased biomass under global change conditions. Note that increased biomass could even be associated with a decreased yield of grain (or sugar, oil, etc.) if one of the consequences of increased CO_2 will be a redistribution of biomass among plant organs (there are indications of relative increases of root growth).

THE HYDROLOGICAL CYCLE AND SOILS

Even a slight increase in surface temperatures will affect evaporation, atmospheric moisture and precipitation (Figure 1.1). While it is generally agreed that rainfall will increase (by an estimated 10 to 15%), two aspects have to be elucidated: how will rainfall intensities be affected, and what are the details of spatial changes. This is still largely a matter of discussion among experts (for details see chapter by Evans).

Based on palaeoclimatic analogies, certain authors predict more favourable rainfall conditions in the present-day Sahel (Petit-Maire, 1992). If the increase in precipitation should be associated with increased rainfall intensities, then the quality and quantity of soil and water resources would decline, for instance through increased runoff and erosion, increased land degradation processes, and a higher frequency of floods and possibly droughts. However:
- the extra precipitation on land, if indeed including present subhumid to semiarid areas, will increase plant growth in these areas, leading to an improved protection of the land surface and increased rainfed agricultural production; in

> **Box 1.2.** Some mechanisms likely to affect biomass production under global change conditions. Note that the ratio between economic yield (e.g., grain, fibre) and biomass may change relative to current conditions

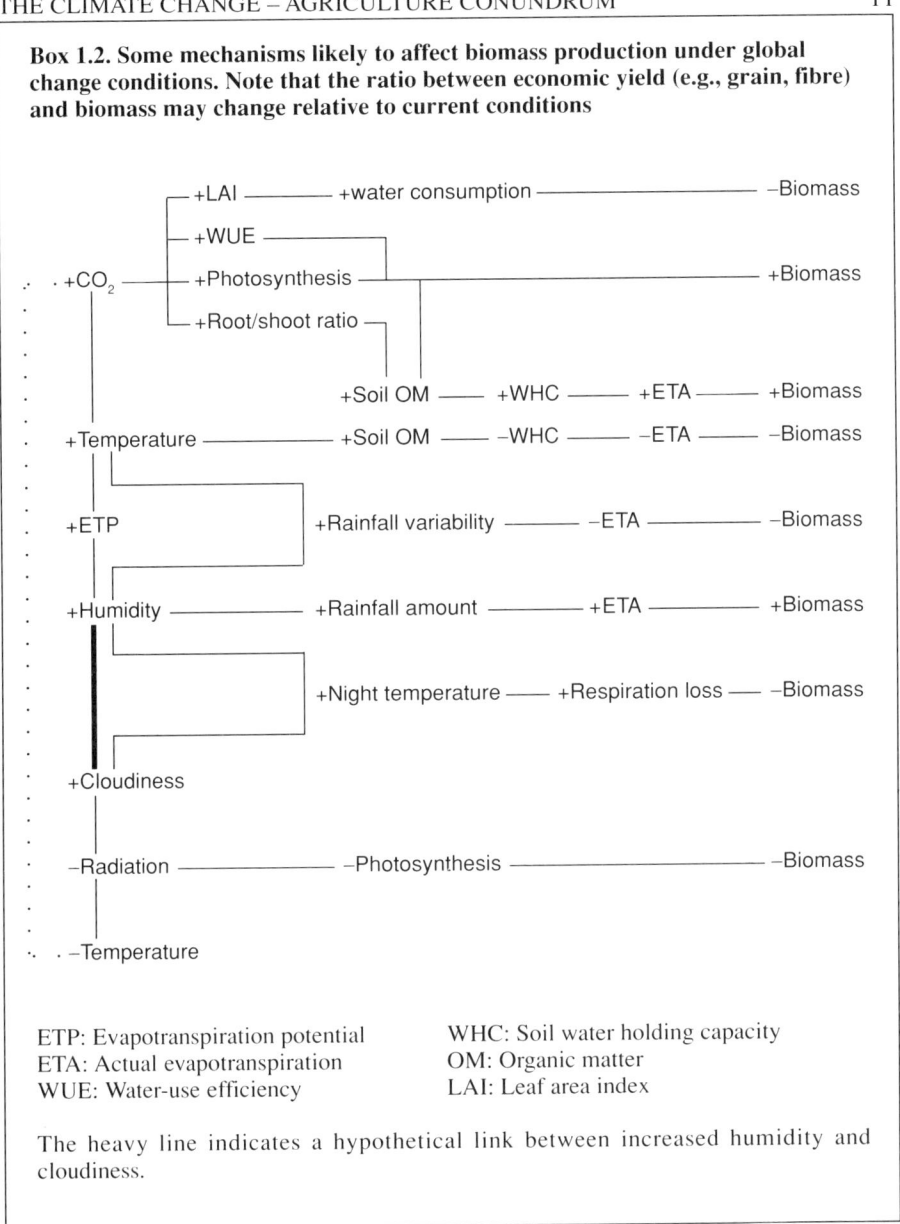

ETP: Evapotranspiration potential
ETA: Actual evapotranspiration
WUE: Water-use efficiency
WHC: Soil water holding capacity
OM: Organic matter
LAI: Leaf area index

The heavy line indicates a hypothetical link between increased humidity and cloudiness.

already humid areas the extra rainfall may, however, impair adequate crop drying and storage;
- the extra precipitation predicted to occur in some regions provides possibilities for off-site extra storage in rivers, lakes and artificial reservoirs (on-farm or at subcatchment level) for the benefit of improved rural water supply and expanded or more intensive irrigated agriculture and inland fisheries;

- the effects on water resources and water apportioning of international river and lake basins can be very substantial, with political overtones.

The greatest risks are often estimated to be associated with increased soil loss through erosion.

Soils, as a medium for plant growth, would be affected in several other ways (for details see chapter by Brinkman and Sombroek):
- increased temperatures may lead to more decomposition of soil organic matter;
- increased plant growth due to the CO_2 fertilization effect may cause other plant nutrients such as N and P to become in short supply; however, CO_2 increase would stimulate mycorrhizal activity (making soil phosphorus more easily available), and also biological nitrogen fixation (whether or not symbiotic). Through increased root growth there would be extra weathering of the substratum, hence a fresh supply of potassium and micronutrients;
- the CO_2 fertilization effect would produce more litter of higher C/N ratio, hence more organic matter for incorporation into the soil as humus; litter with high C/N decomposes slowly and this can act as a negative feedback on nutrient availability;
- the 'CO_2 anti-transpirant' effect would stimulate plant growth in dryland areas, and more soil protection against erosion and lower topsoil temperatures, leading to an 'anti-desertification effect'.

CONCLUSIONS

Global climate change, if it occurs, will definitely affect agriculture. Most mechanisms, and two-way interactions between agriculture and climate, are known, even if not always well understood.

It is evident that the relationship between climate change and agriculture is still very much a matter of conjecture with many uncertainties (see also Rosenzweig and Hillel, 1993); it remains largely a conundrum.

Major uncertainties affect both the Global Circulation Models (GCMs) and the response of agriculture, as illustrated by differences among models, especially as regards effects at the national and subregional levels. In addition, many of the models do not take into consideration CO_2 fertilization and improved water-use efficiency, the effect of cloud cover (on both climate and photosynthesis), or the transient nature of climate change.

It is also worth remembering that enormous knowledge gaps still affect the carbon cycle (with a missing sink of about 2 Gt of carbon), the factors behind the recent near-stabilization of the atmospheric methane concentrations or the unexplained reduced rate of CO_2 increase in recent years, the effect of volcanic eruptions (such as the recent Pinatubo eruption), the effect of any increased cloudiness, etc.

REFERENCES

FAO. 1990. *1989 Production Yearbook No. 43. FAO Statistics Series No. 94*. AGROSTAT-PC, the PC version of the yearbook, was also used.

FAO. 1995a. *World Agriculture: Towards 2010, An FAO Study*. N. Alexandratos (ed.). John Wiley, Chichester, UK, and FAO, Rome. 488 p.

FAO, 1995b. FAOCLIM 1.2, Worldwide agroclimatic data. *FAO Agrometeorology Working Paper Series No. 11*. FAO, Rome. 1 CD-ROM and 66 p.

German Bundestag. (ed.). 1991. *Protecting the Earth*. A status report with recommendations for a new energy policy. Bonn, Deutscher Bundestag, Referat Oeffentlichkeitsarbeit. Vol. 1, 672 p.

Gommes, R. 1993. Current climate and population constraints on world agriculture. In: *Agricultural Dimensions of Global Climate Change*. H.M. Kaiser and T.E. Drennen (eds.). pp. 67-86.

Holmes, R. 1995. Arctic ice shows speed of climate 'flips'. *New Scientist* **145**(1967):13.

Houghton, J.T., Meira Filho, L.G., Bruce, J., Lee, H., Callander, B.A., Haites, E., Harris, N. and Maskell, K. (eds.). 1995. *Climate Change 1994*. Radiative forcing of climate change; and an evaluation of the IPCC IS92 emission scenarios. Cambridge University Press, Cambridge, New York, Melbourne.

IPCC. 1992. *Climate Change 1992*. The supplementary report to the IPCC scientific assessment. J.T. Houghton, B.A. Callandar and S.K. Varney (eds.). Cambridge University Press. Cambridge.

Kaiser, H.M. and Drennen T.E. (eds.). 1993. *Agricultural Dimensions of Global Climate Change*. St. Lucie Press, Delray Beach, Florida. 311 p.

Katz, R.W. and Brown, B.G. 1992. Extreme events in a changing climate: variability is more important than averages. *Clim. Change* **21**: 289-302.

Keeling, C.D., Whorf, T.P., Wahlen, M. and van der Plicht, J. 1995. Interannual extremes in the rate of rise of atmospheric carbon dioxide since 1980. *Nature* **375**: 666-670.

Kukla, G. and Karl, T.R. 1993. Nighttime warming and the greenhouse effect. *Envir. Sci. Technol.* **27**(8): 1468-1474.

Lehman, S. 1993. Ice sheets, wayward winds and sea change. *Nature* **365**: 108-110.

MacKenzie, J.J. and El-Ashry, M.T. 1988. Links between air pollution and damage to crops. In: *Ill Winds: Airborne Pollution's Toll on Trees and Crops*. World Res. Institute. pp. 25-33.

Mitchell, J.F.B., Johns, T.C., Gregory, J.M. and Tett, F.B. 1995. Climate response to increasing levels of greenhouse gases and sulphate aerosols. *Nature* **376**: 501-504.

Paillard, D. and Labeyrie, L. 1994. Role of thermohaline circulation in the abrupt warming after Heinrich events. *Nature* **372**: 162-164.

Petit-Maire, N. 1992. Lire l'avenir dans les archives géologiques. *La recherche* **23**(243): 566-569.

Rahmsdorf, S. 1994. Rapid climate transitions in a coupled ocean–atmosphere model. *Nature* **372**: 82-85.

Rosenzweig, C. and Hillel, D. 1993. Agriculture in a greenhouse world: potential consequences of climate change. *National Geographic Research and Exploration* **9**: 208-221.

Runeckles, V.C. and Krupa, S.V. 1994. The impact of UV-B radiation and ozone on terrestrial vegetation. *Envir. Pollut.* **83**(1-2): 191-213.

Seitz, F. 1994. *Global Warming and Ozone Hole Controversies, A Challenge to Scientific Judgement.* G.C. Marshall Institute, Washington DC. 32 p.

Solbrig, O.T., van Emden, H.M. and Van Oordt, P.G.W.J. 1992. Biodiversity and global change. *IUBS Monograph No. 8*, IUBS Press, Paris. 222 p.

Tans, P.P., Fung, I.Y. and Takahashi, T. 1990. Observational constraints on the global atmospheric CO_2 budget. *Science* **247**: 1431-1438.

UNEP. 1993. *Action on Ozone.* UNEP, Nairobi. 24 p.

Wolfe, D.W. and Erickson, J.D. 1993. Carbon dioxide effects on plants: uncertainties and implications for modelling crop response to climate change. In: *Agricultural Dimensions of Global Climate Change.* H.M. Kaiser and T.E. Drennen (eds.). pp. 153-178.

2. The Effects of Changes in the World Hydrological Cycle on Availability of Water Resources

TERRY E. EVANS
Former Director, Mott MacDonald Co., UK

The development of large water resource systems can take 10 to 20 years from the initiation of preliminary studies to realization of projects, and therefore an assessment of the impact of global warming on water resources is urgently required. The aim of this chapter is to present evidence of hydrological changes which have already occurred, and to evaluate whether, at the present time, our knowledge is sufficient to make predictions of changes in the hydrologic cycle with enough accuracy to be of value in water resource planning.

Methods for evaluating future available water are discussed and examples given of impact studies using outputs from General Circulation Models (GCMs). In this respect, Beran (1986) has emphasized a distinction between hydrology and available water resources. The former deals with evaluating the total resource (naturalized river flow) whilst the latter requires the quantification of the exploitable amount.

The world's agricultural output is heavily dependent on irrigation. Out of a global cropped area of 1 500 million ha, 16% is irrigated. Global warming is likely to have a major impact on the hydrological cycle and consequently on irrigated agriculture. This impact may be more significant than the direct effect of higher temperatures. Even though this may be the case, it is essential to view the impact of climate change in context with other major global changes. Demands on water are multi-sectorial and irrigation is already in competition for potable water, industrial, power generation, and recreational and environmental uses: most of which command a higher price. Within many Near East and African countries current water resources are almost fully exploited and supplies will have to double over the next 20 to 30 years to maintain the current, even unsatisfactory status quo. When the problem of increasing levels of water pollution is added into the equation, a situation arises of 'Water in Crises' (Gleick, 1993).

Unfortunately, in most cases hydrology is treated on a national, rather than a regional, continental or global scale. It is heavily dependent on accurate data collection. With national water resource agencies generally small units within major ministries, and having low priority, funding has fallen to such an extent that in the majority of African countries reliable data are no longer collected (World Bank/UNDP/ADB/

EC/French Government, 1993). The seriousness of the problem has yet to be addressed; although the World Bank is promoting a new 'Integrated Approach to Water Resource Management for Sub-Saharan Africa' and other regions of the world, whilst WMO is attempting to set-up a 'World Hydrological Cycle Observing System (WHYCOS)' (Rodda et al., 1993) and FAO is establishing a programme to develop a worldwide water database with hydrological modelling capability using Geographic Information System (GIS) and remote sensing technology.

In this chapter some of the problems associated with climate change and the hydrological cycle are discussed.

METHODS OF EVALUATION

Different approaches can be used to estimate the effect on water resources of global warming of which the main ones are:
- use of instrumental records;
- palaeoclimatic analogues;
- GCM outputs.

In the first approach the natural variability of climate is used to predict rainfall patterns during sequences of extremes of dry or wet years. Difficulties arise due to the short period of records, although stochastic data generation techniques can be used to extend the database. Over such limited time spans, however, longer-term changes in vegetation, sea surface temperatures, ocean currents, etc., which would be expected to take place with a doubling of atmospheric forcing functions such as CO_2, are not adequately accounted for. Irrespective of this limitation, significant long-term changes in surface runoff have occurred over the past 30 years in a number of major river basins, especially in Africa. Recently, strong scientific evidence has also been produced to suggest that such changes are directly related to global warming triggered by the increase in anthropogenic greenhouse gases (Thomson, 1995; Hadley Centre, 1995). As some of the recorded trends in river flows are of the same order of magnitude as would be expected for a $2 \times CO_2$ scenario, a review of recent changes in the hydrological cycle adds to an understanding of the problems associated with long-term changes in gross water resources (section *Recent changes in global runoff*).

Palaeoclimatic analogues fall into the same category as instrumental records though with obvious drawbacks: estimating past climates is extremely unreliable and fragmentary. Their main benefit lies in being able to cover the range of temperatures likely to be experienced over the next century or two (Table 2.1).

Recent studies suggest that global temperatures during previous ice ages fluctuated widely over time periods as short as decades or centuries (Heinrich cycles) and temperatures were not relatively constant as generally assumed (Maslin, 1993).

There is also no consensus amongst scientists over the prime cause for past temperature changes in the earth's atmosphere, or for the triggering of the ice ages; it

Table 2.1. Past global temperatures

Mean global temperature above present-day values	Date
+ 1.5°C	6 000 BP
+ 2.5°C	125 000 BP
+ 4.0°C	4 000 000 BP

certainly was not anthropogenic. The general accepted viewpoint is that changes in solar radiation received at the earth's surface are the prime instigators (Milankovitch, 1930). However, with current research spanning many disciplines, there has been a proliferation of new theories related to the controlling factors responsible for producing ice ages and postglacial optimums. Current theories range from rates at which the earth's crust is formed, extent of coverage of peat bogs, release of methane from shallow seas due to formation of sea ice, rate of mountain building and resulting erosion and sediments deposited in the oceans, and cyclic changes in iceberg production affecting the north-south ocean 'conveyor belt', possibly due to instability from trapped thermal heat below glaciers, etc. Until the casual factors involved in past climate change are known, it would be unwise to assume that they would have no influence in determining the resulting climate; an inherent assumption in the climate analogue method. Present land use, vegetation, ice cover, atmospheric composition, etc. are all likely to be so different from those experienced during previous postglacial optimums as to make a direct comparison between past and present climates of speculative value only.

GCMs are the only credible tool for predicting climate change and for providing inputs to hydrological models. Unfortunately, at the present time it is not possible to use precipitation outputs from GCMs directly as inputs to hydrological models. The outputs are not sufficiently accurate to simulate daily and monthly sequences (see Table 2.2).

However, the use of GCMs is the only way forward and with improved understanding and better modelling of the systems (especially ocean circulation and the land phase of the hydrological cycle) and larger computer memories, it is only a matter of time before these deficiencies will be remedied.

Table 2.2. Comparisons between observed rainfall over England and estimates from three GCMs (Arnell *et al.*, 1990)

Source	Dec-Feb
Observed	2.5 mm/day
UK Meteorological Office GCM (Bracknell)	1.2 mm/day
National Center for Atmospheric Research GCM (Boulder, Colorado)	3.4 – 4.5 mm/day
Goddard Institute for Space Studies GCM (New York)	3.5 – 4.5 mm/day

GENERAL CIRCULATION MODELS

A GCM model typically has four components: atmospheric, land, ocean and sea ice. As the four models are interactive, production runs must include all four model components. Initially, the atmospheric model was developed to a relatively high level of sophistication, whilst the land phase and the ocean components were very simplistic.

Runoff enters into current GCM model simulations at two points (Rowntree, 1989). Firstly, at the boundary between the atmosphere and the land surface, where flux transfers are converted into surface runoff, and secondly at the boundary between the ocean and land, where inflow hydrographs are required as input to the ocean model. At present neither are simulated with sufficient accuracy. Kite *et al.* (1994) have demonstrated some of the problems associated with GCM modelling of the hydrological cycle, particularly the lack of adequate lateral transfer of water. The problem of inadequate routing of surface runoff is highlighted by studies of Semtner (1984, 1987) where a reduction in freshwater inflow to the Arctic Ocean would have a significant effect on salinity gradient and it was calculated that the outflow from the Kara and Barent seas could halve should there be a total diversion of rivers flowing to these seas. The oceans are therefore quite sensitive to freshwater inputs.

It has been demonstrated that a high degree of 'recycling' of rainfall occurs over land and if, for instance, land evaporation is much reduced or eliminated, then a significant reduction in precipitation results (Shukla and Minz, 1982; Hall and Sarenije, 1993). In order to determine actual evapotranspiration an accurate assessment of surface runoff is required. Current GCMs do not achieve this. It is likely, therefore, that shortcomings in the modelling of the hydrological cycle are reducing the accuracy of outputs from current GCMs, as has been highlighted by Rind *et al.* (1990). The GISS GCM, and others, have been shown to greatly underestimate actual evapotranspiration (ET) through overestimating the reduction in transpiration with increased soil moisture deficits. For example over the USA the ratio of Actual ET/Potential ET for current climatic conditions was estimated by GCM to be 0.23, whereas the actual ratios should lie between 0.6 and 0.9. Such a large discrepancy arises in part due to vegetation cover being inadequately represented in the land phase in GCMs. It was concluded by Rind (1995) that ET loss has been seriously underestimated by GCMs and that global precipitation increases of 9 to 15% (Grotch, 1989) are not compatible with a 30% increase in atmospheric water holding capacity corresponding to a 4°C warming. Either the rainfall estimates are underestimated or much more serious droughts will occur over continents.

Initially, climate change was seen, understandably, as a meteorological problem. The importance of the atmospheric-ocean interface was always known, but only now are results from coupled models becoming available. Similar improvements are required in the hydrological cycle components. The Global Energy and Water Cycle Experiment (GEWEX) of the World Climate Research Programme of ICSU/WMO/IOC should provide the impetus for such improvements. However, until the land phase is adequately modelled by GCMs then there is little hope that its outputs can be

EFFECTS ON AVAILABILITY OF WATER RESOURCES 19

used with confidence in water resource planning. Furthermore, river basin impact studies undertaken to date have been based on results obtained from equilibrium GCMs in which the oceans are modelled simply as a fixed layer(s). Because of this major limitation, they are only used to simulate equilibrium states such as $1 \times CO_2$ or $2 \times CO_2$, neither states of which are, or will be, in equilibrium. All the GCMs are structurally similar and contain the same basic algorithms and will inherently include the same errors. It is rather disturbing, therefore, that such large differences in output values are produced (see Table 2.2 and Figure 2.1). Equilibrium models are relatively stable and do not suffer to the same extent from cold start and drift problems that afflict transient models. Because of the shortcomings of the equilibrium models the reliability of the outputs, particularly precipitation, is seriously compromised. The range in outputs from the different equilibrium models is generally so large that it is often not possible to determine the sign of the change in runoff over large regions or even over some continents. Because of this problem some researchers have used outputs from a number of GCMs and developed levels of probability for decreases (or increases) in temperature and precipitation (Howell and Allan, 1994). Although the results from such a risk analysis approach present a useful guide for water resource planning, when all the other uncertainties are included, such studies are of little practical value in the detailed planning of future water resource development, except to make the planner aware of the increased uncertainty due to climate change.

A significant improvement in recent years has been the development of transient, (ocean-coupled) GCMs. However, the computer time utilized and the costs in running these models are at present exorbitant. As an interim measure techniques have been developed which allow for the rescaling of equilibrium model results and inte-

Figure 2.1. Comparison between percentage change in low and high mean annual precipitation estimates for composite GCM scenarios (1990-2050) (after Hulme, 1994)

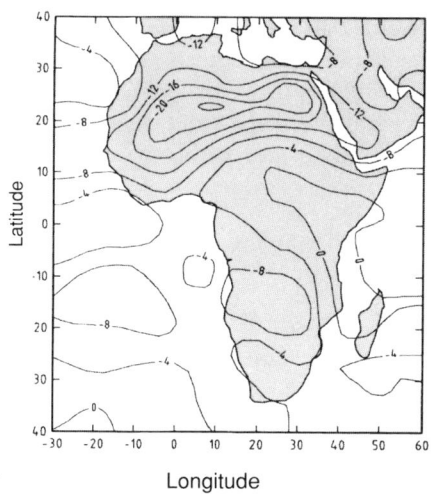
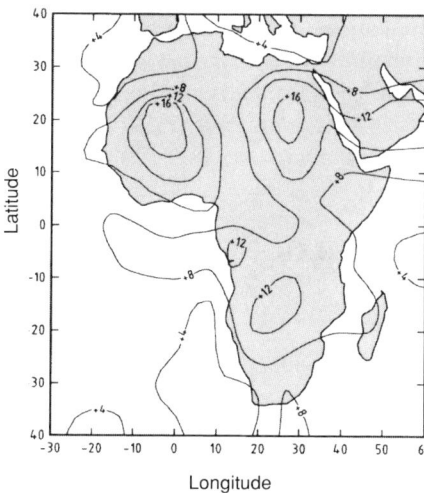

grating them with those of individual transient model runs to provide an indication of temporal changes (Viner and Hulme, 1993). Figure 2.1 shows the range of estimated precipitation changes using the outputs from seven GCMs scenarios (Table 2.3) over the period 1990 to 2050. The low and high rainfall changes are based on the lower and upper 90% confidence limits. Although this approach enables the results from equilibrium models to be modified to provide outputs for different time horizons, the serious drawbacks which apply to the basic equilibrium model results remain.

The first transient model experiment (UKTR) undertaken by the Hadley Centre (Hadley Centre, 1992) determined the climate's response for an increase in CO_2 concentration of 1% per annum over a 75 year period. The northern hemisphere was found to warm twice as rapidly as the southern hemisphere. The rise in temperature of the Antarctic was of a similar magnitude to that for the southern hemisphere as a whole, i.e., very different from previous equilibrium GCM results. In addition, the temperature rise in the transient model experiment was only 60% of that for the equilibrium model, due to the large inertia of the oceans. The change in global rainfall was, however, similar to that produced by early UK Meteorological Office GCMs. Further experiments (Hadley Centre, 1995) have attempted to model historical changes in climate from 1860 to 1990 incorporating for the first time the attenuating effects of sulphate aerosol emissions. The results are encouraging, and the anomalous reversals in global temperature rise in the early 1940s, and the rapid warming from the mid-1970s, are reproduced by the model. Evidence is mounting that the recorded changes in climate over this century are consistent with those expected from increasing emissions of greenhouse gases.

Unfortunately, transient coupled GCMs are not yet capable of reproducing dynamic regional sea surface temperature patterns, and the associated precipitation changes, to the accuracy required to simulate the trends in regional precipitation which have occurred this century. To do this it would be necessary to understand the mechanisms which trigger the Southern Oscillation and to be able to simulate both them and those of the north Atlantic ocean conveyor system, and other significant ocean circulations. It is not possible, therefore, to confirm at the present time, using outputs from transient GCMs, that recorded global precipitation changes are the direct results of global warming. Until such calibrations are effected, then precipitation outputs from GCMs, transient and equilibrium, will remain speculative and should not be used directly in detailed planning of new water development projects.

HYDROLOGICAL MODELS

TYPES

The models under discussion here are concerned with the land phase of the hydrological cycle, and generate river flow and groundwater recharge. The most commonly used is the deterministic model. It is a 'conceptual' type model used in catchment hydrology to simulate at various levels of detail the physical processes involved in

Table 2.3. Some characteristics of the seven GCM experiments used (Hulme, 1994)

Model acronym	Resolution Longitude	Resolution Latitude	Ocean heat transport	ΔT_{eq} (°C)	ΔP_{eq} (%)	W (%)	Reference
1 GISS	10.0°	7.8°	Prescribed	4.2	11	5	Hansen et al. (1984)
2 GFDL	7.5°	4.5°	None	4.0	9	10	Wetherald and Manabe (1986)
3 LLNL	5.0°	4.0°	2 layer ocean	3.8	11	5	W.L. Gates (pers. comm., LLNL)
4 ECHAMI-LSG	5.6°	5.6°	11 layer ocean	1.6	3	10	Cubasch et al. (1992a)
5 OSU	5.0°	4.0°	None	2.8	8	20	Schlesinger and Zhao (1989)
6 UKLO	7.5°	5.0°	Prescribed	5.2	15	15	Wilson and Mitchell (1987)
7 UKHI	3.75°	2.5°	Prescribed	3.5	11	35	Mitchell et al. (1990)

Note: ΔT_{eq} and ΔP_{eq} are sensitivity of models for a $2 \times CO_2$ scenario (except 4). W = weight to derive precipitation model-average, based on global pattern of correlation coefficients.

GISS : Goddard Institute for Space Studies
GFDL : Geophysical Fluid Dynamics Laboratory
LLNL : Lawrence Livermore National Laboratory
ECHAM-LSG : European Centre Hamburg Model – Large Scale Geostrophic
OSU : Oregon State University
UKLO : United Kingdom Meteorological Office – Low
UKHI : United Kingdom Meteorological Office – High

moisture flux transfer: from interception, evapotranspiration, soil moisture changes, surface runoff, infiltration, percolation, interflow, groundwater storage and channel routing. They tend to be lumped processes and are the workhorse of applied catchment hydrology. An essential prerequisite in applying these models is a high level of calibration based on recorded data.

Much research in pure hydrology is aimed at improving the individual processes and obtaining better spatial representation of input data. This, eventually, should reduce the importance of calibration. However, there are pitfalls in using over-sophisticated representation of the hydrological cycle in one or two processes at the expense of over-simplification in others, especially without evaluation and verification on actual catchments (Lockwood, 1985; Schnell, 1984). Similarly, statistical and black box models, often regression based, have little use in impact studies, where many parameters affecting runoff are non-stationary, e.g., evapotranspiration, vegetation, land use, groundwater abstractions, etc.

Simple water balance models (Wigley and Jones, 1985), or models incorporating evapotranspiration based solely on temperature, are also of little value either in water resources impact studies or in generalized estimates of regional changes. There is no short cut to evaluating regional change in water resources without involving all the components of the hydrological cycle at a basin or subbasin scale. Time scales in hydrology are extremely important. For example, in arid and semi-arid regions the annual runoff is dependent on rainfall intensities at an hourly time scale falling over relatively small areas and on percolation within the river bed. Times to peak for catchments of areas of thousands of square kilometres may be only one or two hours. In such catchments the intensity is as crucial to surface runoff as the rainfall amount. Similarly, if annual potential evapotranspiration were to increase by 25% over a catchment with few rainfall days in the year, it would have little impact on the volume of runoff generated. Simple annual and monthly water balance models will work satisfactorily in such conditions. In catchments with significant soil storage, the runoff is governed by available storage and rainfall intensity. In the simple balance equation:

$$Q = P - E$$

where Q = surface runoff, P = precipitation, E = catchment losses.

Q is not independent of E, and if treated as such, will produce significant distortion when calculating changes in Q relative to changing of P and E (Wigley and Jones, 1985).

Also, in most catchments both geology and vegetation cover can be extremely variable. These parameters alone produce large differences in seasonal and annual runoff. For example in the River Canje catchment in Guyana, half the catchment is covered by tropical forest with a closed canopy and underlain by pervious 'White Sands'; whilst the remainder is largely covered by savanna and relatively impermeable sediments. As a result, runoff ratios (Q/P) in the wet season are about 0.15 in the forested area, and 0.85 in the savanna; in the dry season the values are 0.50 and 0.10, respectively (MacDonald, 1965).

INTERFACING GCMS AND HYDROLOGICAL MODELS

There are many ways of using outputs from GCMs to provide inputs into hydrological models. The most obvious is to take the GCM simulations of daily rainfall, temperature and other meteorological variables and feed them directly into a calibrated hydrological model. Unfortunately precipitation outputs from different equilibrium models are extremely variable (see Figure 2.1 and Table 2.2). Also, because of the way clouds and precipitation are modelled, GCMs produce unrealistic simulations of day-to-day rainfall. An alternative approach is to use the 'change' in rainfall, temperature and other climate variables either to produce directly a perturbed time series or to produce stochastically generated sequences of wet or dry states with amounts based on conditional probability distributions. This latter method is a compromise to cover for the inadequacies of the GCM outputs and to produce more realistic looking data sequences. Similarly, regional GCM-generated pressure patterns can be used with a suitable regional multivariate model to generate sequences of rainfall and evapotranspiration (Mott MacDonald, 1992, 1993). However, the pressure patterns cannot be used directly to produce climatic sequences. None of the methods used gives confidence that results will be other than approximate and of limited value in water resource planning.

IMPACT STUDIES OF CLIMATE CHANGE

To date, most of the impact studies undertaken have not used hydrological models which adequately simulate the transient nature of the parameters of the hydrological cycle. Even fewer have attempted to include changes in water demand.

Many impact studies have found that changes in evapotranspiration rates due to global warming have only a marginal effect on surface runoff when compared with changes in precipitation. These conclusions have been questioned. Rind *et al.* (1990) have suggested that most GCMs fail to simulate the land phase satisfactorily and underestimate actual evapotranspiration rates. The effect of increased potential evapotranspiration may significantly reduce surface runoff in the tropics (Rind, 1995). In addition, often changes in temperature alone have been used to estimate changes in evapotranspiration. This again will underestimate actual evapotranspiration (Arnell and Reynard, 1993). Finally, many studies have used simplified hydrological models, often of the water balance type, to analyse changes in surface runoff (Wigley and Jones, 1985). Such methods can easily be demonstrated to be inappropriate.

The Institute of Hydrology (IH) has undertaken an impact study for 21 catchments in England and Wales (Arnell and Reynard, 1993). The wettest scenario produces a general increase in average runoff across England and Wales, whilst the driest produces a reduction. However, the most interesting findings relate to changes in evapotranspiration. Three different potential evapotranspiration (PET) scenarios were investigated: PET changes based on; Case 1 – increased temperatures only; Case 2 – changes in temperature, net radiation, relative humidity and windspeed; Case 3 – Case 2 plus changes in plant stomatal conductance and leaf area. The results are given in Table 2.4.

Table 2.4. Equilibrium climate change scenarios for 2050 (Arnell and Reynard, 1993)

Climate parameter	Annual changes in England and Wales
Temperature	+2.2%
Rainfall	+4%
PET 1	+10%
PET 2	+35%
PET 3	+24%

An important feature of the results is the quite dramatic increase in PET predicted in England and Wales resulting from the inclusion of climatic elements other than temperature. A change in humidity produced the most significant increase based on PET calculations using the Penman-Monteith formula.

Kite (1993) has recognized the importance of land cover changes in water resource impact studies. The Kootenay catchment in Canada was divided into 50 Grouped Response Units (GRUs) and 10 land cover divisions, giving a potential for subdivision to 500 units. The land cover data were obtained from satellite data. Good calibration was obtained using a distributed hydrological model (SLURP). In the $2 \times CO_2$ scenario with changed land cover, the frequency of high flows was found to more than double and evapotranspiration was found, surprisingly, to decrease by 10%.

Over the last century the hydrological cycle over most river basins has been subjected to extensive human changes with the construction of reservoirs, land-use changes, river abstractions, groundwater abstractions, inter-basin diversions, etc. Such changes significantly alter river flows and the hydrology of the basin. For example, the Nile loses 10% of its flow through reservoir evaporation above the Aswan Dam whilst the Zambezi loses the equivalent of almost 20% of its flow at Victoria Falls from evaporation losses from the Kariba reservoir. On a global scale it has been estimated that sea levels would have risen this century by 2 mm per year, rather than the recorded 1 mm per year, had it not been for increases in reservoir storage and river abstractions, mainly for irrigation purposes (Newman and Fairbridge, 1986). It is important, therefore, that existing modifications to natural river flows are evaluated in impact studies as well as likely future changes in demand and supply.

It is evident that forecasts of future water availability produced by climate impact studies to date have limited value in present-day detailed water development designs. Their main value lies in improving impact modelling techniques. Certainly, too much effort has been placed on routine impact studies at the expense of targeted research. Estimates of the effect of climate change on precipitation and evapotranspiration are not sufficiently accurate to provide credible hydrological predictions.

RECENT CHANGES IN GLOBAL PRECIPITATION

RECORDED RAINFALL, 1931-1960 AND 1961-1990

Hulme *et al.* (1992) have prepared maps (see Figures 2.2 and 2.3) showing global precipitation as well as annual and seasonal changes in precipitation between the periods 1931 to 1960 and 1961 to 1990. The two most prominent features are over northern Russia with increased precipitation of up to 20% and the African Sahel, between latitudes 10°N and 30°N, with reduced precipitation between 20 and 50%. Significant changes have occurred over all continents. These recorded changes over the last 60 years are of a similar magnitude to those predicted by GCMs for a $2 \times CO_2$ scenario (see Figure 2.4). How far these recorded changes are related to radiative forcing due to increased greenhouse gases is controversial. The standard response is to state that the natural variability in climate will mask any forcing for two or more decades. This viewpoint can now be questioned following the results of several recent research studies, most notably Thomson (1995). In a major study of annual solar variations at the earth's surface, it has been demonstrated that, due to approximations in using the tropical year (equinox to equinox) in the Gregorian calendar rather than the anomalistic year (perihelion to perihelion), the starting dates of the seasons should be occurring later each year this century. A reversal occurred in 1923 which cannot be explained by changes in incoming solar radiation or by other phenomena such as the occurrence of volcanic eruptions or the Southern Oscillation, but is completely consistent with recorded increases in greenhouse gas emissions. Moreover, were the recorded increases in global temperatures due to greater solar radiation activity, then the amplitudes in seasonal temperature would be widening. The reverse is the case, which is consistent with the greenhouse effect. In addition, UK Meteorological Office GCM (Hadley Centre, February 1995) modelling of historical changes of climate (1960-1990) has reproduced both the anomalous reversals in global temperature and the rapid increases recorded later in the century. The evidence is mounting, therefore, that recorded changes in climate this century are the direct result of increased greenhouse gas emissions. If this is the case, then changes in river flow regime over this century probably present the best evidence available for water resource planning of the transient effect of global warming.

Although there are no transient GCM results to make comparisons between the observed precipitation change recorded between 1931 to 1960 and 1961 to 1990 (Figure 2.3), an interesting comparison can be made between present-day conditions and those predicted in 50 years' time (Figure 2.4). On a global scale the same pattern of change is modelled. The significant exception is that of Africa, north of the equator, which is shown to experience increased rainfall. The Sahelian drought may therefore be ameliorated to some degree. This is consistent with a predicted reversal in hemispheric warming with the north warming much more rapidly than the south; the opposite situation to that seen over the past 30 years. Aridity is predicted to increase in Australia, a trend which has been apparent in recent years, although clearly associated with the Southern Oscillation, a feature not represented at a decadal scale in

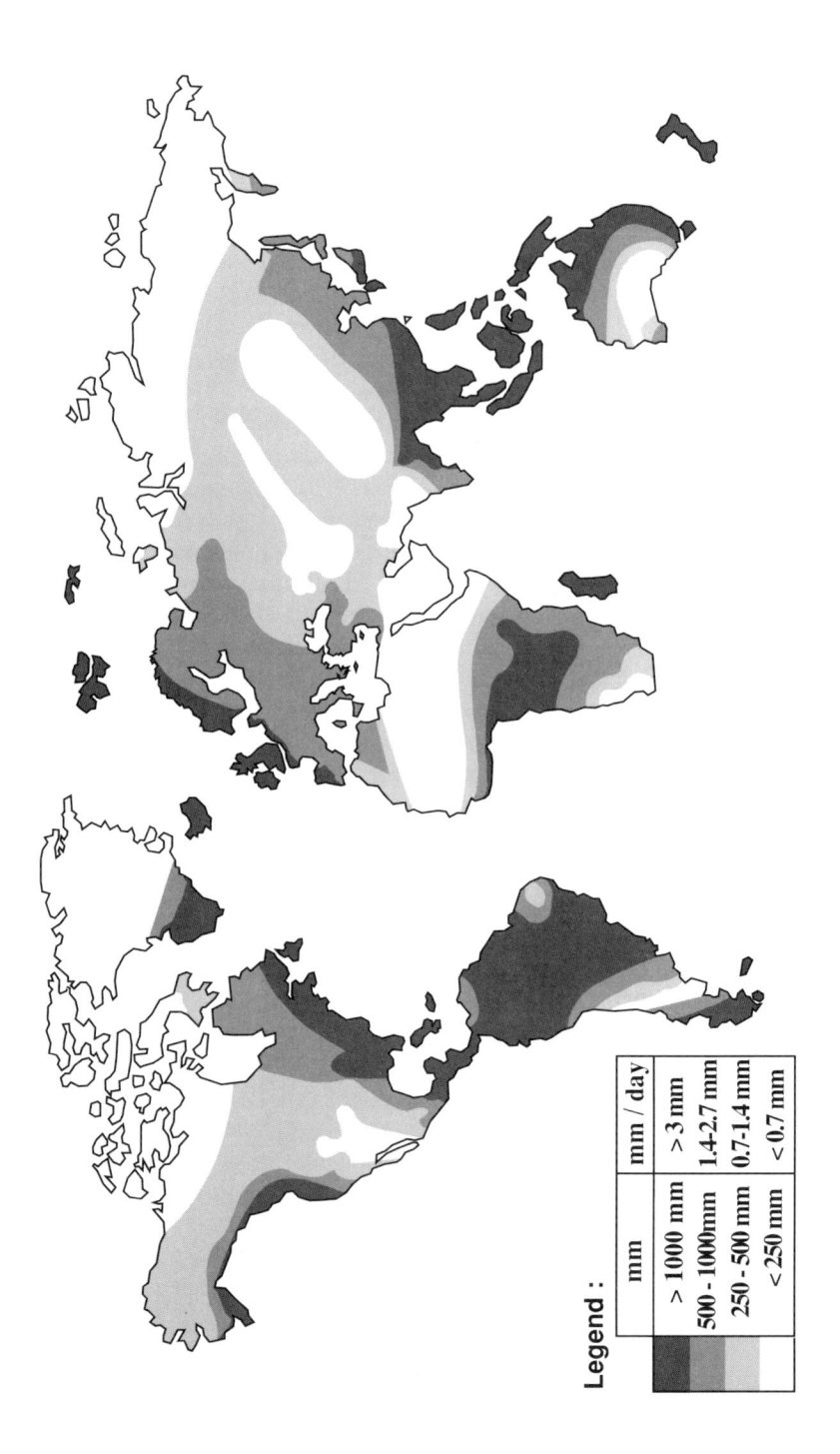

Figure 2.2. Approximate global precipitation

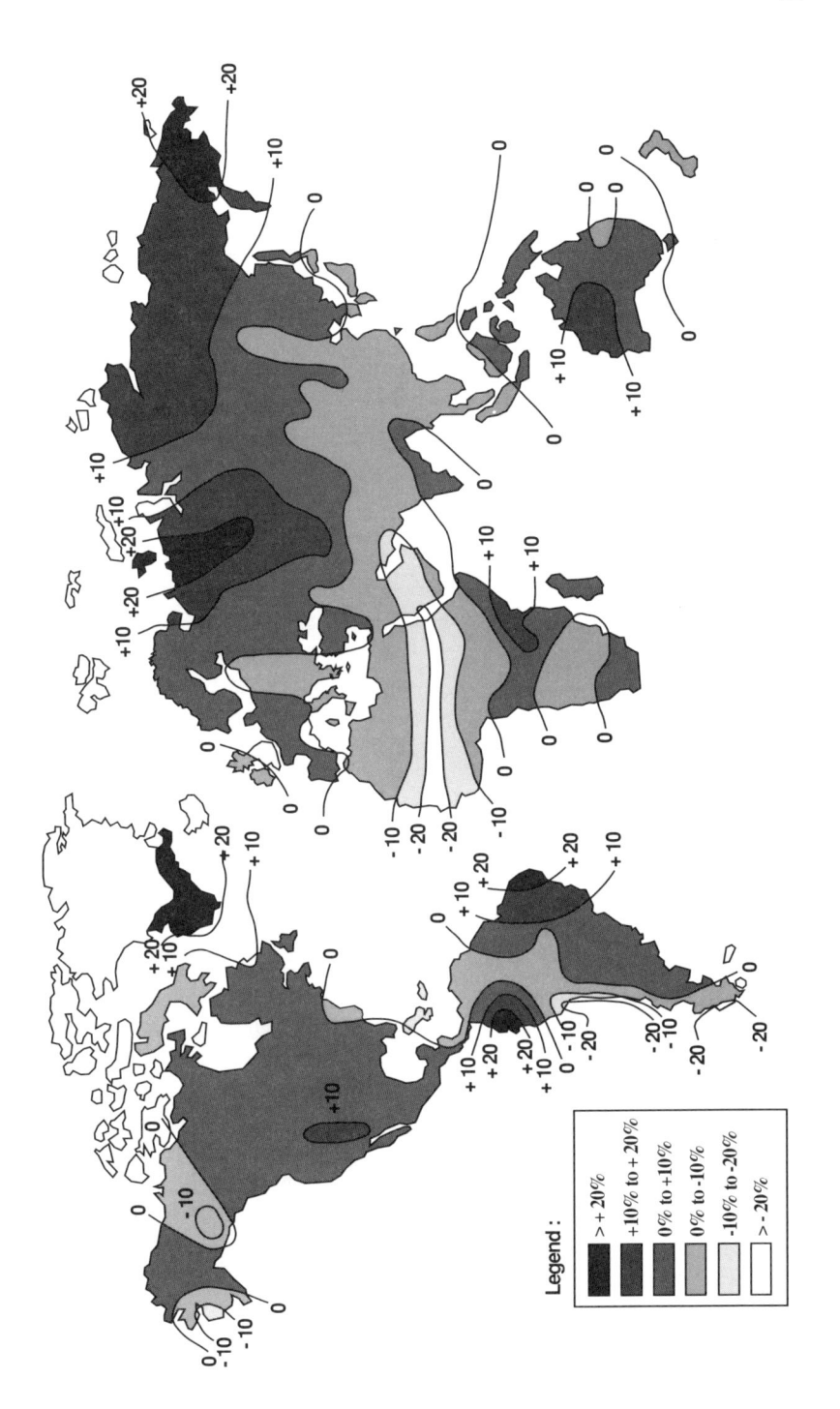

Figure 2.3. Percentage annual precipitation change from 1931-1960 to 1961-1990 (after Hulme et al., 1992)

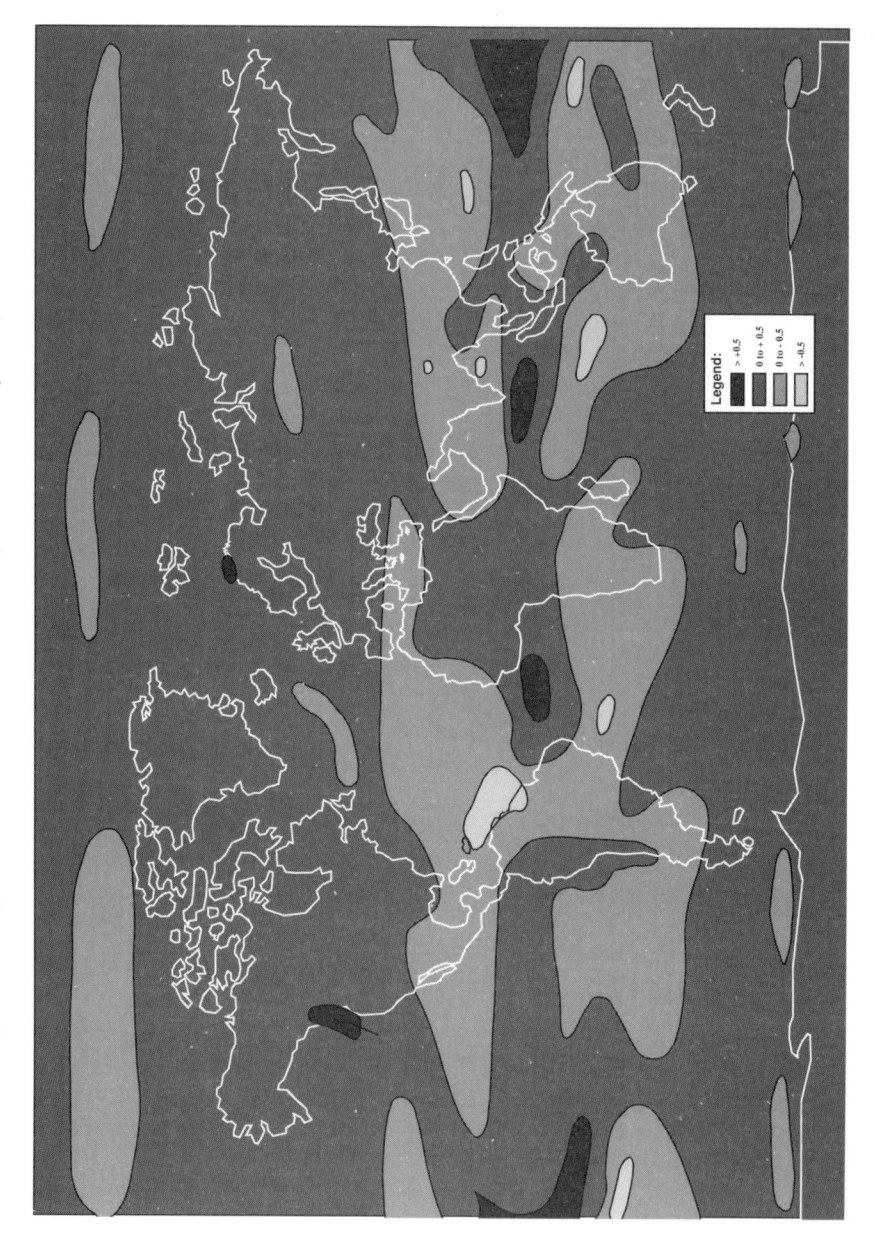

Figure 2.4. Mean annual precipitation changes (mm/d) (1995-2045) (after Hadley Centre, February 1995)

FACTORS AFFECTING AFRICAN HYDROLOGY

The Hurst effect

Hurst (1965) showed that annual Nile floods did not follow the classical theory of a stationary stochastic process. The estimated storage to maximize yield of the Nile was much higher than classical theory indicated, e.g., for a 500-year period Hurst calculated an accumulated departure of flow from the River Nile of 48 standard deviations from the mean compared with the theoretical 28. This problem was a dominant feature of hydrological research in the 1970s. Klemês (1974), in a comprehensive review of research into the Hurst phenomenon, warned of the dangers of deluding ourselves that statistical models such as fractional Brownian noise with its implicit acceptance of infinite memory could be used to explain the phenomenon. A highly successful operational model may turn out to be totally unacceptable from a physical point of view – the Ptolemaic planetary model for example. Klemês also demonstrated that the Hurst phenomenon could be produced using a zero memory process and non-stationary means. Klemês even suggested that operational models of the future may revert to extrapolations based on sound short-term knowledge rather than long-term synthetic data based on 'ignorance and guesses'.

Since the 1970s, interest has waned in trying to explain mathematically why so many long-term natural series fit distributions which are not Gaussian. As the modelling of physical processes has developed, the mechanisms which can lead to persistence of high and low rainfall have become better understood.

Physical feedback processes

The climate of Africa is rather unique. In comparison with the other continents, Africa south of the Atlas Mountains has no major continuous mountain barriers, either longitudinally or latitudinally orientated, to disrupt the circulation. Consequently, its climates and its seasonal variations are less complex and governed to a large extent by the oscillating inter-tropical front which follows the sun on its annual passage between the two hemispheres.

Nicholson (cited in Tyson, 1987) has clearly demonstrated, in a reconstruction of sequences of historical droughts and wet periods of Africa, that the African climate bridges the two hemispheres. The extreme drought years of 1972 and 1984 which devastated the Sahel region from the Atlantic coast to Ethiopia are clearly reflected in the flows of the Rivers Nile, Chari, Zaire, Niger, Zambezi and Orange in these years. Although a high degree of homogeneity does exist between climatic fluctuations in the north and south of the continent, the influences of such large-scale systems as the Southern Oscillation and the Atlantic Oscillation, both in part driven by sea temperature variations, and the monsoons of Asia, do produce spatial anomalies in climate extremes in different areas of Africa.

There is a tendency for an equatorial band across African to react in an opposite direction in terms of precipitation to the rest of Africa. This has been demonstrated clearly by the recent high levels recorded in Lake Victoria, Lake Tanganyika and Lake Malawi. Perhaps the clearest example of this anomaly is that the higher White Nile flows, originating in equatorial Africa, have persisted from 1961, whilst in contrast the Blue Nile has suffered the most serious reductions in discharge recorded this century, or over the past 1 300 years if the Roda Nilometer records are accepted as a reliable source (Howell and Allan, 1994). It seems that for extreme wet and dry years the whole continent can be affected with only minor deviations. Nicholson (1986) isolated four different preferential climatic patterns over the African continent. These modes are:
- northern and southern subtropical areas dry; equatorial regions wet;
- northern and southern subtropical areas wet; equatorial regions dry;
- whole continent dry;
- whole continent wet.

Dry continent and dry subtropical area modes have dominated the continent since 1970.

From a water resource planning viewpoint the major concern is whether or not the drought, which started in the late 1960s in the Sahel, and became dominant in southern Africa throughout the 1980s, will continue. One viewpoint, which has been strongly supported in the past in southern Africa, is the existence of cyclic behaviour (Tyson, 1987). Such fluctuations, and particularly the persistence in climate whereby one dry year seems to be followed by another and one wet year by another wet year, can be explained by biogeophysical short-term feedback processes, the so called 'Joseph' and 'Noah' effects: reduced vegetation leads to increased albedos and increased radiation losses, surface cooling and greater atmospheric stability which reduces rainfall and encourages persistence (Charney, 1975). Similar persistences are produced by lowering soil moisture levels in GCM simulations, indicating the importance of rainfall itself in initiating a significant feedback process, and by the increase in atmospheric particles from sandstorms. Such feedback processes have always been around and explain the Hurst effect and almost certainly explain much of the natural persistence of sequences of wet years and dry years found in the Sahel region, although fluctuations in sea surface temperature also play a role.

Water resource engineers in the past have relied on statistical data, whether stochastically generated or historic, for their estimation of reliability of yield. Long-term trends are generally ignored unless physical explanations can be presented to justify their inclusion. In this respect the Sahelian drought has been with us for almost 30 years and suggests that it has been maintained by causal factors other than normal feedback process. Global warming is an obvious explanation.

Global causes

The search for a causal explanation of the African drought received an impetus by research undertaken in the UK Meteorological Office (Folland *et al*., 1986), which demonstrated a strong correlation between sea surface temperature (sst) anomalies on a global scale with wet and dry periods in the Sahel. After the mid-1960s, a marked cooling of the oceans of the northern hemisphere and simultaneous warming in the southern hemisphere was observed. A reversal occurred around 1970, since when temperatures in both hemispheres have increased. A time series plot of sst differences between oceans of the southern hemisphere and those of the northern hemisphere, and rainfall anomalies for the Sahel, shows a strong negative correlation. The correlation between the July-September sst and Sahel rainfall for the period 1901 to 1984 was -0.62, which is significant at the 99.9% probability level (see Figure 2.5). Numerical equilibrium GCM experiments with prescribed sea temperatures were also undertaken by the UK Meteorological Office which were able to replicate rainfall reductions in the Sahel for recent drought years: 30% reductions over western Sahel, 20% over eastern Sahel and up to 50% over the mountains of southern Sudan and northeast Ethiopia.

The observed and unexpected warming of the southern oceans at a faster rate was thought to be due to a reduction in the heat transfer from southern to northern hemispheres, although the detailed mechanisms of the transfer are still the subject of much research. Alternative scenarios include increased deep water circulation from 1960 to 1970 in the Atlantic and the effect of sulphate aerosols which are dominant the northern hemisphere. If, however, the reduction in heat transfer is related to the north-south conveyor system combined with a slowdown in formation of north Atlantic deep water at high latitudes, owing to a reduction in the extent of sea ice (Street-Perrot and Perrot, 1990), then the Sahelian drought may persist until the greater land mass in the northern hemisphere starts to dominate the effects of a slowdown in ocean transfer and the attenuation effect of sulphate aerosols. However, confirmation will depend on further research developments into detailed coupled transient GCM models, which can be calibrated against recent climates and sea temperatures. It is likely that the results from the equilibrium GCMs will be found wanting once more reliable coupled models have been developed.

A plausible scenario of future global warming links a weakening in carbon sinks and radiation sinks in the polar regions with reduced deep water formation due to reduced heat transfers from the southern hemisphere to the north. It is postulated that reductions to the radiational and CO_2 sinks could give rise to significant positive feedbacks leading to an increase in global warming (Lewis, 1989).

32 GLOBAL CLIMATE CHANGE AND AGRICULTURAL PRODUCTION

Figure 2.5. Sahel rainfall departures and sea surface temperature differences between hemispheres (after Folland *et al.*, 1986). (a) rainfall departures Sahel; (b) sea surface temperature (sst) differences (southern hemisphere, including Indian Ocean, minus northern hemisphere)

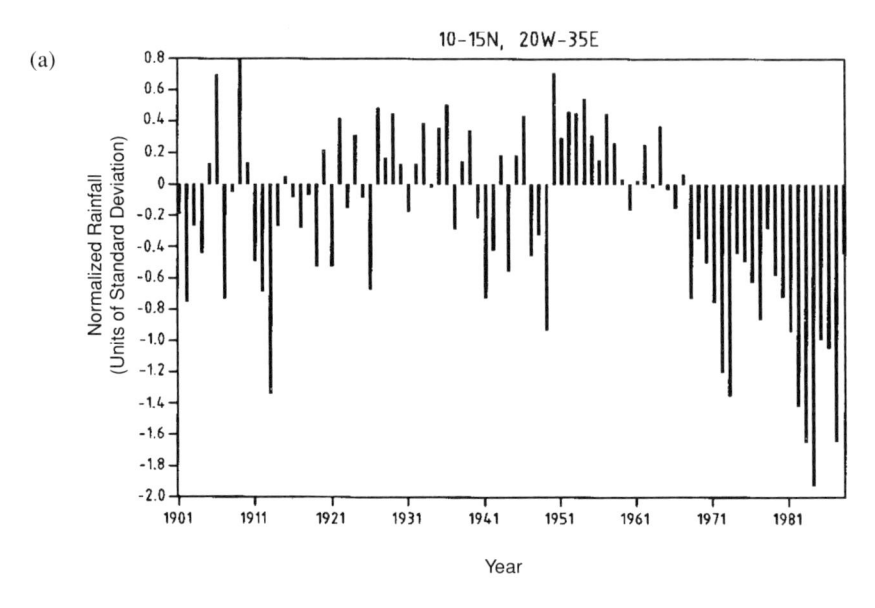

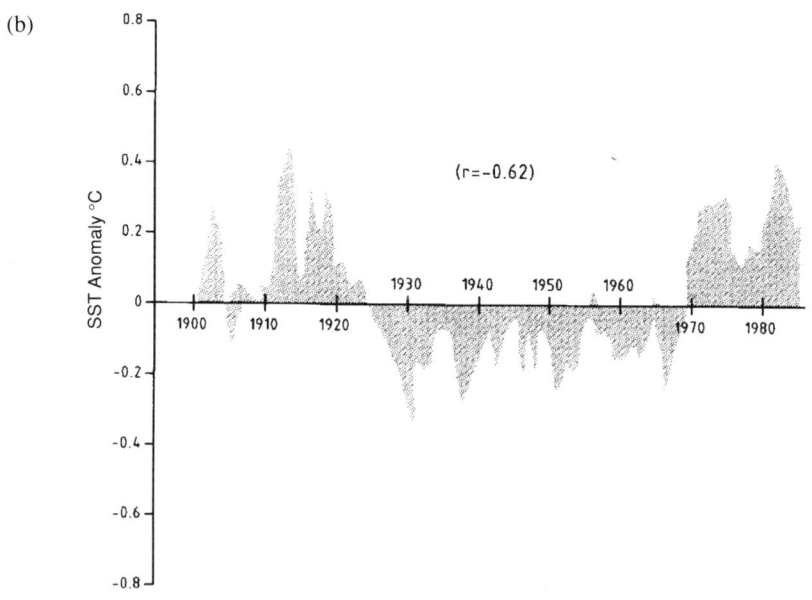

RECENT CHANGES IN GLOBAL RUNOFF

LAKE CHAD

Lake Chad is situated on the border of the Sahara between latitudes 12°N and 14°30'N and longitudes 13°E and 15°30'E. The lake has no outlet to the sea and is the main focus of an internal drainage basin of 2 500 000 km² that collects water from Algeria, Chad, Niger, Nigeria, Cameroon, Sudan and the Central African Republic. Over 90% of the total lake inflow comes from the Chari-Logone river system which rises on the southern margin of the basin and has a combined catchment of about 570 000 km². Lake Chad is located centrally within the Sahel, which is often defined as the 250 to 500 mm rainfall band across Africa below the Sahara desert. Lake Chad is shallow; average depths vary between 1.5 and 5 m, with the result that its surface area is very sensitive to changes in inflows. The lake has a number of unusual features. In spite of being an enclosed basin, it remains a freshwater lake. Its size ranges from 100 km² for the 'Gran Chad' to 10 to 30 km² for the 'Petit Chad'. The water tends to be replaced every year or every other year, with an average annual inflow of about 30 to 40 km³. The saline water in normal times is pushed to the northwest of the lake where its salinity is concentrated and natural seepage to the north maintains the salt balance in equilibrium.

Feasibility studies began in 1972 to investigate the irrigation of lands bordering Lake Chad in Nigeria. As a result, the South Chad Irrigation Project (SCIP) was designed for three-stage implementation. Two stages have been built: Stage I in 1979 and Stage II in 1983. A pump station with a total capacity of 103 m³/s (current capacity 75 m³/s) was constructed to pump water from the lake via a canal. The full scheme has a gross area of 66 000 ha and would abstract approximately 3% of the annual inflow to Lake Chad. The layout of the scheme was based on 4 ha units to provide 16 000 households, or a 100 000 population, with a more secure livelihood. It was planned to produce two crops per year based on rice (50%), cotton (20%) and wheat (60%), with a cropping intensity of 130%.

In the mid-1960s Lake Chad had reached record levels for this century giving rise to speculation that the 'Gran Chad' as observed by explorers of the last century would be re-established. However, this failed to happen and lake levels fell during the 1970s and at the time that the feasibility studies were completed were similar to those experienced during the first half of the century. No one anticipated the drastic fall in rainfall which has occurred since (Figure 2.6).

The sudden onset of the Sahelian drought in the 1970s resulted in a severe contraction in the size of the lake, changing its character to a swampy delta at the mouth of the River Chari, with dire consequences for lake-side communities and the SCIP irrigation scheme.

Lake Chad divided into northern and southern pools in 1973 and by 1976 levels had fallen below the bed of the intake canal. Although water levels recovered during October and November, providing irrigation water for wheat crops planted in November, shortage of water in June to September prevented the cultivation of the

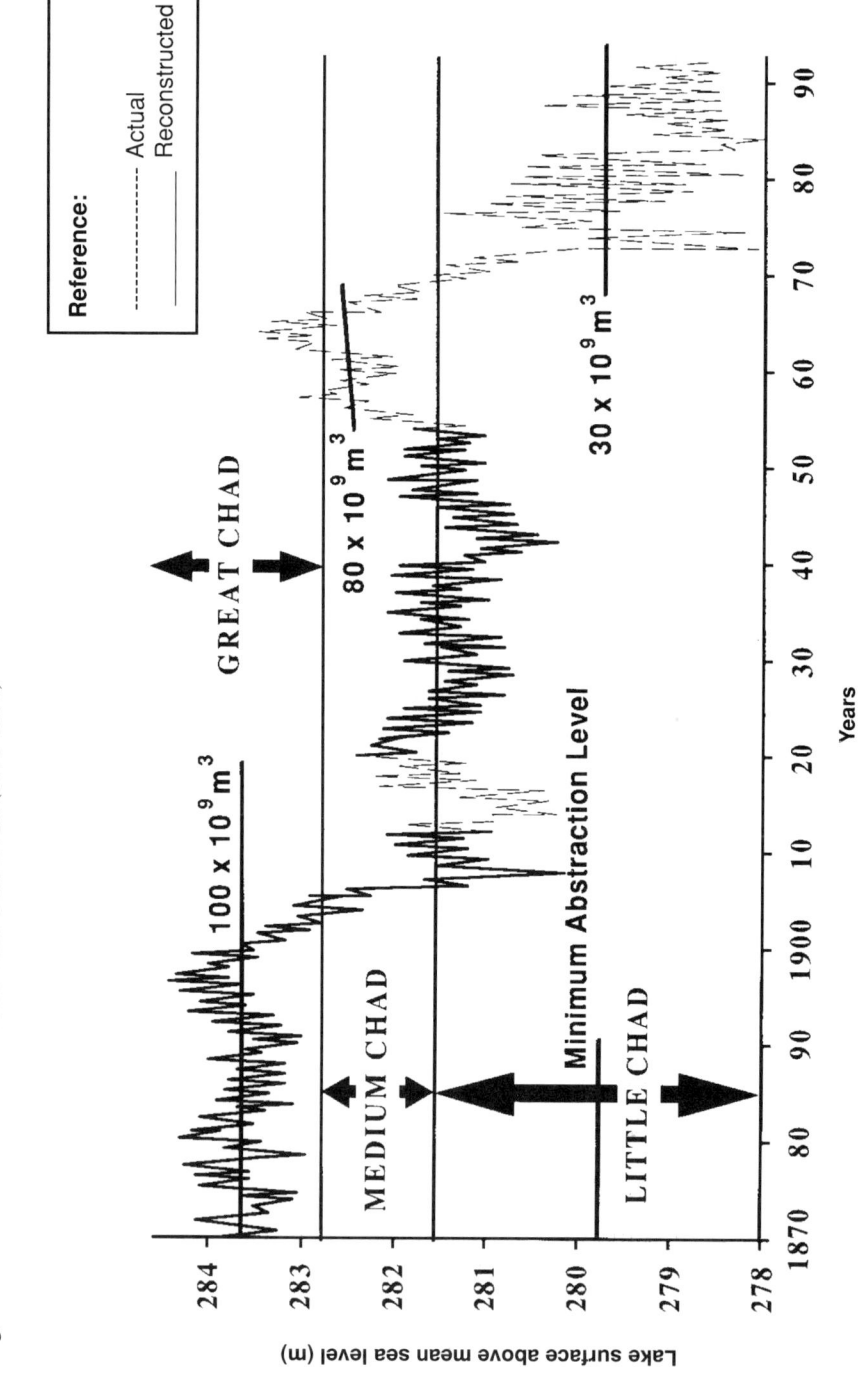

Figure 2.6. Measured and reconstructed Lake Chad levels (1870-1992)

EFFECTS ON AVAILABILITY OF WATER RESOURCES 35

Table 2.5. River Chari annual runoff (km^3) at N'Djamena (1971-1992) (mean flow before 1971 – 40 km^3/annum)

Year	Flow	Year	Flow	Year	Flow
1971	31	1978	29	1986	15
1972	18	1979	21	1987	10
1973	18	1980	25	1988	27
1974	29	1981	19	1989	16
1975	35	1982	21	1990	12
1976	29	1983	17	1991	19
1977	25	1984	7.5	1992	21
		1985	17	Mean annual flow	22

Figure 2.7. Departures from mean annual flow, River Chari at N'Djamena

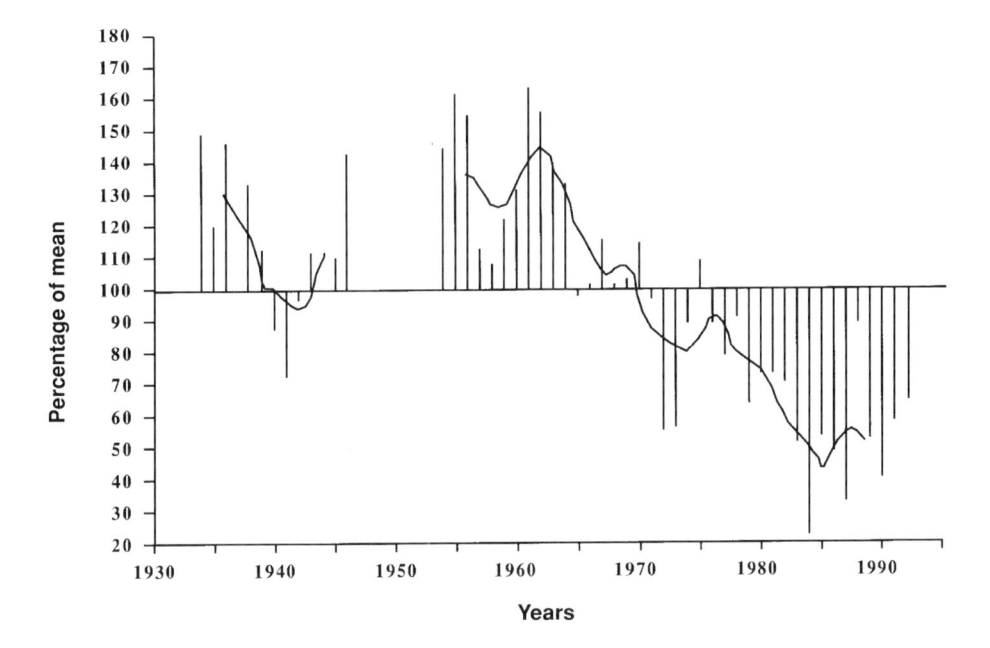

main rice crop. The two pools have remained separated and since 1983 levels have continued to fall and have prevented even winter-season irrigation. The scheme is maintained in good condition awaiting the recovery of the lake.

The reason for the fall in lake levels is simply that flows in the River Chari-Lagone basin failed. The total runoff in 1984 was only around 20% of the long-term mean. Since 1971 flows have been reduced to 50% of the previous long-term mean annual flows of over 40 km^3, although rainfall is estimated to have fallen by only 25%. Of all the major basins in the world, probably Lake Chad has been affected most by climate

36   GLOBAL CLIMATE CHANGE AND AGRICULTURAL PRODUCTION

change. The most ambitious irrigation scheme built in Africa since the mid-1940s has yet to be operated in earnest (Table 2.5 and Figure 2.7)

Falling Lake Chad levels demonstrate the effect of a rainfall reduction of 25% on river flow in a semi-arid region where rainfall gradients are steep and small shifts in the atmospheric circulation can cause large rainfall changes. As a result, proposals have been put forward for a major inter-basin transfer from the River Zaire to the River Chari basin. With an annual runoff of 1 250 km^3, potential supplies are available.

RIVER NILE

To a large extent Egypt has remained unscathed by the drought owing to the large over-year storage in Lake Nasser. However, other factors have contributed to reducing the drought's impact. One of these is the exceptionally high levels in Lake Victoria which have helped maintain higher White Nile flows. The higher levels have resulted from very heavy rainfall in Kenya and Uganda between 1961 and 1963 and above average rainfall since. The higher White Nile flows have helped to compensate for the lower discharges in the Blue Nile, which have been more adversely affected by the Sahelian drought.

The construction of the High Aswan Dam in 1963 provided the complete regula-

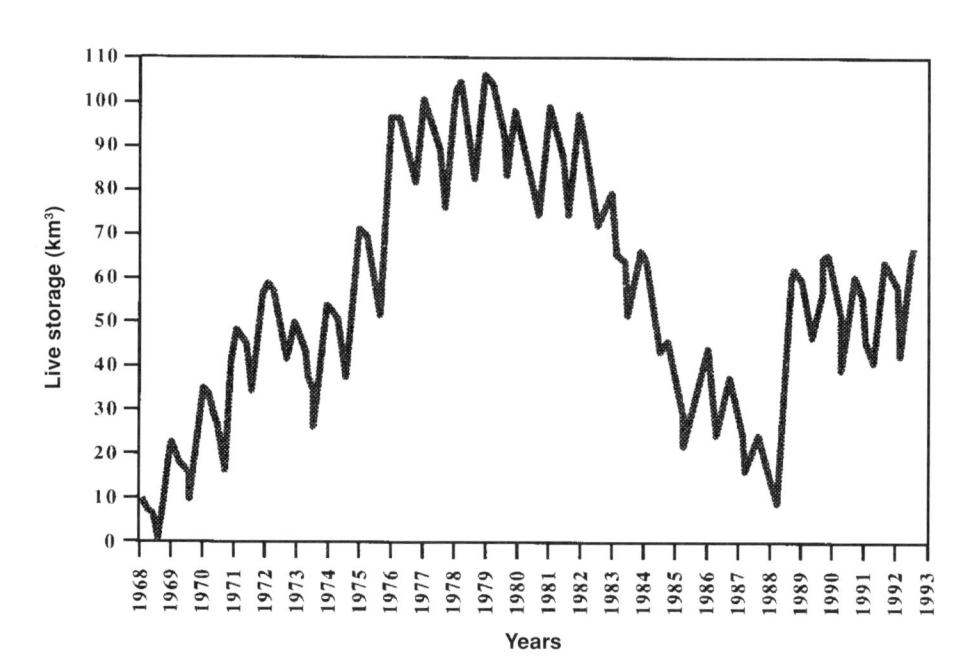

Figure 2.8. Lake Nasser live storage (1968-1992)

tion of the annual Nile flood and also gave sufficient over-year storage to supply Egypt's share under the Nile Waters Agreement (1959) of 55.5 km^3, with a very high level of reliability (96%).

Following the failure of Nile flows in 1984 (a naturalized inflow to Lake Nasser of 59 km^3 was recorded, compared with a mean flow of 84 km^3), there was considerable concern that Lake Nasser's resources would fall to a level where irrigation supplies would have to be reduced with both agricultural and political implications (Figure 2.8).

Effect of Sahel drought on Nile flows

A main feature of the Nile flow series, apart from the high flow period at the end of the last century, is the steep fall in discharge from the mid-1960s. It is more pronounced and persistent than any previous low flow periods. An analysis of Roda Nilometer data also indicates that during the last two centuries the variability of the annual flood far exceeds that observed since records began in AD 622 (Howell and Allan, 1994).

Most previous yield studies of the River Nile have analysed River Nile flows at Aswan. However, the flow records over the past two decades have demonstrated the importance of the different rainfall regimes over the catchments of the Blue and White Nile. Their influence on White and Blue Nile flows can clearly be seen by comparing Figures 2.9 and 2.10. Flows in the White Nile between 1962 and 1985 have increased by 32% or 8 km^3 above the 1912 to 1961 mean. This has occurred at a time when Blue Nile flows have decreased by 9 km^3 over the period 1965 to 1986, or 16% below their 1912 to 1964 mean. It could be inferred that the two rainfall regimes are negatively correlated. However, past records show that the relationship between annual inflow to Lake Victoria and recorded Blue Nile flows at Khartoum is random (MacDonald, 1988). This, however, contradicts results of GCM outputs which suggest that rainfall over the White and Blue Nile catchments may be negatively correlated (Howell and Allan, 1994).

Lake Nasser simulations

River Nile flow sequences were used to investigate the ability of the Nile systems to withstand drought (MacDonald, 1988) should the Sahelian drought prove to be a persistent feature of climate of the region in the coming decades. Two sequences were prepared for severe and moderate droughts (based on flow statistics from 1968 to 1988) and the results are shown in Table 2.6. They indicate that Lake Nasser releases would have to be reduced below 55.5 km^3 (Egypt's allocation under the 1959 Nile Waters Agreement) to 52 km^3 and 49 km^3, respectively, and that these yields would have reliabilities of 87% and 88%, i.e., water shortages would occur on average in 13 and 12 years respectively every 100 years. The effect of the Sahelian drought on River Nile flows and irrigation in Egypt, although problematical, would be nowhere near as severe as for Lake Chad. Even under the most severe drought condi-

Figure 2.9. Departures from mean annual White Nile flows at Mogren (1912-1989)

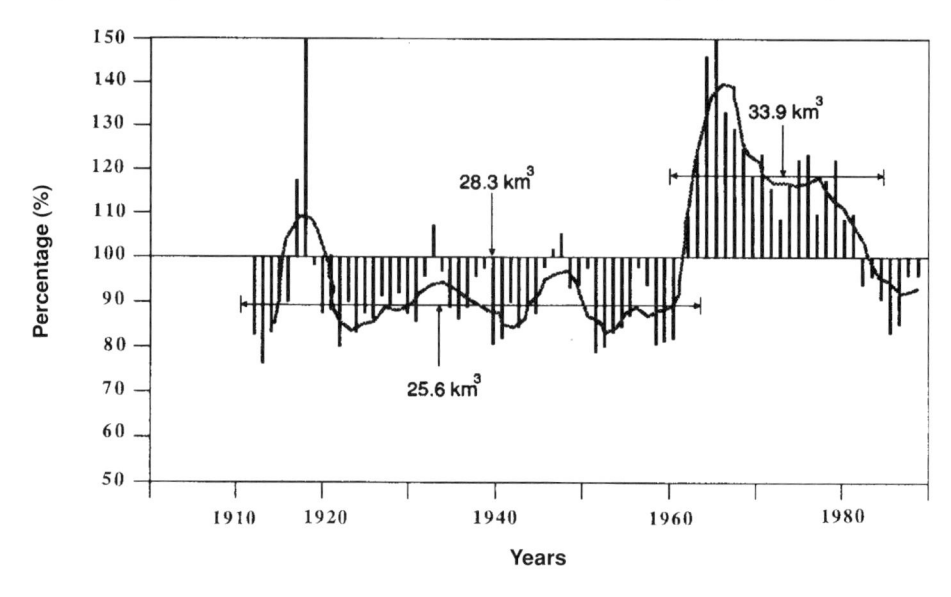

Figure 2.10. Departures from mean Blue Nile flows at Khartoum (1911-1989)

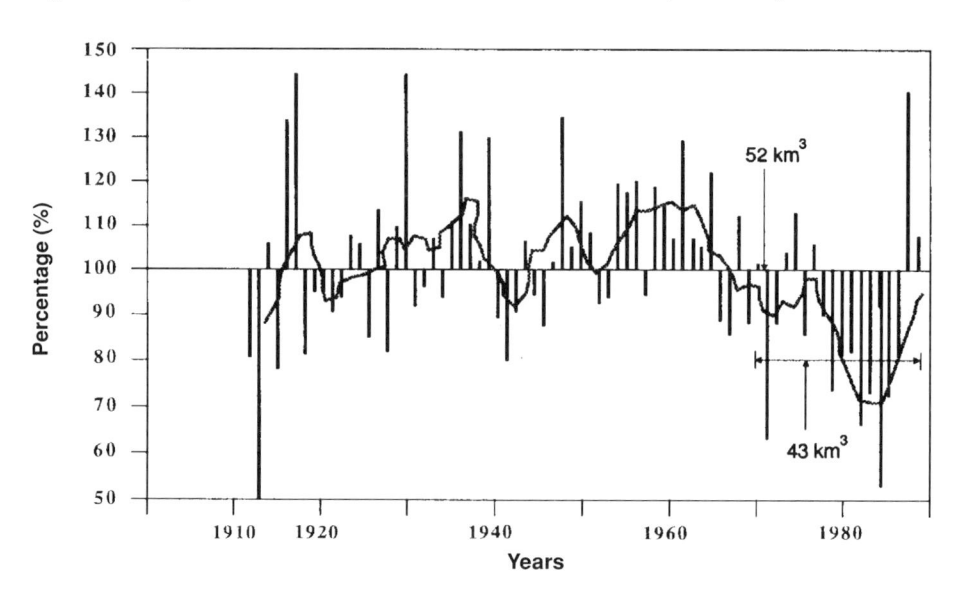

tions the yield from Lake Nasser is predicted to fall only from 55.5 km³ design yield to 48.4 km³, or less than 15%, although the reliability of supply would fall from a design yield of 96 to 88% (Table 2.6)

Table 2.6. Summary of Lake Nasser reservoir simulations

Sequence type	Total number of years	Release from Lake Nasser (km^3)	Number of years with reductions	Reliability yield (%)	Minimum annual release (km^3) 1st lowest	Minimum annual release (km^3) 2nd lowest	Minimum annual release (km^3) 3rd lowest	Average yield obtained (km^3)
Moderate drought sequence	400	52	52	87	39.6	40.1	41.3	51.2
Severe drought sequence	100	49	12	88	38.9	39.8	40.4	48.4

Options do remain, however, for increasing supplies by reducing drainage outflow and operational losses to the sea, but these will largely be needed to meet increased municipal and industrial demands. Conway *et al.* (1996), in a recent study on future water availability in the Nile basin, utilized GCM outputs and modelled runoff incorporating land cover changes. The precipitation changes ranged from +18 mm to +39 mm and Nile flows were calculated to range from a 3% decrease to a 10% increase. These changes are much smaller that those recorded over the last two decades.

The Nile basin is a closed system and in irrigation terms is very efficient (>70%). Improvements in efficiently are therefore limited. Because it is a closed system increased groundwater abstractions do not represent extra water in the long term and provide only short-term temporary benefits. There is potential to save water by selecting less water demanding crops and the use of short-term varieties (5% savings). However, significant enhanced supplies could be obtained from the Sudd and other wetlands in Sudan but these all have major political or ecological constraints. Ignoring the political and ecological implications there is still scope for further water development within the Nile basin. However, when these are set against the population explosion in the region and possible problems of climatic change, the outlook cannot be viewed with optimism.

CASPIAN SEA LEVELS

Background

Another region which has experienced a major change in water resources is in Russia. The River Volga, which rises near the Baltic Sea and discharges into the Caspian Sea, is the largest river in Europe, 3 530 km long. Its catchment area of 1.5 million km^2 makes it one of the major rivers in the world. Its river basin is larger than that of the Zambezi, the Indus or the Ganges. Around 60% of its discharge is derived from melting snow. Three main rivers, the Volga, Ural and Terek, account for 88% of the total river inflow to the Caspian Sea and, of this, over 50% arises from the Volga river. It is connected by canals to Moscow, Volgograd, St. Petersburg and the Arctic Ocean for river traffic and, with hydropower stations and extensive irrigation systems around the north Caspian Sea coast and the Azov Sea, the Caspian basin is crucial to the economy of Russia, Azerbaijan, Kazakhstan and the Ukraine.

The Caspian Sea has been called the greatest salt lake in the world, although its salt content is only one-third of that of the main oceans. It has no natural outlet. It covers an area of 372 000 km^2 with its surface at a level of −28.5 m below mean sea level, whilst the total basin area extends to some 3.7 million km^2. An extensive sedimentary plain surrounds the outlet of the River Volga where the sea is shallow with depths of only 4-6 m. The total capacity of the sea is 66 960 km^3, compared with an annual inflow of around 310 km^3. The annual volume of inflow would raise levels by 0.8 m if evaporation is ignored. Net evaporation (evaporation − rainfall) is therefore close to this figure. Being an enclosed sea, levels fluctuate with inflow and historically have shown a wide variation. Also, developments within the Caspian Sea basin have seriously reduced River Volga flows. Levels fell dramatically from 1930 reaching a low in 1977, although surprisingly since this date levels have started to rise and plans are in hand to mitigate flood damage. However, without predictions of future sea levels and a knowledge of the processes causing the fluctuations, attempts to mitigate the damages resulting from too high or too low levels are likely to be both ineffective and uneconomic.

Water development

The River Volga plays a vital role in the economic life of the states through which it passes. It is used for:
- navigation (151 000 km of waterways connecting the Arctic Ocean to the Caspian Sea, with the Moscow canal, the Volga-Don canal and the Volga-Baltic waterway all with a 4.5 m navigation depth);
- power generation (a cascade of 11 hydroelectric schemes with a capacity of 11 000 MW;
- irrigation (in the Caspian-Azov basins some 7.5 million ha of irrigation have been developed; 4 million ha using sprinkler systems).

Relative water consumption over the years has been compiled and forecasts made for the future based on the construction of the Volga-Don and Volga-Chograi irrigation canals (Table 2.7) (Berezner, 1987).

For withdrawals of 37 km^3 and a residual flow of 266 km^3/year the Caspian Sea was in equilibrium at a level around −28.5 m amsl. This is considered to be about the optimum level for the valuable fisheries. Over the last millennium the highest level was thought to be −22 m recorded in the 1600s, and the lowest levels, −31 m in the 1400s, and −29.0 in 1977, a range of some 9 m (Figure 2.11).

Because of the extensive and dramatic water developments undertaken by the former USSR in the Volga basin since the 1930s, the first step in any evaluation is to naturalize the flow to remove the influences of human activities (Vali-Khodjeni, 1991). But for the massive water developments, the present-day levels of 27.7 m amsl would now have returned to the 1929 high of −26.2 m amsl (Figure 2.12). The larger ab-

Table 2.7. Abstractions from Volga/Caspian Sea, post-1930 (km^3/yr) (after Berezner, 1987)

Water consumption	1980	1990	2000	2010
Water supply	5.6	7.3	9	10.5
Irrigation	18.6	23	33	36
Reservoir evaporation	6	7	8	9
Filling of reservoirs	3	2	0	0
Inflow to Kara-Bogaz-Gol	0	2	2	2
Water consumption in Iran	3.5	4	5	6
TOTAL	**37**	**44**	**57**	**63**

Figure 2.11. Variations in levels in the Caspian Sea (1839-1987) (after Vali-Khodjeni, 1991)

Figure 2.12. Recorded and naturalized levels in the Caspian Sea (1930-1990) (after Vali-Khodjeni, 1991)

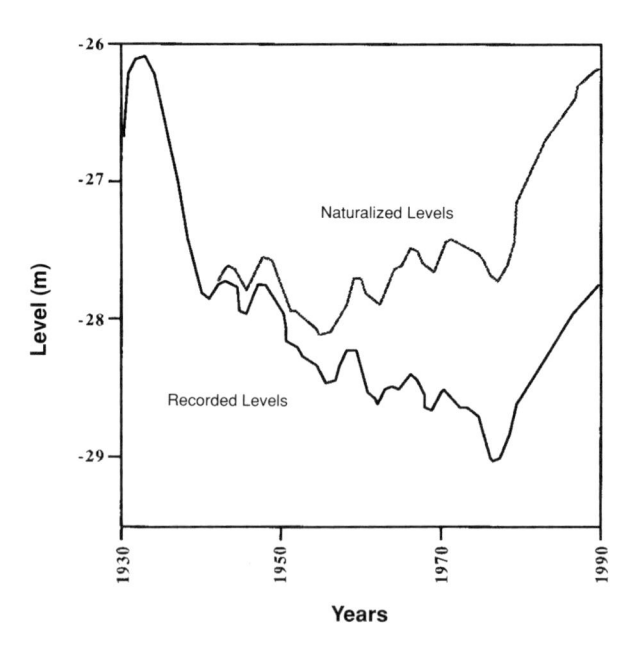

stractions from the Volga river are therefore contributing significantly to reducing sea levels, albeit fortuitously, and so preventing more serious damage to irrigation areas and urban development adjacent to the Caspian Sea (most of which have been developed in recent years). There has been therefore a dramatic reversal from the problems of falling sea levels to one of rising levels and how to contain increasing flood waters.

Between 1931-1960 and 1961-1990 rainfall on the Volga catchment has increased by over 10% whilst naturalized inflow to the Caspian Sea has increased by a similar amount.

The Intergovernmental Panel on Climate Change (IPCC, 1990) suggests increases in precipitation of between 100 and 200 mm over the winter period for a doubling in greenhouse gases. If this prediction proves correct then it would produce a very significant increase in surface runoff in the River Volga which is predominantly snow- melt fed. The additional precipitation will fall mainly as snow on frozen ground. There will therefore be little increase in losses and most of the additional snow will be converted to surface runoff and will increase River Volga discharges substantially. River Volga flows could increase by between 25 and 50%.

If the present change in climate with increased precipitation over Russia continues over the coming decades, serious problems in terms of drainage and flooding of low-lying irrigation lands are likely to be experienced. However, in the longer term the additional water would be extremely beneficial for agricultural development in the Caspian Sea sedimentary basin where 100 000 km^2 of land lies below sea level. Alternatively, surplus water could be used to replenish all of the damage inflicted on the Aral Sea: an annual flow of around 50 km^2 would be required to restore the Aral Sea to its pre-1960 condition. A surplus of 100 km^3/year of freshwater would certainly find customers in the Near East, although the task of transferring it would be formidable.

CONCLUSIONS

1. The effect of climate change on available water resources in most regions, while significant, will be small compared with demands generated by population growth, industrialization, urbanization, land-use changes and improved standards of living. In many countries resources are already fully committed and water will become a scarce commodity.
2. Although agriculture is the most prolific user of the world's water resources, it is more vulnerable to water shortage because of the higher priority given to potable water supply and other high value users.
3. GCMs represent the only plausible method for predicting the effect of global warming on the hydrological cycle. Unfortunately, existing GCMs are not capable of producing realistic precipitation outputs which are required in water resource planning. This is due in part not only to the way clouds and precipi-

tation are modelled, and to coarse resolution, but also to the inadequate representation of the land phase and ocean systems in existing equilibrium GCMs and in the new transient GCMs. For example, lateral water transfer and artificial influences are not taken into account. Because of the interactive relationships between the atmosphere, land and oceans, such simplifications are likely to affect outputs from current GCMs. As a result, even the sign of the change in runoff over regions and some continents cannot be predicted with any confidence.

4. Until transient GCMs can simulate sea surface temperature patterns and regional precipitation changes, observed this century (and the changes are significant), future precipitation changes cannot be predicted with confidence.

5. Predictions from the large number of impact studies undertaken to date to evaluate the effect of climate change on the hydrological cycle can largely be discounted. Their value at present lies in improving impact study techniques rather than assisting in water development planning. However, much of the research has been directed at trying to convert inadequate outputs from GCMs to more realistic values for use in the impact studies rather than trying to target specific problem areas.

6. Deterministic models, which adequately represent the hydrological cycle and are calibrated, should be used in impact studies. Black box, simple water balance and stochastic models have a limited role to play where catchment characteristics are changing rapidly with time.

7. All catchment models used in impact studies should be adequately calibrated. Research into the development of remote sensing algorithms for satellite data and its interfacing with GIS systems, together with better methods of applying small-scale physically based techniques at larger scales, will eventually reduce the present dependence on calibration. The calibration requirement for river basin studies applies equally to the land phase of GCMs.

8. Under conditions of rising temperature, vegetation cover of basins will change significantly and hydrological models used in impact studies should allow for this.

9. Most impact studies have found that surface runoff is influenced more by changes in precipitation than by changes in evapotranspiration. This has arisen in part due to ignoring parameters other than temperature, such as windspeed, net radiation and humidity, in calculating evapotranspiration; realistic values of these are difficult to obtain from GCM outputs. Recent studies indicate that increased evapotranspiration may be more significant than previously thought, although reduced stomatal conductance and reduced leaf area will partially limit its effect.

10. Because the relationships between rainfall and runoff are non-linear, the response of runoff generated by rainfall is magnified. In addition, the rainfall over a large basin is integrated into a single runoff value. As such, flow data represent valuable means of monitoring climate change.

11. Predicted GCM global precipitation increases for $2 \times CO_2$ scenarios range from 8 to 15%, whilst the water holding capacity of the atmosphere for a global 4°C rise will increase by 30%. Under transient conditions the land will warm more rapidly than the oceans, guaranteeing that ocean evaporation (global precipitation) will not keep pace with potential evapotranspiration over the continents. This situation is likely to persist for a century or more. In broad terms then, runoff would appear likely to decrease due to greater aridity and increased frequency of droughts over the next century, providing precipitation forecasts are correct.
12. Recent research supports the claim that changes in climate recorded over the present century are due to increases in greenhouse gases in the atmosphere. If this is the case, then valuable information should be obtained from recorded changes in flow regime in major river basins. It is essential that transient GCMs are capable of simulating these changes.
13. Case studies for the River Nile, Lake Chad and Caspian Sea demonstrate how hydrology is highly susceptible to climate change.
14. Data collection of hydrometric data is essential for operating water resource systems. It is also essential for planning water development. It is unfortunate that in many developing countries data collection has almost been abandoned. Water resource assessment agencies are invariably small units within large sectorial ministries and perceived to be of little value. It is essential that this situation is remedied as a matter of urgency.
15. Hydrology should be treated on a regional, continental and global scale, in the same way, not nationally in isolation. As with climatic parameters, river flows are related on a regional, continental and global scale. All measures which will promote this change should be supported by governments and international agencies.
16. Increased storage in water-critical regions is desirable for agricultural development and sustainability. Worldwide climate change may exacerbate the differences between water-rich and water-poor areas. Water transfer will therefore assume increased urgency and importance.
17. Special water resources design criteria taking into account climate variability and change should be developed particularly for vulnerable areas of the world. Hydrological design for water development projects should continue to be based on the past hydrological records. However, the design should consider both extremes from the past and scenarios from the GCMs in order to achieve the necessary flexibility to accommodate possible future changes.

REFERENCES

Arnell, N.W. and Reynard, N.S. 1993. *Impacts of Climate Change on River Flow Regimes in the United Kingdom.* Institute of Hydrology, NERC, Water Directorate, DOE, July 1993. 130 p.

Arnell, N.W., Brown, R.P.C. and Reynard, N.S. 1990. *Impact of Climatic Variability and Change on River Flow Regimes in the UK.* Report No. 107. Institute of Hydrology, Dec. 1990. 154 p.

Beran, M.A. 1986. The water resources impact of future climatic change and variability. In: *Effects of Changes in Stratospheric Ozone and Global Climate.* Vol. 1. J.G. Titus (ed.). US EPA/UNEP, Washington DC. pp. 299-328.

Berezner, A.S. 1987. Probability prediction of the Caspian Sea level with consideration of the development of water consuming industries in its basin. *Water Resources* **14**(1): 34-40.

Charney, J.G. 1975. Dynamics of deserts and droughts in the Sahel. *Quart. J. R. Met. Soc.* **101**(428): 193-202.

Conway, D., Krol, M., Alcamo, J. and Hume, M. 1996. Water availability in the Nile Basin: the interaction of global, regional and basin scale driving forces. *Ambio* (in press).

Cusbach, W.P., Hasselmann, K., Hoeck, H. *et al.* 1992. Time-dependent greenhouse warming computations with a coupled ocean-atmosphere model. *Cllimate Dynamics,* **8**: 55-69.

Folland, C.K., Palmer, T.N. and Parker, E.D. 1986. Sahel rainfall and world-wide sea temperatures 1901-1985. *Nature* **320**: 251-254.

Gleick, P.H. 1993. *Water in Crises – A Guide to the World's Freshwater Resources.* Oxford University Press. 473 p.

Grotch, S.L. 1989. *A Statistical Intercomparison of Temperature and Precipitation Predicted by Four General Circulation Models.* DOE Workshop on Greenhouse-Gas Induced Climate Change. US Dept. of Energy, Univ. Mass, May 1989.

Hadley Centre. 1992. *Transient Climate Change Experiment.* Hadley Centre, UK Met. Office, Aug. 1992. 20 p.

Hadley Centre. 1995. *Modelling Climate Change, 1860–2050.* Hadley Centre, UK Met. Office, Feb. 1995. 12 p.

Hall, M.J. and Sarenije, H.H.G. 1993. *Climate and Land Use: A Feedback Mechanism?* Conference Water and Environment – Key to Africa's Development, IHE Delft, Report Series 29, June 1993.

Hansen, J., Lacis, A., Rind, D. *et al.* 1984. Climate sensitivity: analysis of feedback effects. In: Climate Processes and Climate Sensitivity, *Geophysical Monograph* **29**: 130-163.

Howell, P.P. and Allan, J.A. 1994. The Nile – sharing a scarce resource. In: *Global Climate Change and the Nile Basin.* N.Hume. Cambridge University Press, Cambridge. pp. 139-162.

Hulme, M. 1994. Regional climate change scenarios based on IPCC emissions projections with some illustrations for Africa. *Area* **26**(1): 33-44.

Hulme, M., Marsh, R. and Jones, P.P. 1992. Global changes in a humidity index between 1931–1960 and 1961–1990. *Clim. Res.* **2**: 1-22.

Hurst, H.E. (1965). *Long-term Storage*. Constable, London.

IPCC [Intergovernmental Panel on Climate Change]. 1990. *Climate Change – The IPCC Scientific Assessment*. Cambridge University Press, Cambridge.

Kite, G.W. 1993. Application of a land class hydrological model to climate change. *Water Resource Research* **29**(7): 2377-2384.

Kite, G.W., Dalton, A. and Dion, K. (1994). Simulation of streamflow in a macroscale watershed using GCM data. *Water Research Paper* **30** (5): 1547-1559.

Klemeš, V. 1974. The Hurst Phenomenon: a puzzle. *Water Resources Research* **10**(4): 675-688.

Lewis, M.R. 1989. The variegated ocean: a view from space. *New Scientist* **1685**(7 Oct. 1989): 37-40.

Lockwood, J.G. 1985. *World Climate Systems*. Edward Arnold, London. 292 p.

MacDonald, Sir M. & Partners. 1965. Survey of Canje Reservoir Scheme, British Guyana. BRG/3, SF4/5. FAO/UN Special Fund. FAO, Rome.

MacDonald, Sir M. & Partners. 1988. Rehabilitation and improvement of water delivery systems in Old Lands, Egypt. Annex A: Water Resources. IBRD/UNDP Project No. EGY/85/012. Washington DC.

Maslin, M. 1993. Waiting for the Polar meltdown. *New Scientist*, 4 Sept. 1993: 36-41.

Milankovitch, M. 1930. *Handbuch der Klimatologie l*. Teil, A. Ed Koppen and Geiger, Berlin.

Mitchell, J.F.B., Manabe, S., Meleshko, V. and Tokioka, T. 1990. Equilibrium climate change – and its implications for the future. In: *Climate Change, the 1990 Scientific Assessment*, J.T. Houghton, G.J. Jenkins and J.J. Ephraums (eds.). Cambridge University Press, Cambridge, UK. pp. 131-172.

Mott MacDonald Ltd. 1992, 1993. *Scenario Planning and Risk Analysis*. Anglian Water Services Ltd, October 1992 and October 1993. Cambridge. UK.

Newman, S. and Fairbridge, W. 1986. The management of sea-level rise. *Nature* **320**: 319-321.

Nicholson, S.E. 1986. The nature of rainfall variability in Africa south of the equator. *J. Climatology* **6**: 515-530.

Rind, D. 1995. The tropics drying out. *New Scientist*, 6 May 1995: 37-40.

Rind, D., Goldberg, R., Hansen, J. *et al*. 1990. Potential evapotranspiration and the likelihood of future drought. *J. Geophysical Res.* **95**(D7): 9983-10004.

Rodda, J.C., Pieyns, S.A., Sehmi, N.S. and Matthews, G. 1993. Towards a world hydrological cycle. *J. Hydrological Sciences* **38**(5): 373-378.

Rowntree, P. 1989. *The Needs of Climate Modellers for Water Runoff Data*. Workshop on Global Runoff Data Sets and Grid Estimation, 10-15 Nov. 1988, Koblenz. World Climate Programme, WMO, June 1989. Geneva.

Schlesinger, M.E. and Zhao, Z.C. 1989. Seasonal climatic change introduced by doubled CO_2 as simulated by the OSU atmospheric GCM/mixed-layer ocean model. *J. Climate*, **2**: 429-495.

Schnell, C. 1984. Socio-economic impacts of a climate change due to a doubling of atmospheric CO_2 content. In: *Current Issues in Climate Research*. Reidel, Dordrecht. pp. 270-287.

Semtner, A.J. 1984. The climate response of the Arctic Ocean to Soviet river diversions. *Climate Change* **6**: 109-130.

Semtner, A.J. 1987. A numerical study of sea ice and ocean circulation in the Arctic. *J. Phys. Ocean.* **17**: 1077-1099.

Shukla, J. and Minz, Y. 1982. Influence of the land surface evapotranspiration on the earth's climate. *Science* **215**: 1077-1099.

Street-Perrot, A.E. and Perrot, R.A. 1990. Abrupt climate fluctuations in the tropics: the influence of Atlantic Ocean circulation. *Nature* **343**: 607-611.

Thomson, D.J. 1995. The seasons, global temperature and precession. *Science* **268**: 59-68.

Tyson, P.D. 1987. *Climatic Change and Variability in Southern Africa*. Oxford Univ. Press, Capetown. 220 p.

Vali-Khodjeni, A. 1991. Hydrology of the Caspian Sea and its problems, hydrology of man-made lakes. Proc. of Vienna Symposium, August 1991. *IAHS Publ. No. 206*. IAHS, Waterloo, Ontario.

Viner, D. and Hume, M. 1993. The UK Met. Office High Resolution GCM Transient Experiment (UKTR). *Tech Note No. 4*. CRU/Hadley Centre, Oct. 1993.

Wetherald, R.T. and Manabe, S. 1986. An investigation of cloud cover change in response to thermal forcing. *Climatic Change*, **10**: 11-42.

Wigley, T.M.L. and Jones, P.D. 1985. Influences of precipitation changes and direct CO_2 effects on stream flow. *Nature* **314**: 149-152.

Wilson, C.A. and Mitchell, J.F.B. 1987. Similated climate and CO_2-induced climate change over Western Europe. *Climate Change*, **10**: 11-42.

World Bank/UNDP/ADB/EC/French Government. 1993. *Sub-Saharan Hydrological Assessment*. July 1993. Washington DC. 17 p.

3. The Effects of Global Change on Soil Conditions in Relation to Plant Growth and Food Production

ROBERT BRINKMAN AND WIM G. SOMBROEK
Land and Water Development Division, FAO, Rome, Italy

The main potential changes in soil-forming factors (forcing variables) directly resulting from global change would be in organic matter supply from biomass, soil temperature regime and soil hydrology, the latter because of shifts in rainfall zones as well as changes in potential evapotranspiration. Soil changes because of a potential rise in sea level resulting from a net reduction in Antarctic ice cap volume and ocean warming are discussed in Brammer and Brinkman (1990) and are summarized at the end of this paper.

The biggest single change in soils expected as a result of these postulated forcing changes would be a gradual improvement in fertility and physical conditions of soils in humid and subhumid climates. Another major change would be the poleward retreat of the permafrost boundary, discussed by Goryachkin and Targulian (1990). Other widespread changes would be in degree rather than in kind. Certain tropical soils with low physico-chemical activity, such as in the Amazon region, may undergo a radical change from one major soil-forming process to another (Sombroek, 1990), as discussed below under *Processes in soils*.

The changes in temperature but particularly in rainfall to be expected as a result of global warming are subject to major uncertainties for several reasons. Different global circulation models do not lead to mutually consistent results (an example for Europe is given by Santer, 1985), and they are not yet adequately verified. Also, the interaction with changes in location and intensity of major ocean currents and resultant possible modifications in sea surface temperatures is still most uncertain, as well as the interaction with possible major changes in cloudiness and land cover and the resulting changes in albedo and actual evapotranspiration.

Indirect effects of climate change on soils through CO_2-induced increases in growth rates or water-use efficiencies, through sea-level rise, through climate-induced decrease or increase in vegetative cover, or a change in human influence on soils because of the changes in options for the farmer, for example, may well each be greater than direct effects on soils of higher temperatures or greater rainfall variability and larger or smaller rainfall totals.

POSSIBLE CHANGES IN FORCING VARIABLES

With these caveats, one could stipulate the following changes in forcing variables as likely to materialize sometime during the next century:

- A gradual, continuing rise in atmospheric CO_2 concentration entailing increased photosynthetic rates and water-use efficiencies of vegetation and crops, hence increases in organic matter supplies to soils.
- Minor increases in soil temperatures in the tropics and subtropics; moderate increases and extended periods in which soils are warm enough for microbial activity (warmer than about 5°C) in temperate and cold climates, parallel to the changes in air temperatures and vegetation zones as summarized by Emanuel *et al.* (1985).
- Minor increases in evapotranspiration in the tropics to major increases in high latitudes caused both by temperature increase and by extension of the growing period.
- Increases in amount and in variability of rainfall in the tropics; possible decrease in rainfall in a band in the subtropics poleward of the present deserts; minor increases in amount and variability in temperate and cold regions. Peak rainfall intensities could increase in several regions.
- A gradual sea-level rise causing deeper and longer inundation in river and estuary basins and on levee backslopes, and brackish-water inundation leading to encroachment of vegetation that accumulates pyrite in soils near the coast.

This paper will not touch upon a possibly increased frequency and severity of cyclonic storms in the present cyclone belts or conceivable poleward widening of these belts because of increased sea surface temperatures, which would also give rise to greater frequencies of high-intensity rainfall events.

EFFECTS OF HIGHER CO_2 ON SOIL FERTILITY, PHYSICAL CONDITIONS AND PRODUCTIVITY

Higher atmospheric CO_2 concentration, as discussed in subsequent chapters, increases growth rates and water-use efficiency of crops and natural vegetation in so far as other factors do not become limiting. The higher temperature optima of some plants under increased CO_2 would tend to counteract adverse effects of temperature rise, such as increased nighttime respiration. The shortened growth cycle of a given species because of higher CO_2 and temperature would be compensated for in natural vegetation by adjustments in species composition or dominance. In agro-ecosystems the choice of longer-duration cultivars or changes in cropping pattern could eliminate unproductive periods that might arise because of the shorter growth cycle of the main crop.

There will be adequate time to adjust to the changes since these are expected to occur over decades, rather than years or days as in all present experimental situations. This chapter deals with the effects of *gradually* rising CO_2 concentrations as

EFFECTS ON SOIL CONDITIONS   51

observed in the recent past and stipulated in simulation models that apply *transient* scenarios.

As summarized in Figure 3.1, the increased productivity is generally accompanied by more litter or crop residues, a greater total root mass and root exudation, increased mycorrhizal colonization and activity of other rhizosphere or soil micro-organisms, including symbiotic and root-zone N_2 fixers. The latter would have a positive effect on N supply to crops or vegetation. The increased microbial and root activity in the soil would entail higher CO_2 partial pressure in soil air and CO_2 activity in soil water, hence increased rates of plant nutrient release (e.g., K, Mg, micronutrients) from weathering of soil minerals. Similarly, the mycorrhizal activity would lead to better phosphate uptake. These effects would be in synergy with better nutrient uptake by the more intensive root system due to higher atmospheric CO_2 concentration. There is no *a priori* reason why the degree of synchrony between nutrient release and demand by crops or natural vegetation would be subject to major changes under high CO_2 conditions.

The greater microbial activity tends to increase the quantity of plant nutrients cycling through soil organisms. The increased production of root material (at similar temperatures) tends to raise soil organic matter content, which also entails the temporary immobilization and cycling of greater quantities of plant nutrients in the soil. Higher C/N ratios in litter, reported by some workers under high CO_2 conditions, would entail slower decomposition and slower remobilization of the plant nutrients from the litter and uptake by the root mat, and would provide more time for incorporation into the soil by earthworms, termites, etc. Higher soil temperatures would counteract increases in 'stable' soil organic matter content but would further stimulate microbial activity.

In all experimental situations, whether chamber-type or free-air enrichment, CO_2 increases are rapid or sudden, often to double ambient concentration, sometimes higher. The consequently rapid increases in soil organic matter dynamics and soil micro-organisms may cause temporary competition for plant nutrients. These temporary effects have on occasion been reported as negative factors affecting plant response to elevated CO_2. However, increased organic matter dynamics and microbial activity in soils are positive for the soil-plant system when CO_2 concentrations rise gradually over decades, as currently and in the recent past. Future experiments could be set up to compensate for the temporary effects caused by the suddenness of the CO_2 increase, for example by artificially higher soil organic matter contents estimated to be near equilibrium with each stepwise higher CO_2 concentration, in a range between 350 and 600 ppm.

Increased microbial activity due to higher CO_2 concentration and temperature produces greater amounts of polysaccharides and other soil stabilizers. Increases in litter or crop residues, root mass and organic matter content tend to stimulate the activity of soil macrofauna, including earthworms, with consequently improved infiltration rate and bypass flow by the greater number of stable biopores. The greater stability and the faster infiltration increase the resilience of the soil against water erosion and consequent loss of soil fertility. The increased proportion of bypass flow also

52   GLOBAL CLIMATE CHANGE AND AGRICULTURAL PRODUCTION

Figure 3.1. Qualitative relationships between gradually increasing atmospheric CO_2 concentration[1], soil characteristics and medium-term processes in soils[2], and biomass or crop productivity[3]

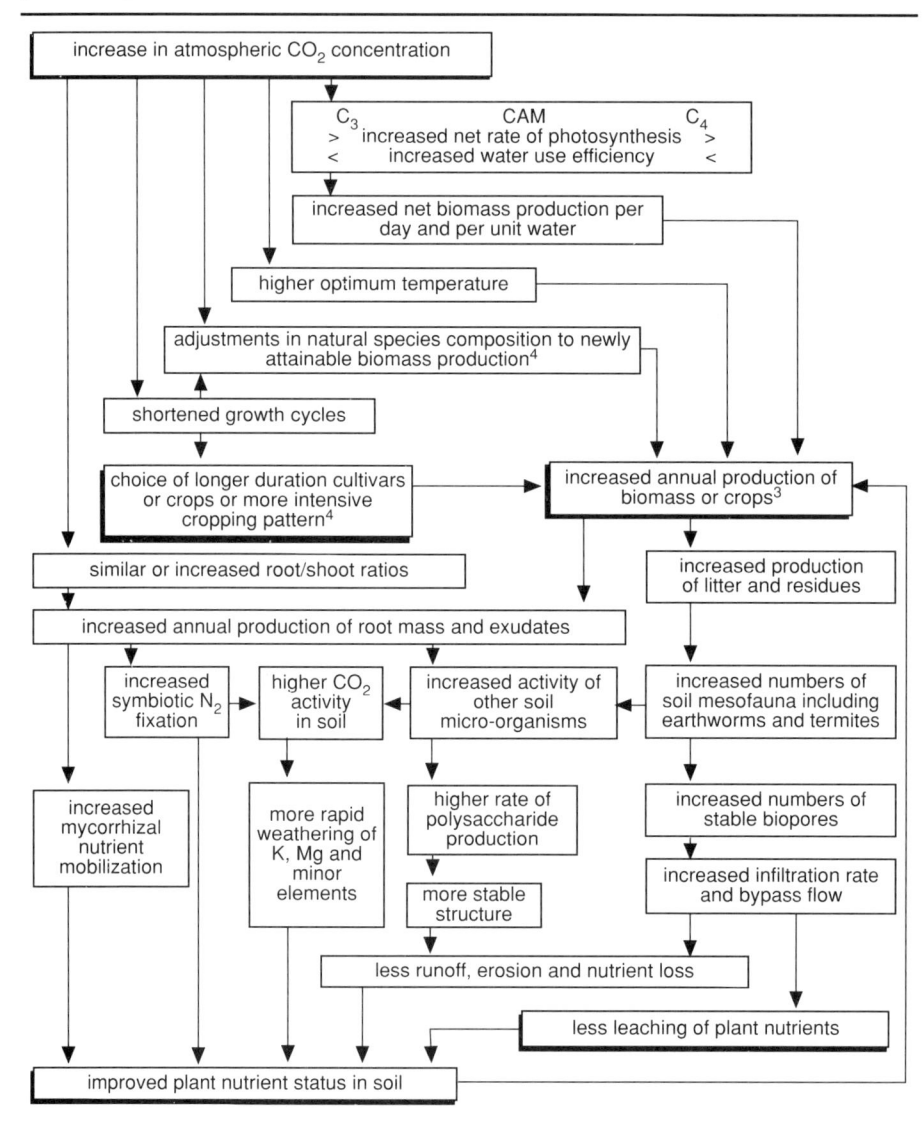

[1] Gradually rising CO_2 as in this century and in *transient* global change scenarios.
[2] Soils with some weatherable minerals at least in the subsoil or substratum within rooting depth.
[3] Extreme weather events may disrupt some relationships in the figure, so any major increase in their frequency or intensity may counteract positive effects shown.
[4] Species composition adjusts, or choices indicated are made to adjust, to the newly attainable biomass or crop production under increased atmospheric CO_2, compensating for shortened growth cycles of existing species or crops. The figure does not include the positive effects of higher temperatures on length of growing periods in temperate or boreal climates.

decreases the nutrient loss by leaching during periods with excess rainfall. This refers to the available nutrients in the soil, including well-incorporated fertilizers or manure, but not to fertilizers broadcast on the soil surface. These are subject to loss by runoff or leaching.

These changes increase the resilience of the soil against physical degradation and nutrient loss by increased intensity, seasonality or variability of rainfall, as well as against some of the unfavourable changes in rate or direction of soil-forming processes discussed in the next sections.

If the partial pressure of CO_2 in the soil air would rise, and that of O_2 decrease to levels impairing root function, part of the benefits indicated would not materialize. The improved gas exchange with the atmosphere through increased numbers of stable biopores would tend to keep CO_2 and O_2 in the soil at 'safe' levels, at least in naturally or artificially well-drained soils. Wetland crops such as rice or jute have their own gas exchange mechanisms and would not be affected; neither would natural wetland vegetation.

The positive effect on weathering rate and plant nutrient availability would occur in soils with significant amounts of weatherable minerals, not in very deeply and strongly weathered or otherwise very poor soils.

EFFECTS OF RAINFALL AND TEMPERATURE CHANGES IN DIFFERENT CLIMATES

In the humid tropics and monsoon climates, increased intensities of rainfall events and increased rainfall totals would increase leaching rates in well-drained soils with high infiltration rates, and would cause temporary flooding or water-saturation, hence reduced organic matter decomposition, in many soils in level or depressional sites. This may affect a significant proportion of especially the better soils in Sub-Saharan Africa, for example. They would also give rise to greater amounts and frequency of runoff on soils in sloping terrain, with sedimentation downslope and, worse, downstream. Locally, there would be increased chances of mass movement in the form of landslides or mudflows in certain soft sedimentary materials, discussed below. Soils most resilient against such changes would have adequate cation exchange capacity and anion sorption to minimize nutrient loss during leaching flows, and have a high structural stability and a strongly heterogeneous system of continuous macropores to maximize infiltration and rapid bypass flow through the soil during high-intensity rainfall.

In subtropical and other subhumid or semi-arid areas, the increased productivity and water-use efficiency due to higher CO_2 would tend to increase ground cover, counteracting the effects of higher temperatures. If there would be locally much less rainfall and increasing intra- and inter-annual variability, these could lead to less dry-matter production and hence, in due course, lower soil organic matter contents. Periodic leaching during high-intensity rainfall with less standing vegetation could desalinize some soils in well-drained sites, cause increased runoff in others, and lead to soil salinization in depressional sites or where the groundwater table is high. Soils most resilient against the effects of such increasing aridity and rainfall variability would have a high structural stability and a strongly heterogeneous system of continuous macropores (the same as in the tropics); hence a rapid infiltration rate, as well as a large available water capacity and a deep groundwater table.

Higher temperatures, particularly in arid conditions, entail a higher evaporative demand. Where there is sufficient soil moisture, for example in irrigated areas, this could lead to soil salinization if land or farm water management, or irrigation scheduling or drainage are inadequate. On the other hand, recent experiments by the Salinity Laboratory, Riverside, California, point to increased salt tolerance of crops under high atmospheric CO_2 conditions (E.V. Maas, pers. comm.; Bowman and Strain, 1987).

In temperate climates, minor increases in rainfall totals would be expected to be largely taken up by increased evapotranspiration of vegetation or crops at the expected higher temperatures, so that net hydrologic or chemical effects on the soils might be small. The negative effect on soil organic matter contents of a temperature rise might be more than compensated by the greater organic matter supply from vegetation or crops growing more vigorously because of the higher photosynthesis, the greater potential evapotranspiration and the higher water-use efficiency in a high-CO_2 atmosphere. The temperate zone would thus be likely to have the smallest changes in soils, even in poorly buffered ones, directly caused by the effects of global change. A minor and probably slow, but very visible, change could be a reddening of presently brown soils where increased periods with high summer temperatures would coincide with dry conditions, so that the iron oxide haematite would be stable over the presently dominant goethite. This mineralogical change might decrease the intensity and amount of phosphate fixation. An overview of such changes, with emphasis on temperate climate zones, is given by Buol *et al.* (1990).

In boreal climates, the gradual disappearance of large extents of permafrost and the reduction of frost periods in extensive belts adjoining former permafrost are expected to improve the internal drainage of soils in vast areas, with probable increases in leaching rates. The appreciable increase in period when the soil temperature is high enough for microbial activity would lead to lower organic matter contents, probably not fully compensated by increased primary production through somewhat higher net photosynthesis and the longer growing period. Paradoxically, the extent of soils subject to periodic reduction could well increase in level areas, in spite of the greater leaching, because of increased periods when the soils are water-saturated but also sufficiently warm for microbial activity. Soils most resilient against such effects, including the leaching of nutrients and periodic soil reduction, would have similar characteristics as the most resilient ones in other climates: adequate cation exchange capacity and anion sorption to minimize nutrient loss during leaching flows, a high structural stability and a strongly heterogeneous system of continuous macropores to maximize rapid bypass flow during periods with excess meltwater.

PROCESSES IN SOILS

The most rapid processes of chemical or mineralogical change under changing external conditions would be loss of salts and nutrient cations where leaching increases, and salinization where net upward water movement occurs because of increased evapotranspiration or decreased rainfall or irrigation water supply. The clay mineral composition and the mineralogy of the coarser fractions would generally change little, even over centuries. Exceptions would be the transformation of X-ray amorphous material into the clay mineral halloysite when a volcanic soil previously under perennially moist conditions becomes subject to periodic drying, or the gradual dehydration of goethite to haematite in soils subject to higher temperatures or severe drying, or both. Changes in the surface properties of the clay fraction, while generally slower than salt movement, can take place much faster than changes in bulk composition or crystal structure. Such surface changes have a dominant influence on soil physical and chemical properties (Brinkman, 1985, 1990).

Changes in the clay mineral surfaces or the bulk composition of the clay fraction of soils are brought about by a small number of transformation processes, listed below (Brinkman, 1982). Each of these processes can be accelerated or inhibited by changes in external conditions due to global change.

- hydrolysis by water containing carbon dioxide, which removes silica and basic cations;
- cheluviation, which dissolves and removes especially aluminium and iron by chelating organic acids;
- ferrolysis, a cyclic process of clay transformation and dissolution mediated by alternating iron reduction and oxidation, which decreases the cation exchange capacity by aluminium interlayering in swelling clay minerals;
- dissolution of clay minerals by strong mineral acids, producing acid aluminium salts and amorphous silica;
- reverse weathering, i.e., clay formation and transformation under neutral to strongly alkaline conditions, which may create, e.g., montmorillonite, palygorskite or analcime.

Hydrolysis and cheluviation may be accelerated by increased leaching rates. Ferrolysis may occur where soils are subject to reduction and leaching in alternation with oxidation: in a warmer world, this may happen over larger areas than at present, especially in high latitudes and in monsoon climates. Dissolution by strong acids would occur, e.g., where sulphidic materials in coastal plains are oxidized with an improvement of drainage; however, a rise in sea level would reduce the likelihood of this occurring naturally. Reverse weathering could begin in areas drying out during global warming, and would continue in most presently arid areas.

These processes would influence the surface properties of the clay fraction only over a period of centuries, even with the changes envisaged as a consequence of global warming. By contrast, direct human action can vastly accelerate some of these processes as is evident, for example, from the severe effects of acid rain on sandy soils in parts of Europe (Van Breemen, 1990) or from the extremely rapid ferrolysis in soils seasonally inundated by water level fluctuations in the Volta lake in Ghana (Amatekpor, 1989).

Not only the speed of soil formation can be accelerated by human action, but also, albeit much more locally, its very nature or direction. In most places, the natural soil-forming processes are not fundamentally changed, but there are certain threshold situations, generally with fragile soils, where even a small change in external conditions may cause a major, and adverse, change from one dominant soil-forming process to another. The four examples summarized below (from Sombroek, 1990) illustrate a change from hydrolysis to cheluviation (Ferralsols to Podzols); irreversible hardening of the subsoil; clay illuviation forming a dense subsoil in originally homogeneous, porous Ferralsols; and salinization.

The yellowish sandy Ferralsols and Ferralic Arenosols of Eastern Amazonia, Kalimantan and the Zaire basin may rapidly change into Podzols or Albic Arenosols (giant Podzols) with even small increases in total rainfall or stronger seasonality, or increased input of acidic ('poor') organic matter. An increase in effective rainfall due to climate change may cause a major increase in the extent of Podzols formed from present-day yellowish sandy Ferralsols where, presently, Podzols occur in patches within the Ferralsols area (Lucas *et al.*, 1987; Dubroeucq and Volkoff, 1988).

The imperfectly drained loamy Plinthosols on the flat interfluves of Western Amazonia would change into shallow, droughty soils with an irreversibly hardened subsoil if subject to drying out with climate change.

The deep reddish, porous loamy to clayey Ferralsols of the transition zones between forest and savanna in Eastern Africa, stable under the present vegetation, may be leached so far that a denser subsoil with washed-in clay is formed below an unstable topsoil with little organic matter, as already observed where the land was cleared several decades ago; the same may happen over more extensive areas under a sparser vegetation brought about by a somewhat drier climate.

The silty Fluvisols in the broad river valleys of the Sudano-Sahelian zone of West Africa, such as the interior delta of the Niger river, may become saline or sodic upon even minimal change in precipitation and flooding regimes – as exemplified by current human actions with the same soil-hydrological implications (Sombroek and Zonneveld, 1971).

SOME PROPERTIES OF CLAY SURFACES

Soils with a naturally high structural stability, for example Ferralsols and Nitisols, occupy sizeable areas in the tropics. The former are widespread, among others, in northern South America, the latter in East Africa. The clay fraction in such soils generally has oxidic surfaces: mainly iron(III) and aluminium oxides or hydroxides, while the bulk of the clay fraction may have different compositions. The oxidic surfaces could form from parent materials with moderate or high iron contents under long-continued hydrolysis by water (containing carbon dioxide).

At the other extreme are soils with a very low structural stability, or with a severe hazard of failure under load or shock (so-called quick clays). The surfaces of the clay minerals in these soils are generally covered by amorphous, gel-like material with a high silica content (McKyes et al., 1974). Such material may have originated in earlier periods when the soils were strongly saline and, presumably, subject to processes of reverse weathering. Examples are the quick clays of the Champlain Sea sediments in Ontario and Quebec and of parts of Scandinavia. Such soils are most likely to generate high proportions of runoff and suspended sediment, but also most liable to mudflows once sloping sites are water-saturated to appreciable depth. Some Andosols are thixotropic and have similarly low stability because of their similar composition, derived from volcanic materials (tuff).

Most soils fall somewhere between these extremes. Vertisols, for example, have moderate or low structural stability, and clay surfaces that are mainly silica, but with generally small amounts of amorphous coating. In Planosols, if formed by ferrolysis, the clay fraction in the upper, eluvial horizons has been partly decomposed with a residue of amorphous silica, but the remaining smectite or illite has been interlayered with aluminium hydroxide polymers, which has decreased the swell-shrink potential and the cation exchange capacity of the clay fraction. Concurrently, parts of the free iron oxides have been reduced and leached out. The net effect of these changes generally is a decrease in structural stability.

RESILIENCE AGAINST PHYSICAL AND CHEMICAL SOIL DEGRADATION

As discussed, most soils do not have a high intrinsic resilience against physical soil degradation by, for example, high-intensity rainfall. In natural conditions in humid climates, it is the complete soil cover near ground level combined with the perforating activity of the soil fauna that makes the soil-vegetation system resilient against physical degradation.

In the Rhine river plain in the Netherlands, for example, most of the originally calcareous alluvial soils have been decalcified within a millennium or so. Only in small areas on the highest levees of that age that have continually remained under forest, soils are still calcareous, and even have lime pseudomycelia (filaments) indicative of less humid soil conditions, and abundant vertical macropores produced by earthworms. In these soils, faunal activity is high because of the adequate litter supply; the resulting macropores remain open, protected against rain impact by litter and undergrowth; and excess water from heavy rain passes to the substratum through the macropores without leaching lime from most of the soil mass.

Research to increase resilience of soils against any adverse effects of climate change could be done in conjunction with that on soil resilience against direct adverse human impacts. Until site-specific management procedures have been elaborated, soil and crop (including trees and pasture) management should aim to maintain soil cover and organic matter supply to soil biota, while minimizing mechanical disturbance by heavy traffic, cultivation or excessive grazing intensity. Such kind of management may also help to conserve plant nutrients (in soils not flooded for wetland cultivation) since the stable, heterogeneous system of biopores produced by the soil fauna would favour bypass flow of any excess moisture and thus decrease leaching through the soil mass.

A single management recipe would not be generally applicable in different conditions. Minimizing damage by certain termite species harming crop performance may necessitate a period without residues on the soil, for example; or crop residues may be needed for feed or fuel. Management methods for wetland need to be developed that make optimum use of any increased potential productivity, while minimizing secondary effects such as increased CH_4 or N_2O emission from the reduced soil. Such factors, and others, should be taken into account in designing an optimum management strategy for any specific natural and cultural environment.

RESILIENCE AGAINST SOIL REDUCTION (ANOXIC CONDITIONS)

Soil reduction, which would limit land suitability for dryland crops, or strong reduction, which would be liable to produce toxins even for wetland crops, may take place once the soil is water-saturated long enough for microbial action to exhaust the oxygen remaining in the soil when water-saturation started. Another necessary condition is the presence of sufficient readily decomposable organic matter as an energy source for the microbial activity. In most soils, during reduction the redox status is stabilized at an Eh about 100-200 mV near neutrality by the Fe^{2+} - $Fe(OH)_3$ equilibrium, except where the content of readily decomposable organic matter is very high or the content of free iron(III) oxides very low. In such cases, negative Eh values may occur, and toxic hydrogen sulphide or low-molecular organic compounds – including methane – may be formed.

Resilience against soil reduction in practice depends on the drainage conditions, since most soils have sufficient organic matter for reduction to start within about a week after water-saturation. Soils most resilient against reduction in conditions of increased rainfall variability and incidence of high-intensity rainfall have similar properties as those resilient against the negative effects of other perturbations: high infiltration rate, high structural stability and a permanent heterogeneous system of tubular macropores, good external drainage.

SOIL REACTION (pH)

Most soils would not be subject to rapid pH changes resulting from climate change. Exceptions might be found in potential acid sulphate soils, extensive in some coastal plains and estuaries, if they become subject to increasingly long dry seasons. Even though most of such soils are clays with moderate or high cation exchange capacity, the amounts of acid liberated in such soils upon oxidation generally exceed this rapid buffering capacity. Therefore, pH values may temporarily reach 2.5 to 3.5 and a small part of the clay fraction may be decomposed as indicated under *Processes in soils*, above. This then buffers the pH generally between 3.5 and 4 in the long run. Depending on the efficiency with which the excess acid formed can be leached out, the period of extreme acidity and aluminium toxicity may last between less than a year and several decades.

In calcareous soils, soil reaction may range between about 8.5 and 7 depending on the partial pressure of CO_2 in the soil; this range is maintained against leaching of basic cations by the different soil processes as long as a few per cent of finely distributed lime remain. Buffering in non-calcareous soils is less strong, but depends on the cation exchange capacity at soil pH. In soils with variable-charge surfaces of the clay fraction, this decreases with acidification.

It should be noted that the simple modelling of accelerated $CaCO_3$ leaching under a doubled atmospheric CO_2 concentration generally does not hold true. In most soils, the ongoing decomposition of organic matter maintains CO_2 concentrations in the soil air far above atmospheric concentration even now, and $CaCO_3$ solubility is determined by the partial pressure of CO_2 in soil air and its activity in soil water, rather than in the atmosphere. Leaching of lime is thus positively related to rate of organic matter decomposition, negatively to gas diffusion rate, and positively to amount of water percolating through the soil.

In conditions where leaching is accelerated by climate change, it would be possible to find relatively rapid soil acidification after a long period with little apparent change, as has been the case – but after a shorter latent period – in some soils in Europe that have been subject to acid rain for several decades. The soil might in fact be steadily depleted of basic cations, but a pH change may start, or may become more rapid, once certain buffering pools are nearly exhausted. Such non-linear and time-delayed effects have been discussed in the context of soil and water pollution by Stigliani (1988); they are also expected to occur in various ways at different times after increased temperatures and changed rainfall patterns will have been operative.

EFFECTS OF A RISING SEA LEVEL ON SOILS IN COASTAL AREAS

The probable effects on soil characteristics of a gradual eustatic rise in sea-level will vary from place to place depending on a number of local and external factors, and interactions between them (Brammer and Brinkman, 1990). In principle, a rising sea level would tend to erode and move back existing coastlines. However, the extent to which this actually happens will depend on the elevation, the resistance of local coastal materials, the degree to which they are defended by sediments provided by river flow or longshore drift, the strength of longshore currents and storm waves, and on human interventions which might prevent or accelerate erosion.

In major deltas, such as those of the Ganges-Brahmaputra and the major Chinese rivers, sediment supplies delivered to the estuary will generally be sufficient to offset

the effects of a rising sea level. Such deltaic aggradation could decrease, however, under three circumstances:
- where human interventions inland, such as large dams or successful soil conservation programmes, drastically reduce sediment supply to the delta: e.g., the construction of the Aswan high dam in 1964 has led to coastal erosion and increased flooding of lagoon margins in the Nile delta (Stanley, 1988);
- where construction of embankments within the delta interrupts sediment supply to adjoining backswamps, exposing them to submergence by a rise in sea level: e.g., embankments along the lower Mississippi river have cut off sediment supplies to adjoining wetlands which formerly offset land subsidence occurring due to compaction of underlying sediments (Day and Templet, 1989);
- where land subsidence occurs due to abstraction of water, natural gas or oil: e.g., as is presently happening in Bangkok and in the northern part of the Netherlands.

In coastal lowlands which are insufficiently defended by sediment supply or embankments, tidal flooding by saline water will tend to penetrate further inland than at present, extending the area of perennially or seasonally saline soils. Where *Rhizophora* mangrove or *Phragmites* vegetation invades the area, that would over several decades lead to the formation of potential acid sulphate soils. Impedance of drainage from the land by a higher sea level and by the correspondingly higher levels of adjoining estuarine rivers and their levees, will also extend the area of perennially or seasonally reduced soils and increase normal inundation depths and durations in river and estuary basins and on levee backslopes. In sites which become perennially wet, soil organic matter contents will tend to increase, resulting eventually in peat formation. On the other hand, where coastal erosion removes an existing barrier of mineral soils or mangrove forest, higher storm surges associated with a rising sea level could allow sea-water to destroy existing coastal eustatic peat swamps, which might eventually be replaced by freshwater or saltwater lagoons.

The probable response of low-lying coastal areas to a rise in sea level can be estimated in more detail on the basis of the geological and historical evidence of changes that occurred during past periods when sea level was rising eustatically or in response to tectonic or isostatic movements: e.g., around the southern North Sea (Jelgersma, 1988); in the Nile delta (Stanley, 1988); on the coastal plain of the Guyanas (Brinkman and Pons, 1968); in the Musi delta of Sumatera (Brinkman, 1987). Contemporary evidence is available in areas where land levels have subsided as a result of recent abstraction of water, natural gas or oil from sediments underlying coastal lowlands. Further studies of such contemporary and palaeoenvironments are needed together with location-specific studies in order to better understand the change processes, identify appropriate responses and assess their technical, ecological and socio-economic implications (e.g., Warrick and Farmer, 1990).

CONCLUSIONS

Some major and widespread soil changes expected as a result of any global change are positive, especially the gradual increases in soil fertility and physical qualities consequent on increased atmospheric CO_2. The increased productivity and water-use efficiency of crops and vegetation, and the generally similar or somewhat higher rainfall indicated by several global circulation models, not fully counteracted by higher evapotranspiration, would be expected to lead to widespread increases in ground cover, and consequently better protection against runoff and erosion.

Major but less widespread soil changes, including greater biological activity and increased extent of periodic reduction in soils, would be expected where permafrost would disappear. In unprotected low-lying coastal areas, gradual encroachment of *Rhizophora* mangroves or *Phragmites* following more extensive brackish-water inundation may give rise to the formation of potential acid sulphate soil layers after several decades. Deeper and longer-duration flooding of basins and levee backslopes in adjacent river and estuary plains could lead to more extensive reducing conditions and increased organic matter contents, and locally to peat formation.

Other changes due to climate change (temperature and precipitation) are expected to be relatively well buffered by the mineral composition, the organic matter content or the structural stability of many soils. However, decreases in cover by vegetation or annual or perennial crops, caused by any locally major declines in rainfall not compensated by CO_2 effects, could lead to soil structure degradation and decreased porosity, as well as increased runoff and erosion on sloping sites and by the concomitant more extensive and rapid sedimentation. Changes in options available to land users because of climate change may have similar effects.

In certain fragile soils, the nature of the dominant soil-forming process may change for the worse with increased, decreased or more strongly seasonal rainfall.

In most cases, changes in soils by direct human action, on-site or off-site (whether intentional or unintended), are far greater than the direct climate-induced effects. Soil management measures designed to optimize the soil's sustained productive capacity would therefore be generally adequate to counteract any degradation of agricultural land by climate change. Soils of nature areas, or other land with a low intensity of management such as semi-natural forests used for extraction of wood and other products, are less readily protected against the effects of climate change but such soils, too, are threatened less by climate change than by human actions – off-site, such as pollution by acid deposition, or on-site, such as excessive nutrient extraction under very low-input agriculture.

To armour the world's soils against any negative effect of climate change, or against other extremes in external circumstances such as nutrient depletion or excess (pollution), or drought or high-intensity rains, the best that land users could do, would be:
- to manage their soils to give them maximum physical resilience through a stable, heterogeneous pore system by maintaining a closed ground cover as much as possible;
- to use an integrated plant nutrient management system to balance the input and

offtake of nutrients over a cropping cycle or over the years, while maintaining soil nutrient levels low enough to minimize losses and high enough to buffer occasional high demands.

An analogous philosophy, at lower levels of external inputs, could be formulated for extensive grazing land and production forest, whether planted or managed natural forest.

Human action and management has been emphasized in these conclusions because most of the world's land is used and, to different degrees, managed rather than under natural conditions.

ACKNOWLEDGEMENTS

Critical comments by R. Gommes, F.O. Nachtergaele and D.W. Sanders are gratefully acknowledged.

REFERENCES

Amatekpor, J.K. 1989. The effect of seasonal flooding on the clay mineralogy of a soil series in the Volta lake drawdown area, Ghana. *Land Degradation and Rehabilitation* **1**: 89-100.

Bowman, W.D. and Strain, B.R. 1987. Interaction between CO_2 enrichment and salinity stress in the C4 non-halophyte *Andropogon glomeratus* (Walter) BSP. *Plant Cell Environ.* **10**: 267-270.

Brammer, H. and Brinkman, R. 1990. Changes in soil resources in response to a gradually rising sea-level. Chapter 12. In: Scharpenseel *et al.* (eds.). 1990. pp. 145-156.

Brinkman, R. 1982. Clay transformations: aspects of equilibrium and kinetics. In: *Soil Chemistry. B. Physicochemical Models. Developments in Soil Science 5B.* G.H. Bolt (ed.). 2nd ed. Elsevier, Amsterdam. pp. 433-458.

Brinkman, R. 1985. Mineralogy and surface properties of the clay fraction affecting soil behavior and management. In: *Soil Physics and Rice.* T. Woodhead (ed.). International Rice Research Institute, Los Baños, Philippines. pp. 161-182.

Brinkman, R. 1987. Sediments and soils in the Karang Agung area. In: *Some Aspects of Tidal Swamp Development with Special Reference to the Karang Agung Area, South Sumatra Province.* R. Best, R. Brinkman and J.J. van Roon. Mimeo. World Bank, Jakarta. pp. 12-22.

Brinkman, R. 1990. Resilience against climate change? Soil minerals, transformations and surface properties, Eh, pH. In: Scharpenseel *et al.* (eds.). 1990. pp. 51-60.

Brinkman, R. and Pons, L.J. 1968. A pedo-geomorphological classification and map of the Holocene sediments in the coastal plain of the three Guyanas. *Soil Survey*

Paper No. 4. Soil Survey Institute (Staring Centre), Wageningen. 40 p, separate map.

Buol, S.W., Sanchez, P.A., Kimble, J.M. and Weed, S.B. 1990. Predicted impact of climatic warming on soil properties and use. *American Soc. Agron. Special Publ.* **53**: 71-82.

Day, J.W. and Templet, P.H. 1989. Consequences of sea level rise: implications from the Mississippi delta. *Coastal Management* **17**: 241-257.

Dubroeucq, D. and Volkoff, B. 1988. Evolution des couvertures pédologiques sableuses à podzols géants d'Amazonie (Bassin du haut Rio Negro). *Cahiers ORSTOM, Série Pédologie* **24**(3): 191-214.

Emanuel, W.R., Shugart, H.H. and Stevenson, M.P. 1985. Climatic change and the broad-scale distribution of terrestrial ecosystem complexes. *Climatic Change* **7**: 29-43.

Goryachkin, S.V. and Targulian, V.O. 1990. Climate-induced changes of the boreal and subpolar soils. In: Scharpenseel *et al.* (eds.). pp. 191-209.

Jelgersma, S. 1988. A future sea-level rise: its impacts on coastal lowlands. In: *Geology and Urban Development. Atlas of Urban Geology*. Vol. 1. UN-ESCAP. pp. 61-81.

Lucas, Y., Boulet, R., Chauvel, A. and Veillon, L. 1987. Systèmes sols ferrallitiques-podzols en région amazonienne. In: *Podzols et podzolisation.* D. Righi and A. Chauvel (eds.). AFES-INRA, Paris. pp. 53-68.

McKyes, E., Sethi, A. and Yong, R.N. 1974. Amorphous coatings on particles of sensitive clay soils. *Clays and Clay Minerals* **22**: 427-433.

Santer, B. 1985. The use of general circulation models in climate impact analysis – a preliminary study of the impacts of a CO_2-induced climatic change on West European agriculture. *Climatic Change* **7**: 71-93.

Scharpenseel, H.W., Schomaker, M. and Ayoub, A. (eds.) 1990. Soils on a warmer earth. Effects of expected climatic change on soil processes, with emphasis on the tropics and subtropics. *Developments in Soil Science 20*. Elsevier, Amsterdam. xxii + 274 p.

Sombroek, W.G. 1990. Soils on a warmer earth: the tropical regions. In: Scharpenseel *et al.* (eds.). 1990. pp. 157-174.

Sombroek, W.G. and Zonneveld, L.S. 1971. Ancient dune fields and fluviatile deposits in the Rima-Sokoto river basin (N.W. Nigeria). Stiboka (Staring Centre). *Soil Survey Paper No. 5*. Wageningen. 109 p.

Stanley, D.J. 1988. Subsidence in the Northeastern Nile delta: rapid rates, possible causes, and consequences. *Science* **240**: 497-500.

Stigliani, W.M. 1988. Changes in valued 'capacities' of soils and sediments as indicators of nonlinear and time-delayed environmental effects. *Environmental Monitoring and Assessment* **10**: 245-307.

Van Breemen, N. 1990. Impact of anthropogenic atmospheric pollution on soils. In: *Climate Change and Soil Processes*. UNEP, Nairobi. pp. 137-144.

Warrick, R. and Farmer, G. 1990. The greenhouse effect, climatic change and rising sea level: implications for development. *Trans. Inst. Br. Geogr. N.S.* **15**: 5-20.

4. The CO_2 Fertilization Effect: Higher Carbohydrate Production and Retention as Biomass and Seed Yield

L. HARTWELL ALLEN, JR.
US Department of Agriculture, Agricultural Research Service, University of Florida, Gainesville, Florida, USA

JEFF T. BAKER AND KEN J. BOOTE
Agronomy Department, University of Florida, Gainesville, Florida, USA

The rise in atmospheric carbon dioxide (CO_2) concentration from about 280 µmol/mol before the industrial revolution to about 360 µmol/mol currently is well documented (e.g., Baker and Enoch, 1983; Keeling *et al.*, 1995). The consensus of many studies of the effects of elevated CO_2 on plants is that the CO_2 fertilization effect is real (see Kimball, 1983; Acock and Allen, 1985; Cure and Acock, 1986; Allen, 1990; Rozema *et al.*, 1993; Allen, 1994; Allen and Amthor, 1995). However, the CO_2 fertilization effect may not be manifested under conditions where some other growth factor is severely limiting, such as low temperature (Long, 1991). Also, plants grown in some conditions, where limitations of rooting volume (Arp, 1991), light, or other factors restrict growth, have not shown a sustained response to elevated CO_2 (Kramer, 1981).

The main objectives of this chapter are to assess the direct effects of rising atmospheric CO_2 and indirect effects of potential climate changes on crop growth and yield. The approach will be to (1) provide a general overview of CO_2 effects on plant growth processes; (2) analyse some specific experimental data on crop plant responses to elevated CO_2 and climate change factors; (3) summarize recent reviews of plant responses to elevated CO_2; and (4) discuss some crop modelling assessments of rising CO_2 and climate change factors on agricultural productivity based on predictions of global climate change models. Also, some adaptations for improving crop productivity in a higher CO_2 world will be suggested.

OVERVIEW OF CO_2 EFFECTS ON PLANT GROWTH PROCESSES

Most of the following discussion of CO_2 effects on plants applies to species with the C_3 photosynthetic pathway and not necessarily to species with the C_4 pathway. Other aerial, non-biotic environmental factors that affect plant growth and development are light and temperature. Plant photosynthetic rates generally increase linearly with light across relatively low ranges of light intensity, and then the rates decelerate until they reach an asymptotic maximum. Because of crowding and shading of many leaves, most crop canopies do not reach light saturation at full sunlight; that is, they would be able to respond to light levels well beyond full solar irradiance. Likewise, crop photosynthetic rates respond to increasing levels of CO_2, but then level off at higher concentrations (around 700 µmol/mol or greater, depending upon species and other factors). However, leaf photosynthesis usually increases with temperature up to some maximum value, and then declines. Furthermore, temperature affects not only photosynthesis, but also respiration, growth, development phases and reproductive processes.

Elevated CO_2 may have some effects on crop phenology, although stages of development are governed primarily by temperature, time and photoperiod. If dates of planting were to be changed because of the greenhouse effect, then phenological timing of plants could be affected. For example, higher temperatures could decrease yields by decreasing the duration of the grain-filling period or changes in photoperiod could shorten or lengthen the vegetative stage.

The CO_2 fertilization effect begins with enhanced photosynthetic CO_2 fixation. Non-structural carbohydrates tend to accumulate in leaves and other plant organs as starch, soluble carbohydrates or polyfructosans, depending on species. In some cases, there may be feedback inhibition of photosynthesis associated with accumulation of non-structural carbohydrates. Increased carbohydrate accumulation, especially in leaves, may be evidence that crop plants grown under CO_2 enrichment may not be fully adapted to take complete advantage of elevated CO_2. This may be because the CO_2-enriched plants do not have an adequate sink (inadequate growth capacity), or lack capacity to load phloem and translocate soluble carbohydrates. Improvement of photoassimilate utilization should be one goal of designing cultivars for the future (Hall and Allen, 1993).

In the process of growth, photoassimilates are allocated to the vegetative shoots, root system or reproductive organs. In some cases, more photoassimilate of CO_2-enriched plants is partitioned to the root system than to the shoots. Above ground, more photoassimilate usually goes into stems and supporting structures than into leaves. This phenomenon may not be an inherent response to elevated CO_2, but may be a by-product of the larger size of plants often found in CO_2-enriched atmospheres, especially by species that produce branch stems along the aerial mainstems (Allen *et al.*, 1991).

Reproductive biomass growth as well as vegetative biomass growth are usually increased by elevated CO_2. However, the harvest index, or the ratio of seed yield to above-ground biomass yield, is typically lower under elevated CO_2 conditions (Allen, 1991; Baker *et al.*, 1989), which may also be evidence of the lack of capacity to utilize completely the more abundant photoassimilate.

In many cases, both the amount and the carboxylation activity of ribulose 1,5-bisphosphate carboxylase-oxygenase enzyme (rubisco) is decreased in leaves of plants grown under elevated CO_2. This acclimation phenomenon may produce 'downregulation of photosynthesis'; however, this is not universally the case. For example, there is little evidence of this downregulation response in soybean (*Glycine max* L. Merr.), a C_3 legume. In fact, photosynthetic capacity per unit leaf area of soybean is increased under CO_2 enrichment. Leaves often develop an additional layer of mesophyll cells. Also, more structural carbohydrate may be produced in leaves, as well as stems, of CO_2-enriched plants.

We can obtain clues about reasons that some plants downregulate and others do not in response to elevated CO_2 by focusing on the capabilities of soybean. This plant has (a) symbiotic N_2 fixation; (b) the capacity to form additional layers of palisade cells in the leaf tissue; (c) the capacity to shunt much of the photoassimilate into relatively inert starch rather than soluble sugars during photosynthesis; (d) a relatively strong leaf and stem sink during vegetative development; and (e) a strong seed-fill sink during reproductive development. Plants which lack these capacities, either inherently or because of growth in limiting environments, are more likely to demonstrate some degree of downregulation of photosynthesis (Allen, 1994).

The carbon:nitrogen ratio of leaves of plants is usually increased under CO_2 enrichment. Plants may acclimate to elevated CO_2 by requiring less rubisco and photosynthetic apparatus, which would lead to lower nitrogen contents. The overall change in C:N ratios is governed both by increases in structural and non-structural carbohydrates, and by decreases in protein content. However, seed nitrogen content is little affected (Allen *et al.*, 1988).

Specific respiration rates may be reduced by both short-term exposure to elevated CO_2 and long-term growth at elevated CO_2 (Amthor, 1995). However, the long-term effect may be similar when respiration rates are reported on a per unit nitrogen basis.

In climate change scenarios, temperatures are predicted to increase following the rise of CO_2 and other greenhouse-effect gases. Carbon dioxide × temperature interactions have been observed for vegetative growth (i.e., the CO_2 fertilization effect is greater at warmer temperatures than at cooler temperatures). Temperature increases in a higher CO_2 world could increase overall biomass productivity for vegetative crops (pastures and forages) both by extending the length of the growing season in temperate regions, and by the interaction of CO_2 × temperature in stimulation of vegetative growth. However, CO_2 × temperature interactions appear to be very small or negligible for reproductive processes (seed set and seed yield) although there may be more initial flowers formed by greater amounts of branching or tillering that is stimulated by CO_2 enrichment (Baker and Allen, 1993a).

Precipitation changes may occur along with other climatic change effects. In general, predictions from crop models show that increased CO_2 should increase productivity of C_3 plants, but the associated predictions of temperature rise will be detrimental. Not surprisingly, changes in precipitation patterns (decreases of rainfall during growth period) could be more detrimental for crop production than changes in temperature.

Under elevated CO_2, stomatal conductance in most species will decrease which may result in less transpiration per unit leaf area. However, leaf area index of some crops may also increase. The typical 40% reduction in stomatal conductance induced by a doubling of CO_2 has generally resulted in only a 10% (or less) reduction in crop canopy water use in chamber or field experimental conditions. Actual changes in crop evapotranspiration will be governed by the crop energy balance, as mitigated by stomatal conductance, leaf area index, crop structure and any changing meteorological factors.

Water-use efficiency (WUE) (ratio of CO_2 uptake to evapotranspiration) will increase under higher CO_2 conditions. This increase is caused more by increased photosynthesis than it is by a reduction of water loss through partially closed stomata. Thus, more biomass can be produced per unit of water used, although a crop would still require almost as much water from sowing to final harvest. If temperatures rise, however, the increased WUE caused by the CO_2 fertilization effect could be diminished or negated, unless planting dates can be changed to more favourable seasons.

Several assessments of impacts of climate change on crop productivity have been published. Progress has been made on integrating the impacts on individual countries and on economic and social interactions. For the most part, these assessments project more favourable climates for agriculture in northern latitudes and less favourable climates in the tropical and subtropical zones. However, the crop modelling predictions are dependent on the scenarios of outputs of General Circulation Models (GCMs) applied to the greenhouse effect. Thus, the dependency chain of assessments follows: Climate Change Scenarios —> Crop Model Prediction and Agricultural Production Systems (with and without available mitigation and adaptation response strategies) —> National Scenarios of Economics and Well-being of Farmers, Agricultural Commerce, and Consumers —> Country-by-Country and Global Interaction Scenarios of World Trade (food and all other commodities), Population Dynamics and Economic Well-being, and Impacts on Social Systems.

As the world continues to consume fossil fuels, CO_2 concentrations will continue to rise. Other greenhouse-effect gases, such as methane, nitrous oxides, chlorofluorocarbons and chlorofluorocarbon substitutes, and perhaps tropospheric ozone, will likely rise also. The CO_2 fertilization effect on plants will increase and climate changes may occur because of the combined increase of all greenhouse-effect gases. Global agriculture could adapt to gradual regional climate changes, but sudden changes would be more serious. Adaptation and/or mitigation actions could include the following:

1. Selection of plants that can better utilize carbohydrates which are produced when plants are grown at elevated CO_2.
2. Selection of plants that produce less structural matter and more reproductive capacity under CO_2 enrichment. (This applies for seed crop plants, not necessarily vegetative biomass plants.)
3. Search for germplasms that are adapted to higher day and night temperatures, and incorporate those traits into desirable crop production cultivars to improve flowering and seed set.
4. Change planting dates and other crop management procedures to optimize yields under new climatic conditions, and select for cultivars that are adapted to these changed agricultural practices.

5. Shift to species that have more stable production under high temperatures or drought.
6. Determine whether more favourable N:C ratios can be attained in forage cultivars adapted to elevated CO_2.
7. Where needed, and where possible, develop irrigation systems for crops.

SPECIFIC RESPONSES OF CROPS TO ELEVATED CO_2

This section will focus on responses of two C_3 crops to elevated CO_2: soybean, a symbiotic nitrogen-fixing legume representative of the pulses; and rice (*Oryza sativa* L.), a globally important cereal food crop. Findings from seven crop cycles of both soybean and rice grown in sunlit chambers at Gainesville, Florida, USA, will be emphasized (Baker and Allen, 1993a,b). The chambers feature both large rooting volumes of real soil and light from the sun. Some studies included responses to sub-ambient as well as superambient CO_2 concentrations (e.g., Allen *et al.*, 1991; Baker and Allen, 1993a,b, 1994). (The rice experiments have been designated chronologically as RICE I, II, III, IV, V, VI and VII.)

SOYBEAN

Photosynthetic rates

Soybean photosynthetic rates of both leaves and the whole canopy have *always* been increased by elevated CO_2. Jones *et al.* (1984) reported midday maximum canopy photosynthetic rates of 60 and 90 μmol $CO_2/m^2/s$ at 70 days after planting (DAP) for soybean grown at CO_2 concentrations of 330 and 800 μmol/mol, respectively. The leaf area index (LAI) was 6.9 and 9.0, respectively. Midday maximum canopy photosynthetic rates in other studies were 40 and 75 μmol/m^2/s for soybean grown at 320 and 640 μmol/mol CO_2 (Jones *et al.*, 1985c) and 40 and 80 μmol/m^2/s for soybean grown at 330 and 800 μmol/m^2/s (Jones *et al.*, 1985a,b).

Valle *et al.* (1985) found that midday maximum photosynthetic CO_2 uptake rates of soybean leaves ranged from 30 to 50 μmol/m^2/s and 15 to 25 μmol/m^2/s on plants grown at 660 and 330 μmol/mol CO_2, respectively. Allen *et al.* (1990) reported that, at all light levels, leaf photosynthetic rates increased linearly with CO_2 concentration across the range of 330 to 800 μmol/mol.

Valle *et al.* (1985) used a Michaelis-Menten type of rectangular hyperbola to summarize photosynthetic responses of soybean leaves vs. CO_2 concentration. The plants had been grown at 330 and 660 μmol/mol of CO_2 and then exposed to a wide range of CO_2 for a short period.

$$Y = (Y_{max} \times [C])/([C] + K_m) + Y_i \qquad (4.1)$$

where Y is photosynthetic rate in μmol $CO_2/m^2/s$; [C] is CO_2 concentration in μmol/mol; Y_i is the y-axis intercept at zero [C], the apparent respiration rate, in μmol CO_2/

m^2/s; Y_{max} is the response limit of $(Y - Y_i)$ at very high [C], the asymptotic photosynthetic rate, in μmol $CO_2/m^2/s$; K_m is the value of [C] where $(Y - Y_i) = Y_{max}/2$, the apparent Michaelis-Menten constant, in μmol/mol; and Γ_c is the calculated [C] intercept at zero Y, the CO_2 compensation point, μmol/mol (not shown in this equation). The average parameters for responses at 330 and 660 μmol/mol are given in Table 4.1. There was no obvious downregulation of soybean leaf photosynthesis in response to elevated CO_2; in fact, photosynthetic capacity was increased. Leaf quantum yield increased from 0.05 to 0.09 in the soybean leaves exposed to CO_2 of 330 and 660 μmol/mol, respectively (Valle et al., 1985).

Campbell et al. (1988) showed that soybean leaf photosynthetic rates were higher for plants grown at 660 than at 330 μmol/mol CO_2 when measured at common intercellular CO_2 concentrations. Furthermore, Campbell et al. (1988) measured rubisco activity and amount in leaves of soybean grown in CO_2 concentrations of 160, 220, 280, 330, 660 and 990 μmol/mol. They found that rubisco activity was almost constant at 1.0 μmol CO_2/min/mg soluble protein across this CO_2 treatment range. Leaf soluble protein was nearly constant at about 2.4 g/m^2 with 55% being rubisco protein. Specific leaf weight increased across the 160 to 990 μmol/mol CO_2 concentration range, so that the rubisco activity on a leaf dry weight basis decreased.

Campbell et al. (1990) showed that photosynthetic capacity of soybean canopies grown at 330 and 660 μmol/mol CO_2 were similar for short exposures at concentrations below 500 μmol/mol, but above this concentration, canopies grown at 660 μmol/mol had a slightly higher photosynthetic capacity. Thus, soybean did not lose photosynthetic capacity as did some other plant species (see Allen, 1994). In other studies, however, both photosynthetic rate and rubisco activity of soybean declined during long-term CO_2 enrichment (Thorne and Koller, 1974; Delucia et al., 1985).

Respiration

Whole-canopy pre-dawn respiration rates at 40 to 60 DAP were 2-3 and 4-5 μmol/m^2/s for soybean grown at CO_2 levels of 320 and 640 μmol/m^2, respectively, and a constant day/night temperature of 25°C (Jones et al., 1985c). After cross-switching two of the treatment chambers at DAP 52, the subsequent canopy photosynthetic rates and respiration rates quickly adjusted to their new CO_2 exposure conditions.

Table 4.1. Average asymptotic maximum photosynthetic rate (Y_{max}) with respect to y-intercept parameter (Y_i), apparent Michaelis-Menten constant for CO_2 (K_m), and CO_2 compensation point (Γ_c) for leaves grown at two CO_2 treatments and subjected to different short-term CO_2 levels. Condensed from Valle et al. (1985)

Growth CO_2 treatment	Y_{max} μmol/m^2/s	K_m μmol/mol	Y_i μmol/m^2/s	Γ_c μmol/mol
330	51.8	359	-7.8	63
660	126.6	1 133	-4.6	42

Means of Y_{max} and Y_i were significantly different, p = 0.05, by a t-test.

Thus, pre-dawn respiration rates were closely connected to the previous CO_2 fixation rates.

Partitioning

Growth of plants under elevated CO_2 results in changes in partitioning of photoassimilates to various plant organs over time (Table 4.2). In soybean, elevated CO_2 generally promoted greater carbon (dry matter) partitioning to the supporting structure (stems, petioles and roots) than to the leaf laminae during vegetative stages of growth (Allen et al., 1991). During reproductive stages, there tended to be lower relative partitioning to reproductive growth (pods) by plants under elevated CO_2.

Table 4.2. Soybean plant components as a percentage of total dry matter grown at subambient and superambient concentrations of CO_2 in 1984. Condensed from Allen et al. (1991).

Component	_____ CO_2 concentration, µmol/mol _____					
	160	220	280	330	660	990
			13 DAP[1] (V2 Stage)			
Root, %	11.3	9.8	9.3	12.4	9.8	10.5
Cotyledon, %	19.5	15.6	13.9	13.8	11.2	9.4
Stem, %	22.8	25.0	25.8	25.5	26.3	26.7
Leaf, %	46.1	49.5	50.7	48.2	52.5	53.2
			34 DAP (V8 Stage)			
Root, %	6.4	7.1	6.9	7.6	8.9	7.9
Stem, %	23.4	24.4	27.6	25.9	28.2	29.3
Petiole, %	12.2	13.1	14.6	14.4	15.7	15.7
Leaf, %	57.6	55.3	50.8	52.2	47.2	46.9
			66 DAP (R5 Stage)			
Stem, %	16.5	19.6	25.0	21.7	25.2	26.8
Petiole, %	9.9	11.3	12.9	12.7	14.1	13.6
Leaf, %	35.3	35.3	35.4	32.2	31.5	29.8
Pod, %	38.2	33.6	26.5	33.3	28.9	29.6
			94 DAP (R7 Stage)			
Stem, %	10.2	12.8	14.9	14.3	16.3	19.5
Petiole, %	5.5	6.2	6.0	6.7	7.0	6.5
Leaf, %	17.4	16.3	14.9	14.4	11.9	10.9
Pod, %	66.7	64.7	64.1	64.6	64.7	64.0

[1] Days after planting.

Growth rates

During the linear phase of vegetative growth after full ground cover is reached, the growth rates of plants exposed to a range of CO_2 concentrations varied from 5.0 to 20.7 g/m²/d for exposures from 160 to 990 µmol/mol (Allen et al., 1991). The total final dry weight ranged from 12.88 to 39.12 g/plant, and final seed weight ranged from 5.77 to 17.85 g/plant for CO_2 treatments ranging from 160 to 660 µmol/mol.

Carbohydrates

Soybean accumulates non-structural carbohydrates, particularly starch, under CO_2 enrichment. Allen et al. (1988) grew soybean under CO_2 treatments of 330, 450, 600 and 800 µmol/mol. Sucrose, reducing sugars, and total soluble sugars of leaves remained somewhat constant throughout the day, but starch increased steadily at a rate of about 6 g/kg dry matter/hour. Average total soluble sugars increased from 24 to 36 g/kg dry weight and starch increased from 85 to 204 g/kg dry weight across the range of 330 to 800 µmol/mol. Elevated non-structural carbohydrates in CO_2-enriched soybean plants were confirmed by Baker et al. (1989) and Allen et al. (1995). The concentrations also varied across the life cycle of the plants.

Nitrogen

Nitrogen content decreased from 50 to 37 g/kg dry weight over the range of 330 to 800 µmol/mol. When the nitrogen content was adjusted to remove the effect of total non-structural carbohydrate, the relative changes were smaller, 55 to 48 g N/kg dry matter, across the 330 to 800 µmol/mol range (Allen et al., 1988).

Yield

Soybean seed yield was always increased by elevated CO_2. Allen et al. (1987) summarized the photosynthetic, biomass and seed yield responses of several experiments with the equation (4.1) rectangular hyperbola model using data normalized to responses obtained at 330 µmol/mol. The values of K_m, Y_{max} and Y_i parameters for relative photosynthetic rates were 279 µmol/mol, 3.08 and −0.68, respectively, for relative biomass yield were 182 µmol/mol, 3.02 and −0.91, respectively, and for relative seed yield were 141 µmol/mol, 2.55 and −0.76, respectively. This model was used to project yields across several ranges of atmospheric CO_2 concentration increases (Table 4.3). For a doubling of CO_2, this model predicted a 32.2% increase in soybean grain yield and a 42.7% increase in biomass. The ratio of these two numbers, 1.322/1.427 = 0.926, gives the fraction of the harvest index expected under doubled CO_2 in comparison with ambient CO_2.

Table 4.3. Percentage increases of soybean midday photosynthetic rates, biomass yield, and seed yield predicted across selected carbon dioxide concentration [CO$_2$] ranges associated with relevant benchmark points in time. Adapted from Allen et al. (1987)

Period of time (years)	[CO$_2$]-Midday Initial	[CO$_2$]-Midday Final	Biomass photo- synthesis	Seed yield	Biomass yield
	(Nmd/mol)		(% increase over initial [CO$_2$])		
IA-1700[1]	200	270	38	33	24
1700-1973	270	330	19	16	12
1973-2073?[2]	330	660	50	41	31

[1] IA, the Ice Age about 13 000 to 30 000 years before present. The atmospheric CO$_2$ concentrations that prevailed during the last Ice Age, and from the end of the glacial melt until pre-pioneer/pre-industrial revolution times, were 200 and 270 µmol/mol, respectively.
[2] The first world energy 'crisis' occurred in 1973 when the CO$_2$ concentration was 330 µmol/mol. This CO$_2$ concentration is used as the basis for many CO$_2$ doubling studies. The CO$_2$ concentration is expected to double sometime within the 21st century.

Carbon dioxide and temperature

The CO$_2$ fertilization effect appears to be enhanced under elevated temperatures, at least up to a point. Idso et al. (1987) and Kimball et al. (1993) showed that the growth modification factor (or biomass growth modification ratio) due to a 300 µmol/mol enrichment was 0.08 per °C (average daily temperature) across the range of 12 to 34°C. However, for soybean, Allen (1991) calculated a biomass growth modification ratio response to temperature of −0.031 per °C for seed biomass yield and −0.026 per °C for total biomass accumulation. Elevated temperatures tended to shorten the grain-filling period of this crop.

Soybean seed yield tended to decrease slightly with temperature over the day/night range of 26/19 to 36/29°C (Table 4.4). The number of seed per plant increased slightly with increase of both CO$_2$ and temperature. Mass per seed decreased sharply with increasing temperature. Although CO$_2$ enrichment resulted in increased seed yield and above-ground biomass, harvest index was decreased with both CO$_2$ and temperature (Baker et al., 1989). The data of Table 4.4 show no tendency for the growth modification factor to increase with temperature for either seed yield or biomass accumulation.

Subsequent experiments with soybean substantiate these data although it was found that seed yields dropped sharply for day/night temperatures of 40/30°C and above, and biomass yields were maintained up to 44/34°C before rapidly failing (Deyun Pan et al., unpublished).

The sensitivity of the growth modification factor data of Idso et al. (1987) and Kimball et al. (1993) may have been affected by other factors at Phoenix, Arizona. Apparently, the growth vs. temperature data were obtained throughout the year, and the findings may have been impacted by factors such as total solar radiation, photoperiod, stages of development, or other conditions caused by the changing seasons. Of course, these other environmental factors are all part of the complex of

Table 4.4. Seed yield, components of yield, total above-ground biomass and harvest index of soybean grown at two CO_2 concentrations and three temperatures in 1987 (adapted from Baker et al., 1989)

CO_2 conc. (μmol/mol)	Day/night temperature (°C)	Grain yield (g/plant)	Seed/plant (no./plant)	Seed mass (mg/seed)	Above-ground biomass (g/plant)	Harvest index
330	26/19	9.0	44.7	202	17.1	0.53
330	31/24	10.1	52.1	195	19.8	0.51
330	36/29	10.1	58.9	172	22.2	0.45
660	26/19	13.1	58.8	223	26.6	0.49
660	31/24	12.5	63.2	198	27.6	0.45
660	36/29	11.6	70.1	165	26.5	0.44
		F-values				
CO_2 conc.		12.3**	11.4**	2.5*	NA	NA
Temperature		0.0 NS	8.4**	106.2**	NA	NA
CO_2 × Temperature		2.0 NS	0.1 NS	11.2**	NA	NA

*, ** Significant at the 0.05 and 0.01 probability levels, respectively.
NS = not significant; NA = not available.

plant responses to climatic conditions. However, a better test of temperature effects alone would be CO_2 enrichment throughout the season or life cycle of plants under natural conditions, but with consistent temperature differences (cooler and warmer).

Evapotranspiration and water-use efficiency

The direct effect of increasing temperatures across the range of 28 to 35°C appears to increase transpiration rate about 4 to 5% per °C, based on both experimental and modelling studies (Allen, 1991). This is in close agreement with the rise in saturation vapour pressure of about 6% per °C. Allen et al. (1985) showed that the increase in WUE of soybean contributed by increased photosynthesis under elevated CO_2 was much greater than the increase in WUE contributed by decreased transpiration. The fractional increase in WUE attributable to increased photosynthesis and decreased transpiration were about 0.8 and 0.2, respectively (based on data of Jones et al., 1985b).

RICE

Leaf and canopy photosynthesis

Leaf photosynthetic rates, in high light, of cv. IR72 rice, determined in 1992 (RICE VI experiment) for plants grown at three day/night temperature regimes, 32/23, 35/26 and 38/29°C (but measured at ambient temperature), averaged 18.8 and 30.4 μmol $CO_2/m^2/s$ for 330 and 660 μmol/mol CO_2 treatments, respectively (Allen et al., 1995).

THE CO_2 FERTILIZATION EFFECT

This increase caused by elevated CO_2 is about 60%. Leaf A/C_i curves obtained from both CO_2 treatments at the growth temperature of 35/26°C are shown in Figure 4.1. There were essentially no differences in A between the two CO_2 treatments at each specific C_i level. Similar pairs of curves (not shown) were obtained for the other temperature treatments. Each pair of leaf A/C_i curves were similar within each temperature regime. Thus, the leaf A/C_i curves gave no indication of a change in photosynthetic capacity. Extrapolations of the response curves to zero values of A gave an intercept (leaf CO_2 compensation point) of about 60 µmol/mol CO_2.

Following canopy closure, the rice canopy net photosynthetic rate (Pn) vs. photosynthetic photon flux density (PPFD) responses were linear and did not approach light saturation in any of the experiments, probably because of the erect leaf orientation of cv. IR30 and high plant populations. Canopy Pn vs. PPFD at 60 DAP for the six CO_2 treatments of the RICE II experiment (Baker and Allen, 1993b; Allen *et al.*, 1995) gave values of about 34, 50, 60, 80, 85 and 90 µmol $CO_2/m^2/s$ at high light (1600 µmol photons/m^2/s) for treatments of 160, 250, 330, 500, 660 and 900 µmol/mol.

Linear canopy Pn responses to PPFD on day 60 of the RICE IV experiment were similar among all temperature treatments (25/18, 28/21, 31/24, 34/27 and 37/30°C)

Figure 4.1. Photosynthesis CO_2 assimilation rate (A) as a function of intercellular CO_2 (C_i) for single, attached, fully expanded leaves of rice plants grown at CO_2 concentrations of 330 (open symbols) or 660 (filled symbols) µmol/mol and dry bulb air temperatures (day/night) of 35/26°C. Measurements were made on 29 October 1992 at an irradiance of 1 200 to 1 300 µmol (photons)/m^2/s. Adapted from Allen *et al.* (1995)

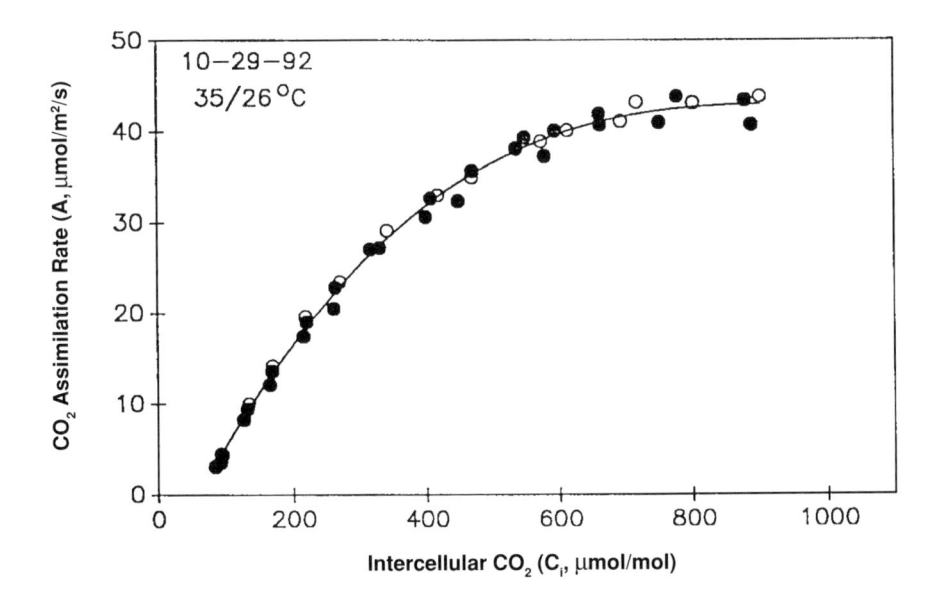

that were exposed to 660 μmol/mol CO_2; responses of the 330 μmol/mol treatment were about 25% less. Other studies of canopy Pn have shown only small differences across wide ranges of temperature; e.g., cotton (Baker et al., 1972) and soybean (Jones et al., 1985a). There may be two reasons for the lack of a clear canopy photosynthetic response to air temperature across the 25 to 37°C range. Evaporative cooling may lower foliage temperature below air temperature increasingly with increasing air temperature and increasing vapour pressure deficit (Allen, 1990; Pickering et al., 1995). The summed response of the photosynthetic rates of leaves to temperature at all exposures of light may broaden the temperature response for the whole canopy photosynthetic rates (Pickering et al., 1995).

Baker and Allen (1993b) fitted whole-day canopy Pn data to a rectangular hyperbola of the form of equation (4.1) for DAP 61 of the RICE II experiment. The parameters were 70.8 μmol/mol, 3.96 and −2.21, for K_m, Y_{max} and Y_i, respectively, with a relative Pn response ceiling of 1.75 at infinite [C]. The calculated percentage increase in response of whole-day canopy Pn at 660 vs. 330 μmol/mol was 36%.

Photosynthetic acclimation to CO_2 at the canopy level

To test for acclimation of canopy photosynthetic capacity, Baker et al. (1990b) used rice that was grown at 160, 250, 330, 500, 660 and 900 μmol/mol CO_2. For half-day periods (mornings) on 62, 63 and 64 DAP, common CO_2 setpoints of 160, 330 or 660 μmol/mol were imposed on each chamber on those respective days. Within each of these short-term CO_2 exposure comparisons, Baker et al. (1990b, 1996b) and Baker and Allen (1993a) showed that the short-term canopy net photosynthetic rate decreased with increasing long-term CO_2 treatment (Figure 4.2). The relative effects of the short-term CO_2 exposure were greatest for the lowest short-term CO_2 concentration. For example, when compared at a common short-term CO_2 exposure of 160 μmol/mol, the canopy photosynthetic rate of the chamber containing the 900 μmol/mol long-term CO_2 growth treatment was only about one-third of that of the 160 μmol/mol long-term CO_2 growth treatment (Figure 4.2). However, a large part of this apparent acclimation effect may be attributed to the greater respiration rates of the plants that had been grown under elevated CO_2 (Boote et al., 1994).

Rubisco protein percentage was used as evidence of leaf acclimation to a wide range of CO_2 concentrations (Baker and Allen, 1994; Allen et al., 1995), shown in Table 4.5 for 34 DAP soybean (Campbell et al., 1988) and 75 DAP rice (Rowland-Bamford et al., 1991). For rice, rubisco activity expressed on a leaf area basis decreased by 66% across the 160 to 900 μmol/mol long-term CO_2 treatments (Rowland-Bamford et al., 1991). A major cause of this decline in rubisco activity was a 32% decrease in the amount of rubisco protein relative to other soluble protein (Rowland-Bamford et al., 1991).

Although Rowland-Bamford et al. (1991) reported decreases in both amount and activity of rubisco with increasing CO_2 in rice plants grown across the range of 160 to 900 μmol/mol, the leaf A/C_i response curves gave no indication of a downward acclimation of photosynthetic capacity across the smaller range of 330 to 660 μmol/

Figure 4.2. Comparison of rice canopy net photosynthetic rate (Pn) vs. long-term [CO_2] acclimation treatment for rice canopies grown at subambient (160, 250), ambient (330), and superambient (500, 660 and 900 μmol CO_2/mol air) [CO_2] treatments in 1987. The Pn estimates were obtained during a short-term [CO_2] change-over study where the [CO_2] was maintained during the morning hours in all six long-term [CO_2] treatments at 160, 330 and 660 μmol/mol on days 62, 63 and 64 days after planting, respectively. The Pn was estimated from linear regression equations of Pn vs. photosynthetic photon flux density (PPFD) with PPFD set to 1500 μmol photons/m²/s. The vertical bars represent 95% confidence intervals. Adapted from Baker *et al.* (1990b, 1996b) and Allen *et al.* (1995)

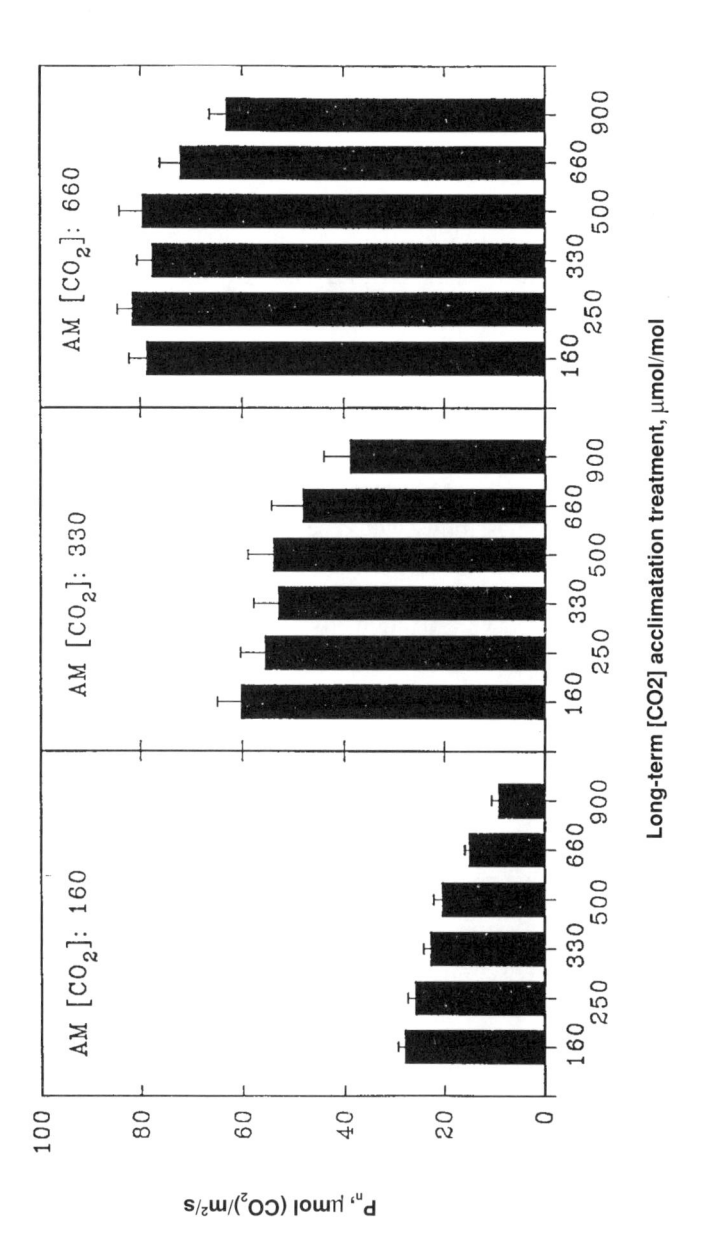

Table 4.5. For soybean, leaf blade soluble protein expressed on a leaf blade area basis and percentage rubisco protein expressed on a leaf blade soluble protein basis for 34-day-old soybean plants grown under a wide range of CO_2 concentrations. Adapted from Campbell *et al.* (1988) and Baker and Allen (1994). For rice, leaf nitrogen content expressed on a leaf area basis and percentage rubisco protein expressed on a leaf soluble protein basis for 75-day-old rice plants grown under a wide range of CO_2 concentrations. Adapted from Rowland-Bamford *et al.* (1991) and Baker and Allen (1994)

Soybean			Rice		
CO_2 growth concentration (µmol/mol)	Leaf soluble protein (g/m^2)	Rubisco protein (%)	CO_2 growth concentration (µmol/mol)	Leaf nitrogen protein (µmol/m^2)	Rubisco protein (%)
160	2.5	56	160	95	62
220	3.2	54	250	90	59
280	2.6	–	330	81	54
330	2.3	57	500	62	49
660	2.3	54	660	78	43
990	2.3	55	900	64	42

mol CO_2. One possibility would be that rubisco is not as limiting for photosynthesis under CO_2 enrichment as would be expected. More work is needed, under a range of CO_2 treatments, to explore the interaction effects of sink capacity, nitrogen nutrition, and other internal CO_2-fixation processes on photosynthetic behaviour and crop yield.

Development stages

In rice, two distinctly different rates of leaf appearance for vegetative and reproductive phases of growth were found (Baker *et al.*, 1990, 1996; Baker and Allen, 1994), as has been observed many times before (Yoshida, 1977; Vergara, 1980). Leaf appearance rates (leaves per day) are about twice as great in the vegetative phase as in the reproductive phase.

In the 'late' planted (23 June 1987) RICE II experiment, the number of mainstem leaves at panicle initiation and the final number of mainstem leaves decreased across CO_2 treatments of 160 to 500 µmol/mol and remained similar from 500 to 900 µmol/mol. Panicle initiation and boot stage occurred about 12 days earlier in the superambient CO_2 treatments compared with the 160 µmol/mol treatment (Baker *et al.*, 1990a; Baker and Allen, 1993b). Therefore, plant developmental rate was clearly accelerated with increasing CO_2 up to about 500 µmol/mol.

Temperature effect data were assembled from treatments ranging from 25/18/21 to 40/33/37°C (day/night/paddy water temperatures) in RICE I to RICE V experiments. Air temperature greatly influenced leaf appearance rate (Figure 4.3-A), developmental rate, and total growth duration, whereas, across the limited range of 330 to 660 µmol/mol the effects were comparatively small (Baker and Allen, 1993b; Baker *et al.*, 1992a,b, 1995). No consistent differences in phyllochron interval (days per leaf appearance) between 330 and 660 µmol/mol were observed, whereas phyllochron

THE CO_2 FERTILIZATION EFFECT

Figure 4.3. (A) Mainstem Haun scale growth units vs. days after planting for rice plants in two different CO_2 and temperature regimes from the RICE III experiment. (B) Phyllochron interval vs. temperature treatment for all five rice experiments. 'Vegetative' data are plotted against paddy water temperature and 'reproductive' data are plotted against average day/night temperature adjusted for thermoperiod. Adapted from Baker et al. (1995) and Allen et al. (1995)

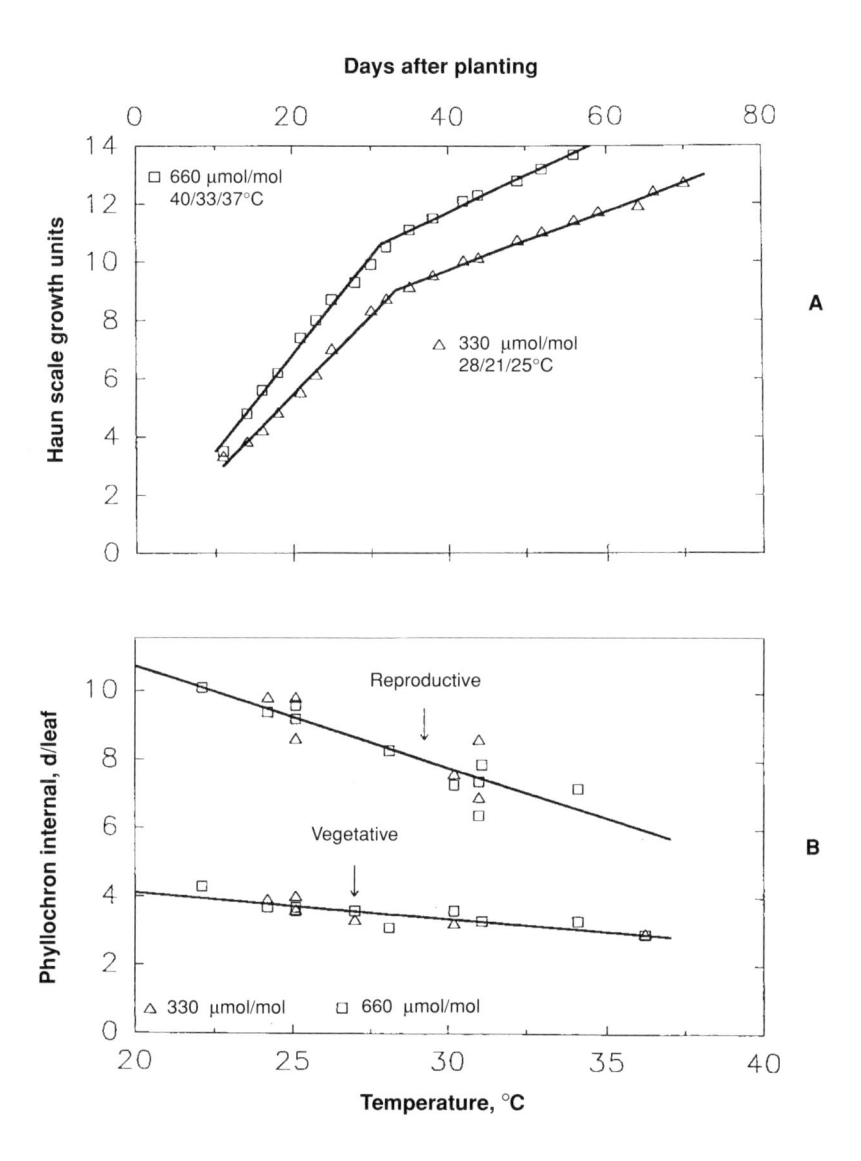

interval increased with increasing temperature across the range of 25/18/21 to 40/33/37°C, especially for the reproductive phase (Figure 4.3-B). In RICE III, IV and V experiments, anthesis ranged from 0 to 6 days earlier in the 660 vs. 330 µmol/mol treatments (Baker et al., 1995). Also, the number of days to anthesis was shortened by approximately 10 days across the temperature treatment range from 25/18/21 to 34/27/31°C for the RICE IV experiment.

Growth and yield

The yield and yield component data of the RICE I and RICE II experiments with CO_2 concentrations spanning 160 to 900 µmol/mol are given in Table 4.6. A rectangular hyperbola (equation (4.1)) was fitted to the seed yield data. Values of $K_m = 284$ µmol/mol, $Y_{max} = 2.24$ and $Y_i = -0.13$ were obtained for relative grain yield with asymptotic relative yield response ceiling $(Y_{max} + Y_i) = 2.11$ at infinite [C]. The calculated percentage increase in response at 660 vs. 330 µmol/mol was 44% for grain yield (Baker and Allen, 1993b).

Across a relatively wide temperature range from 25/18/21 to 37/30/34°C, Baker et al. (1992b) found a broad temperature optimum for biomass production in the mid-temperature ranges. Plants grown at 40/33/37°C were near the upper temperature limit for survival. High temperature spikelet sterility of rice is induced almost exclusively on the day of anthesis (Satake and Yoshida, 1978) when temperatures greater than 35°C for more than one hour induce a high percentage of sterility (Yoshida, 1981). In the 40/33/37°C treatments, plants in the 330 µmol/mol CO_2 chamber died during internode elongation whereas plants in the 660 µmol/mol chamber produced small, abnormally shaped panicles that were sterile (Baker et al., 1992a). Therefore, elevated CO_2 may slightly increase the maximum temperature at which rice plants can survive.

At both CO_2 levels, grain yield was highest in the 28/21/25°C treatment followed by a decline to zero yield in the 40/33/37°C treatment (Table 4.7). The CO_2 enrichment from 330 to 660 µmol/mol increased yield by increasing the number of panicles per plant, whereas the number of filled grains per panicle and individual seed mass were less affected. Temperature effects on yield and yield components were highly significant. The number of panicles per plant increased while the number of filled grains per panicle decreased sharply with increasing temperature treatment. Individual seed mass was stable at moderate temperatures but tended to decline at temperature treatments above 34/27/31°C. Final above-ground biomass and harvest index were increased by CO_2 enrichment while harvest index declined sharply with increasing temperature. Notably, there were no significant CO_2 × temperature interaction effects on yield, yield components, or final above-ground biomass (Table 4.7).

At each CO_2 concentration, polynomial regression equations were fitted to the rice seed yield (Y) vs. temperature (X) data of Table 4.7. A third-degree polynomial $[Y = -239.226 + 25.557*X - 0.848*X^2 + 0.00896*X^3$ $(R^2 = 0.91, S_{y.x} = 1.36)]$ provided the best fit for the 660 µmol/mol CO_2 treatments (Baker and Allen, 1993a). However, a second-degree polynomial $[Y = -1.196 + 1.042*X - 0.0278*X^2$ $(R^2 = 0.91,$

Table 4.6. Grain yield, components of yield, total above-ground biomass and harvest index of rice subambient and superambient CO_2 concentration experiments conducted in 1987. Adapted from Baker et al. (1988), Baker and Allen (1993b, 1994), Baker et al. (1995, 1996).

CO_2	Temperature[1]	Grain yield	Panicle/ plant	Filled grain	Grain mass	Biomass	Harvest index	Experiment RICE No.
μmol/mol	°C	Mg/ha	no./plant	no./panicle	mg/seed	g/plant		
160	31/31/27	3.4c[2]	3.6c	24.8a	17.0a	4.0c	0.36a	I & II
250	31/31/27	4.1c	4.8bc	20.8a	18.2a	5.1bc	0.34a	I & II
330	31/31/27	4.8bc	5.7ab	21.0a	17.6a	6.3ab	0.34a	I & II
500	31/31/27	6.8[3]	7.3	23.0	18.1	9.8	0.30	II only
660	31/31/27	6.6ab	6.5ab	25.0a	17.8a	8.4a	0.35a[4]	I & II
900	31/31/27	7.3a	7.4a[4]	24.8a	17.8a	8.2a	0.39a	I & II

[1] Daytime dry bulb air temperature/nighttime dry bulb air temperature/paddy water temperature.
[2] Values followed by the same letter in the same column are not significantly different by Duncan's Multiple Range Test (p = 0.05).
[3] Plants in 500 μmol/mol CO_2 treatment of the RICE I experiment experienced hydrogen sulphide toxicity in the root zone which inhibited seed fill and reduced grain yield to levels below those measured in subambient treatments. These data were not used in this analysis. The values in the 500 μmol/mol CO_2 treatment row are from the RICE II experiment only.
[4] Indicates significant difference between planting dates as determined by the t-test at the 0.05 level of confidence.

Table 4.7. Grain yield, components of yield, total above-ground biomass and harvest index for five separate rice experiments. Adapted from Baker and Allen (1993a) and Baker et al. (1996)

CO_2	Temperature[1]	Mean air temperature[2]	Grain yield	Panicle/ plant	Filled grain no./panicle	Grain mass	Biomass	Harvest index	Experiment No.
µmol/mol	°C	°C	Mg/ha	no./plant		mg/seed	g/plant		
330	28/21/25	24.2	7.9	5.1	34.5	17.4	7.3	0.47	RICE III
	28/21/25	25.1	6.6	3.9	39.6	17.5	6.5	0.44	RICE IV
	28/21/25	25.1	8.0	4.0	47.5	18.5	8.1	0.43	RICE V
	31/31/27	31.0	5.2	5.9	23.0	17.1	5.5	0.42	RICE I
	31/31/27	31.0	4.3	5.4	19.0	17.9	7.2	0.26	RICE II
	34/27/31	30.2	4.2	7.7	15.2	16.2	5.6	0.43	RICE III
	40/33/37	36.2	0.0	-.-	-.-	-.-	-.-	-.-	RICE III
660	25/18/21	22.1	8.4	4.4	46.0	18.2	7.8	0.47	RICE IV
	28/21/25	24.2	8.4	5.0	37.7	18.0	7.9	0.46	RICE III
	28/21/25	25.1	10.4	4.2	58.3	18.3	8.9	0.50	RICE IV
	28/21/25	25.1	10.1	4.4	54.2	19.0	9.3	0.47	RICE V
	31/31/27	31.0	6.8	6.9	25.1	17.2	7.5	0.40	RICE I
	31/31/27	31.0	6.4	6.0	24.8	18.4	9.3	0.29	RICE II
	34/27/31	30.2	4.8	6.5	18.5	16.7	6.3	0.32	RICE III
	34/27/31	31.1	3.4	7.5	12.9	16.3	8.1	0.18	RICE IV
	37/30/34	34.1	1.0	8.0	3.0	14.2	7.1	0.06	RICE IV
	40/33/37	36.2	0.0	-.-	-.-	-.-	-.-	-.-	RICE III
					F-values				
CO_2			4.2 *	4.3 *	0.4 NS	0.1 NS	13.1 **	4.6 *	
Temperature			51.8 **	22.6 **	51.1 **	27.0 **	3.2 *	32.8 **	
CO_2 × temperature			1.9 NS	1.7 NS	1.3 NS	0.3 NS	0.2 NS	1.9 NS	

[1] Daytime dry bulb air temperature/nighttime dry bulb air temperature/paddy water temperature.
[2] Mean air temperature is the average of day and night temperature adjusted for thermoperiod.
*,** Significant at the 0.01 and 0.05 probability levels, respectively. NS = Not significant.

THE CO$_2$ FERTILIZATION EFFECT

$S_{y.x} = 1.36$)] fits the data best for the 330 μmol/mol treatments. Figures 4.4 and 4.5 show the effect of temperature on seed yield and final biomass, respectively, from a number of experiments based on the tables and figures shown in Baker and Allen

Figure 4.4. Rice seed yield vs. weighted mean day/night air temperature for plants grown to maturity in CO$_2$ concentrations of 330 and 660 μmol/mol in five separate experiments

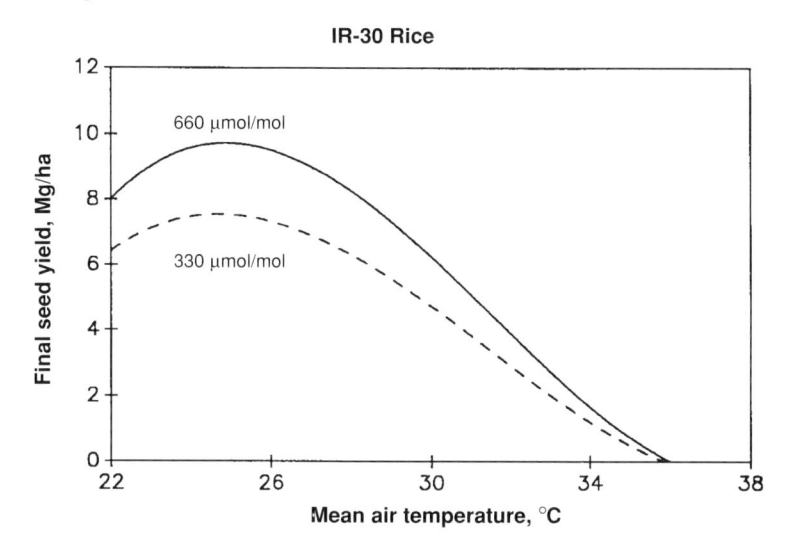

Figure 4.5. Second-degree polynomial fit of rice final biomass yield vs. weighted mean day/night air temperature for plants grown to maturity in CO$_2$ concentrations of 330 and 660 μmol/mol in five separate experiments

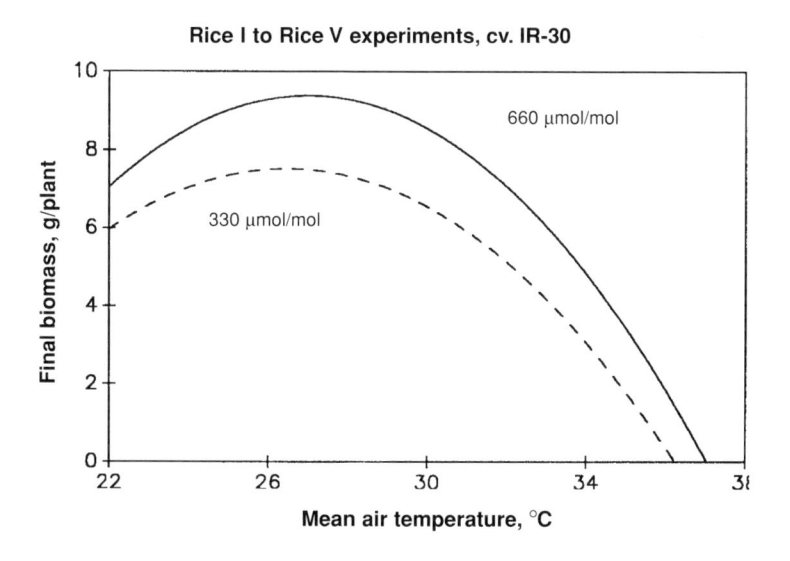

(1993a). While future increases in atmospheric CO_2 should benefit rice yields, large negative effects are likely if temperatures also rise. The figures show that vegetative productivity is maintained at higher temperatures than is reproductive growth.

Canopy dark respiration rates

Plant respiration rates may be decreased by both short- and long-term exposure to high CO_2 concentrations (Bunce, 1990; Amthor, 1991); however Baker et al. (1992c) found that nighttime canopy dark respiration rates [R_d, µmol (CO_2)/m² (ground area)/s] increased for rice exposed to daytime CO_2 ranging from 160 to 900 µmol/mol. Similar to photosynthetic rates, R_d increased with CO_2 exposure from 160 to 500 µmol/mol but levelled off somewhat across the 500 to 900 µmol/mol range. The R_d of the ambient and superambient CO_2 treatments reached a broad maximum around 30 to 50 DAP whereas the broad maximum of the subambient treatments occurred later around 50 to 70 DAP. The maximum values of R_d were about 6, 8, and 9 µmol/m²/s for the 160, 250 and 330 µmol/mol CO_2 treatments and about 11 to 12 µmol/m²/s for the three superambient CO_2 treatments.

Specific respiration rate [R_{dw}, µmol (CO_2)/s/kg (total above-ground dry matter)] decreased exponentially with DAP at all CO_2 exposures and was higher in the subambient (160 and 250 µmol/mol) than in the ambient (330 µmol/mol) and superambient (500, 660 and 900 µmol/mol treatments). At each CO_2 exposure level, the patterns of R_{dw} with DAP were very similar to the patterns of plant tissue nitrogen concentration with DAP (Baker et al., 1992c). Furthermore, R_{dw} was linearly related to total above-ground plant tissue nitrogen concentration [mg (N)/g (DW)] across the range of CO_2 exposures and the six dates of plant sampling ($R_{dw} = -13.0 + 0.952$ * [N]; $r^2=0.91$, $P=0.01$). Baker et al. (1992c) concluded that the CO_2 treatments affected R_{dw} mainly by altering the protein content of the plant tissue. Another explanation is that elevated CO_2 increased the amount of structural and non-structural carbohydrates in the plant tissues, so that a larger proportion of dry matter was sequestered in non-protein materials. This study showed no respiration acclimation to long-term CO_2 enrichment (indirect acclimation) of rice that could not be explained by nitrogen concentration of the plant tissues.

Evapotranspiration and water-use efficiency

Both plant transpiration and direct evaporation from the floodwater surface contribute to water use (evapotranspiration, ET) of rice growing in the SPAR chambers. Diurnal trends of ET followed diurnal patterns of solar irradiance (Baker et al., 1990b; Allen et al., 1995). After canopy closure, maximum ET rates and total daytime water losses were about 35 and 30% greater, respectively, from rice grown at 160 µmol/mol compared to rice grown at 900 µmol/mol CO_2 (Table 4.8). The ET rates were similar when all chambers were exposed for one-half day to the same CO_2 concentration (data not shown, Baker et al., 1990b), which demonstrates the effect of the exposure level of CO_2 on stomatal control of transpiration.

Temperature also has a large effect on diurnal ET rates (Baker and Allen, 1993a; Allen et al., 1995) and daytime water use (Table 4.8), mediated primarily through vapour pressure deficit of the air. Solar irradiance also has a large effect, probably through both the energy inputs to the canopy and through the stomatal opening response to light (directly or indirectly). Midday maximum ET rates were about 75 and 35% higher for rice grown at 40/33/37°C and 34/27/31°C, respectively, compared to rice grown at 28/21/25°C (Baker and Allen, 1993a; Allen et al., 1995).

Water-use efficiency: effects of CO_2

Daytime totals of CO_2 uptake, ET, and calculations of WUE vs. CO_2 treatment are shown in Table 4.8. Stomatal conductance decreases with increasing CO_2 concentration which can cause a reduction of both leaf and whole canopy transpiration. However, CO_2 enrichment may also increase canopy leaf surface area for transpiration, thereby offsetting some of the water savings (Jones et al., 1985b; Allen et al., 1985). The leaf area index of the RICE II experiment ranged from 7.6 to 10.8 across

Table 4.8. Comparison of total daytime responses of rice canopies grown season-long in various CO_2 concentration (RICE II, 23 August 1987, 61 days after planting) and temperature (RICE IV, 10 September 1989, 58 days after planting) treatment regimes. Daytime dry bulb air temperature/ nighttime dry bulb air temperature/paddy water temperature:dewpoint temperature treatments are given in the second column. Adapted from Baker et al. (1990b) and Baker and Allen (1993a,b, 1994).

CO_2	Temperature[1]	Total daytime photons	Total daytime CO_2 uptake	Total daytime H_2O loss	Water-use efficiency[2]
μmol/mol	°C	mol(Photons)/m²	mol(CO_2)/m²	mol(H_2O)/m²	mmol/mol
Effects of carbon dioxide: RICE II, 23 August 1987, 61 DAP					
160	31/31/27:18	39.4	0.74	608.5	1.22
250	31/31/27:18		1.13	643.4	1.76
330	31/31/27:18		1.34	611.0	2.19
500	31/31/27:18		1.80	553.6	3.25
660	31/31/27:18		1.79	536.2	3.34
900	31/31/27:18		1.83	469.5	3.90
Effects of temperature: RICE IV, 10 September 1989, 58 DAP					
330	28/21/25:12.0	26.4	0.87	612.7	1.42
660	25/18/21:10.5		0.83	359.3	2.30
660	28/21/25:12.0		1.13	475.8	2.37
660	31/24/28:13.5		0.83	491.3	1.69
660	34/27/31:15.0		0.89	736.4	1.20
660	37/30/34:16.5		0.97	909.4	1.06

[1] Daytime dry bulb air temperature/nighttime dry bulb air temperature/paddy water temperature: dewpoint temperature.
[2] mmol (CO_2)/mol (H_2O).

the CO_2 treatments from 160 to 900 μmol/mol. Calculations of WUE increased with increasing CO_2 (Table 4.8) due to the decline in ET across this CO_2 range and the increase in Pn with CO_2 up to the 500 μmol/mol treatment.

Water-use efficiency: effects of temperature

In the 28°C air temperature treatment, CO_2 enrichment from 330 to 660 μmol/mol resulted in 30% increase in daytime total CO_2 uptake and 22% decrease in ET with a concomitant increase of WUE from 1.42 to 2.37 μmol (CO_2)/mmol (H_2O), an increase of 67% (Table 4.8). Daytime ET was more than doubled from the 25 to the 37°C daytime temperature treatment. WUE declined from the 28 to the 37°C daytime treatment due to the sharp increase in ET and the relatively stable Pn across this temperature range. If WUE were based on grain yield, it would decrease drastically with increasing temperature not only because of increasing ET, but also because of sharply decreasing seed production.

Rising atmospheric CO_2 is likely to benefit rice production by increasing photosynthesis, growth and grain yield while reducing water use and increasing WUE. In warm areas of the world, however, possible future global warming may result in substantial yield decreases because of the sensitivity of flowering and seed set to high temperatures and the possibility of water shortages that may result from increased evapotranspiration.

SUMMARY OF COMPREHENSIVE REVIEWS

Several recent symposia proceedings and reviews leave little doubt that crop plants can respond well to elevated CO_2 (Rozema et al., 1993; Woodwell and Mackenzie, 1995; Wittwer, 1995). Poorter (1993) compiled information from 156 plant species and found that doubling CO_2 provided an average growth increase of 37%. The distribution of weight ratios of CO_2-enriched and control plants is shown in Figure 4.6. Poorter's compilation showed a 41 and a 22% increase for C_3 and C_4 plants, respectively. As a group, C_3 herbaceous crop plants responded more than wild herbaceous species (58 vs. 35%). Furthermore, the fast growing wild species responded more strongly than slow-growing wild species (54 vs. 23%).

Poorter (1993) imposed two restrictions on his compilations that may have led to larger than expected responses to elevated CO_2. Firstly, plants grown in competition were not included. Secondly, only vegetative stages of plants were compared since compiled data were selected prior to flowering.

A number of studies have shown that vegetative growth responses may be greater than reproductive (seed yield) responses. Therefore, the compiled data of Poorter may give an impression of greater response than would be observed throughout the life cycle. Secondly, crops in field conditions usually are grown in dense populations where they compete for space and light. Under more realistic field conditions, crop plants are likely to respond as a community rather than individual plants, wherein

Figure 4.6. Distribution of the biomass growth modification ratio (weight ratio) of CO_2-enriched plants (600 to 720 μmol/mol) in comparison with control treatments (300 to 360 μmol/mol). The bar graph was created from averages of all the weight ratios of 156 species selected from the literature. Source: Poorter (1993)

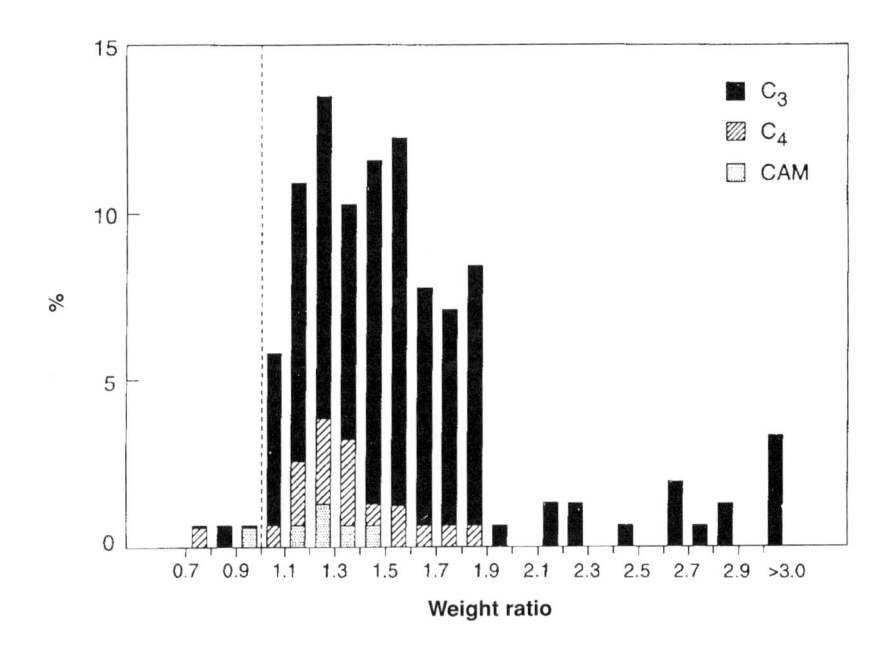

light (solar radiation) becomes a limiting factor for growth. Under these conditions, elevated CO_2 cannot promote horizontal expansion and greater light capture. Although the actual field responses may be less, the CO_2 fertilization effect is clearly well-established.

The CO_2 fertilization effect for forest species has also become firmly established. Wullschleger et al. (1994) estimated the biotic growth factor for 58 controlled-exposure studies of forest tree species which included 398 observations. Their frequency distribution of relative growth response of trees grown at elevated CO_2 vs. ambient CO_2 is given in Figure 4.7. They also found a mean response ratio of 1.32. Of the 398 observations, 51 showed a relative growth response less than 1, and 31 showed responses greater than 2. However, under competitive conditions, tree response may be much less (Bazzaz et al., 1995).

Kimball et al. (1993) discussed the data of Idso et al. (1987) which related the growth modification factor (or biomass growth modification ratio) caused by a 300 μmol/mol increase in CO_2 concentration above ambient. Their data are reproduced as Figure 4.8. Under Phoenix, Arizona, conditions, the growth modification factor across

Figure 4.7. Frequency distribution of the log-transformed biomass growth modification ratio ($\log_{10}$ elevated/ambient CO_2 exposure total dry mass ratios) of 73 tree species. Source: Wullschleger et al. (1994)

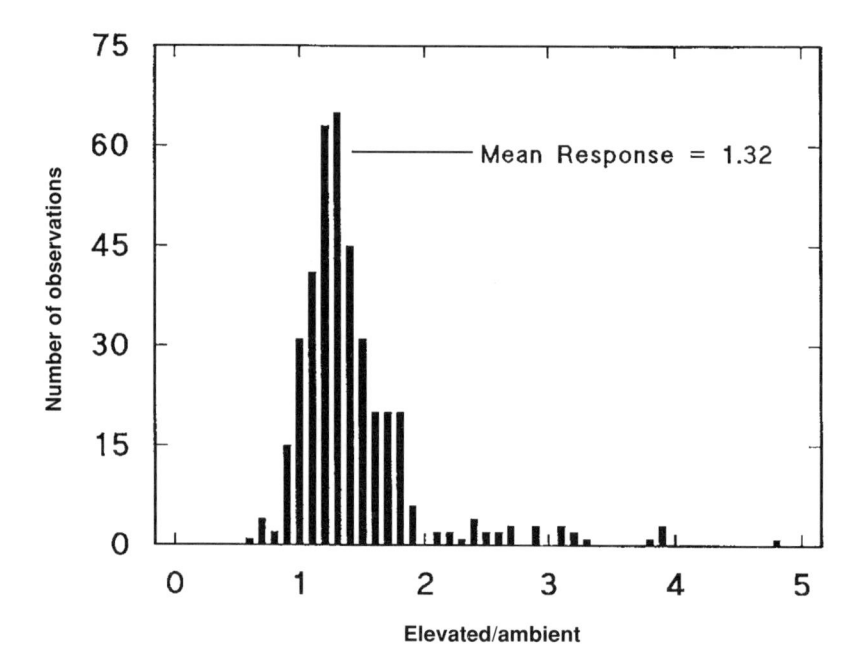

a mean temperature range of 12 to 34 °C was rather linear. Thus, the *vegetative* response to CO_2 should be enhanced at increased temperatures. As discussed before, this growth modification factor may not apply for reproductive growth responses (seed yield) of crops (e.g., rice).

CROP MODELLING: PREDICTIONS FOR THE FUTURE

MODELLING CROP RESPONSES TO CO_2 AND CLIMATE CHANGES

Peart et al. (1989) and Curry et al. (1990a,b, 1995) predicted growth and yield responses of soybean and maize to doubled-CO_2 climate change scenarios of the southeastern USA. Their simulations used 30 years of baseline weather data (1951-1980) from 19 sites in 11 states. Predicted climate changes of the Goddard Institute for Space Studies (GISS) model (Hansen et al., 1988) and the Geophysical Fluid Dynamics Laboratory (GFDL) model (Manabe and Wetherald, 1987) were used to change temperatures, precipitation and solar radiation, month-by-month, for the 30 years of baseline data at each site. These modified baseline weather data provided GISS and

THE CO_2 FERTILIZATION EFFECT

Figure 4.8. Biomass growth modification ratio (growth modification factor or relative increase in biomass growth) resulting from a 300 μmol/mol increase in CO_2 concentration above ambient (almost doubled) versus mean daily air temperature for the plants indicated in the legend. Source: Kimball *et al.* (1993)

CO_2 × Temperature interaction

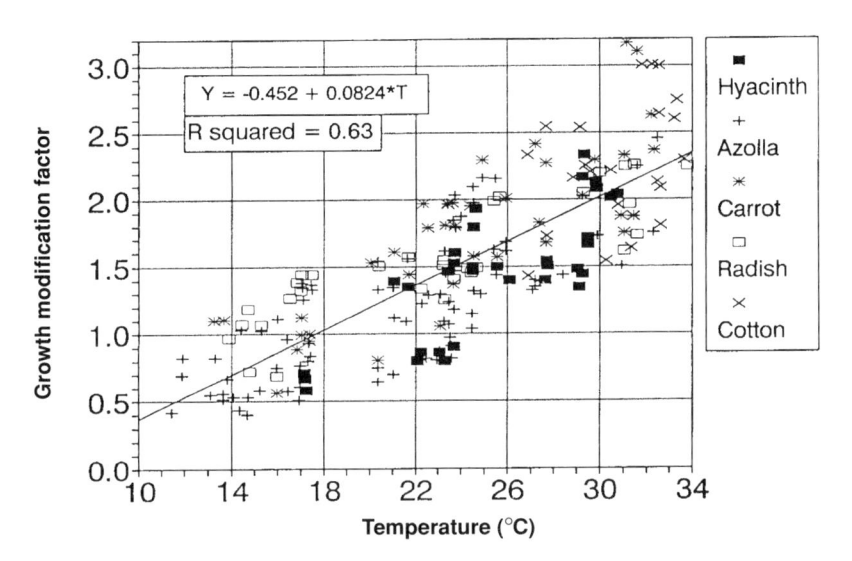

Y = -0.452 + 0.0824*T
R squared = 0.63

Legend: Hyacinth, Azolla, Carrot, Radish, Cotton

GFDL climatic change scenarios (Smith and Tirpak, 1989). At each site, prevailing cropped soil types, planting dates, and cultivars were used in the simulations.

Baseline, GISS and GFDL scenario temperature and rainfall data that were used by Peart *et al.* (1989) were condensed for two representative locations; Columbia, South Carolina, and Memphis, Tennessee (Table 4.9). Crop yield responses to climate change were simulated under four conditions: with or without direct CO_2 fertilization effects, and under rainfed or optimum irrigation culture. The Penman-Monteith equation, which contains a term for canopy conductance, was used to compute the effects of elevated CO_2 on canopy transpiration. Simulations were run for doubled-CO_2 conditions with a crop photosynthetic enhancement factor of 1.35 for soybean (a C_3 plant) and 1.10 for maize (a C_4 plant). No simulations were run with CO_2 fertilization effects only (without climate change effects).

Simulated soybean seed yields were averaged over 30 years and 19 sites (Table 4.10). Under rainfed conditions with climate change effects only, simulated soybean yields under the GFDL scenario were reduced 71% compared to the baseline climate, whereas the average yields were reduced only 23% under the GISS scenario. Yields under the GFDL scenario were severely impacted because of the rainfall reductions predicted by this model (Table 4.9).

Table 4.9. Summary of 30-year mean (1951-1980) baseline (BASE) weather data of two locations with GISS and GFDL scenarios. Condensed from Peart et al. (1989)

	BASE			GISS			GFDL		
	Prec. (mm)	T_{max} (°C)	T_{min} (°C)	Prec. (mm)	T_{max} (°C)	T_{min} (°C)	Prec. (mm)	T_{max} (°C)	T_{min} (°C)
Columbia, SC								
APR-SEP; Jul	684	33.3	21.2	817	35.2	23.1	471	38.2	26.1
OCT-MAR; Jan	561	13.5	0.7	602	15.8	3.0	571	15.6	2.8
TOTAL; Mean	1 245	24.1	10.7	1 419	26.6	13.2	1042	27.3	13.9
Memphis, TN								
APR-SEP; Jul	656	33.1	22.6	748	35.4	24.9	549	36.0	25.5
OCT-MAR; Jan	654	9.0	-0.6	565	12.0	2.4	758	11.3	1.7
TOTAL; Mean	1 310	22.0	11.1	1 313	25.4	14.5	1307	24.8	13.9

Under optimum irrigation conditions, average yields under both the GISS and GFDL scenarios were reduced 18 to 19%. However, in spite of higher temperatures, the irrigated yields under the GISS and GFDL scenarios were about 25% greater than the baseline climate scenario without irrigation.

When CO_2 fertilization effects were included with climate change effects (Table 4.10), simulated average yields of the GISS climate scenario were increased 11% under rainfed conditions, whereas yields under the GFDL climate scenario were still decreased (−52%). Under optimum irrigation with CO_2 fertilization effects, yields under both GISS and GFDL scenarios were increased 13 to 14%.

Yields of maize were simulated for doubled-CO_2 fertilization effects at only four weather station locations (Charlotte, North Carolina; Macon, Georgia; Memphis, Tennessee; Meridian, Mississippi). For climatic change effects alone, predicted maize yields declined only 6% in the GISS scenario, but declined 73% in the GFDL scenario (Table 4.11). Although irrigation increased predicted crop yields, the GISS and GFDL climate scenarios gave yield decreases of 18 and 27%, respectively, with respect to the irrigated baseline weather. The yield reduction of the GFDL scenario with respect to the GISS scenario was 10%, attributable to slightly higher temperatures. Including the CO_2 fertilization effects with climatic change scenarios had little effect on the predicted yields of maize because it is a C_4 plant.

In the Great Lakes and Corn Belt Area, simulations by Ritchie et al. (1989) showed that higher temperature would have the greatest effect of the climate change factors on predicted yields of soybean and maize, mainly through decreases in the crop life cycle. For the southern part of this region, yield reductions were greater for the GFDL scenario. Predicted yields increased for the northernmost stations because temperatures and growing season duration became more favourable. Average irrigation water requirements in this region increased about 90% for the GISS and GFDL scenarios.

Rosenzweig (1989) modelled maize and wheat yields in the Great Plains under

THE CO$_2$ FERTILIZATION EFFECT

Table 4.10. Doubled CO$_2$ soybean yield simulations (SOYGRO) for the southeastern USA

BASE	GISS Model		GFDL Model		Model
Yield (kg/ha)	Yield (kg/ha)	% Diff.	Yield (kg/ha)	% Diff.	% Diff.
Climate Change Effects Only, Rainfed					
2 497	1 929	−23	733	−71	−62
Climate Change Effects Only, Irrigated					
3 837	3 158	−18	3 092	-19	−2
CO$_2$ Fertilization plus Climate Change Effects, Rainfed					
2 497	2 780	+11	1 206	−52	−57
CO$_2$ Fertilization plus Climate Change Effects, Irrigated					
3 837	4 350	+13	4 393	+14	+1

Adapted from Peart *et al.* (1989).

GISS and GFDL scenarios. Yields decreased in the Southern Great Plains because higher temperatures shortened the life cycle of the crops. Where precipitation was predicted to decrease, irrigation requirements increased. In a separate modelling study, Allen and Gichuki (1989) predicted a 15% increased requirement for irrigation for

Table 4.11. Doubled CO$_2$ maize yield simulations (CERES-Maize) for Charlotte NC, Macon GA, Meridian MS, and Memphis TN

BASE	GISS Model		GFDL Model		Model
Yield (kg/ha)	Yield (kg/ha)	% Diff.	Yield (kg/ha)	% Diff.	% Diff.
Climate Change Effects Only, Rainfed					
8 468	7 926	−6	2 289	−73	−71
Climate Change Effects Only, Irrigated					
13,899	11 455	−18	10 257	−26	−10
CO$_2$ Fertilization plus Climate Change Effects, Rainfed					
8 577	8 136	−5	2 224	−74	−73
CO$_2$ Fertilization plus Climate Change Effects, Irrigated					
14 052	11 545	−18	10 363	−26	−10

Adapted from Peart *et al.* (1989).

this region, with greater requirements for alfalfa because its growing season was increased, and lower requirements for maize and winter wheat because their growing seasons were decreased. The CO_2 fertilization effect offset the adverse effects of climate change at some locations of Ritchie *et al.* (1989) and Rosenzweig (1989).

Dudek (1989) predicted productivity changes of several Californian vegetable, fruit and nut crops in response to GISS and GFDL scenarios for doubled CO_2. Without CO_2 fertilization effects, statewide average yield changes, depending on the crop, would be −8 to −34% for the GISS scenario and −6 to −31% for the GFDL scenario. With CO_2 fertilization effects, predicted statewide yield changes ranged from about +17 to −12% for the GISS scenario and about +21 to −8% for the GFDL scenario.

Dudek (1989) used a California Agriculture and Resource Model (CARM) to further predict economic and market impact of the productivity changes. Production declined, in general, under the scenarios of climate change without CO_2 fertilization effects; however, commodity prices generally increased. Under the CO_2 fertilization plus climate change effects scenarios, predicted impacts on commodity prices were much less.

The effect of temperature on the phenology of crop plants plays a critical role. One crucial need is for more detailed research on responses of plants to temperature, and temperature × CO_2 interactions, as inputs to crop models. More modelling studies are also needed on different planting dates as an adaptive strategy. Cultivars need to be designed for future climatic conditions (Hall and Allen, 1993). Factors that should be considered are: extension of the grain-filling period and perhaps shortening the duration of vegetative growth (which would also improve harvest index); ability to flower and set seed at higher temperatures; photoperiod and thermoperiod adaptive interactions; selection for positive photosynthetic acclimation where negative photosynthetic acclimation has been observed; and capability of utilizing photoassimilates (storage carbohydrates) more effectively. These factors need to be integrated into whole-plant physiology under real-world conditions.

ADAPTATIONS AND EVAPOTRANSPIRATION REQUIREMENTS

The simulated crop yield responses to climatic changes provided by Peart *et al.* (1989) and Curry *et al.* (1990a,b, 1995) manifest two main points: (a) the serious adverse impact of inadequate rainfall scenarios on crop production coupled with rising temperature scenarios, and (b) the importance of beneficial CO_2 fertilization effects in the face of elevated temperatures. However, the climate change for an effective doubling of CO_2 may occur at CO_2 concentrations less than those used in this simulation, if radiatively active trace gases other than CO_2 play a large role in the greenhouse effect. In that case, the direct CO_2 effects would be somewhat lower than shown in the example of Tables 4.10 and 4.11 for an equivalent climate change. All of the simulations assumed that climatic changes would occur simultaneously with increasing concentrations of CO_2 and other trace greenhouse-effect gases. If global warming lags the increases of atmospheric CO_2, then some beneficial effects of CO_2 fertilization are likely to occur before the full impact of climate change is manifested. However, Broecker (1987) and others caution that climate changes have not always been gradual during interglacial periods of the Pleistocene Era. There is clear evidence of relatively rapid climatic oscillations in the northern hemisphere during the previous interglacial period 110 000 to 140 000 years before present based on Greenland Ice-core Project (GRIP) records (Anklin et al., 1993). These oscillations produced cold periods that were as severe as the preceeding glacial period.

Much of the reduction in soybean yields reported by Peart et al. (1989) and Curry et al. (1990a,b, 1995) was caused by decreases in the length of the grain-filling period under higher temperatures. Changes in management practices, such as changing planting dates or selection of other cultivars, may help to prevent some of the potential reductions in yield. In the future, plant breeders may need to adapt combinations of temperature tolerance and photoperiod responses into new germplasm. In situations where non-structural carbohydrates accumulate as a CO_2 fertilization effect response, new germplasm needs to be developed that can make better use of the photoassimilate source.

Irrigation is not likely to be a panacea for climate change. Predicted irrigation requirements for soybean in the southeastern USA were increased 33 and 134% under the GISS and GFDL scenarios, respectively, based on simulations of Peart et al. (1989) and Curry et al. (1990a,b). However, under the GFDL scenario, water resources would become scarce, and may not be readily available for crops. Some areas of the USA may have to adapt by irrigating less land area. Increasing temperatures and decreasing precipitation for the USA as predicted by the GFDL model would have a serious negative impact overall on agricultural productivity although producers in favourable regions may benefit from scarcity-mediated higher prices (Adams et al., 1990).

ASSESSING INTERNATIONAL IMPACTS OF RISING CO_2 AND CLIMATE CHANGE

Several assessments have been conducted on the impacts of rising CO_2 and global climatic changes on crop production patterns and economics responses within national or regional zones (Adams et al., 1990; Crosson, 1993; Parry et al., 1988; Smith and Tirpak, 1989) and in various countries around the world (Rosenzweig and Iglesias, 1994; Rosenzweig et al., 1995). Furthermore, progress has been made on predicting impacts of climate change using world trade models (Rosenzweig and Parry, 1993, 1994).

The studies conducted by Rosenzweig and colleagues used existing crop models and climate change scenarios from three GCMs (GISS, GFDL and United Kingdom Meteorological Office Model (UKMO), Wilson and Mitchell, 1987) to predict changes in crop yields for a number of countries around the world. The doubled-CO_2 climate change scenarios (temperature, rainfall and evaporation changes) were based on the climate change potential expected from increases of all greenhouse-effect gases. These climate changes are expected to occur well before CO_2 concentration has actually doubled. Therefore, the CO_2 levels for fertilization effects were estimated to be

555 µmol/mol rather than 660 µmol/mol. Thus, the CO_2 fertilization effect photosynthetic ratios for four crops (soybean, wheat, rice and maize) were taken as 1.21, 1.17, 1.17 and 1.06, respectively.

The impact of climate change scenarios was more severe in the tropical latitudes than in the mid- or high-latitudes. For example, averaged over all three GCM scenarios, wheat production changes predicted in Brazil, India, China and Canada were −47, −43, −11 and −20%, respectively, without CO_2 effects, and −28, −13, +8 and +16% with CO_2 fertilization effects (calculated from data of Rosenzweig and Parry, 1993; Rosenzweig and Iglesias, 1994).

Simulations of soybean yields throughout the USA were a part of this international study (Curry *et al.*, 1995). They found that aggregated yields were 2.42, 2.80, 2.37 and 1.31 Mg/ha for the BASE, GISS, GFDL and UKMO scenarios, respectively. Temperatures were about 4.0°C higher than BASE for the GISS and GFDL scenarios, but were about 5.2°C higher for the UKMO model.

Refinements and improvements in prediction methodology will continue, but these assessments provide the best currently available insights into the CO_2 fertilization and climate change effects on global crop productivity.

SUMMARY AND CONCLUSIONS

Elevated CO_2 increases the size and dry weight of most C_3 plants and plant components. Relatively more photoassimilate is partitioned into structural components (stems and petioles) during vegetative development in order to support the light-harvesting apparatus (leaves). The harvest index tends to decrease with increasing CO_2 concentration and temperature. Selection of plants that could partition more photoassimilates to reproductive growth should be a goal for future research. As more is learned about the effects of anticipated climate changes on crops, more effort should be directed to exploring biological adaptations and management systems for reducing these impacts on agriculture and humanity. Whether regional climates become drier or wetter with global warming remains to be seen.

ACKNOWLEDGEMENTS

Supported in part by the US Department of Energy Interagency Agreements DE-AI02-93ER61720, DE-AI05-88ER69014, and DE-AI01-81ER60001; and by US EPA Interagency Agreement DW12934099 with the US Department of Agriculture, Agricultural Research Service. This work was conducted in cooperation with the University of Florida at Gainesville. Florida Agricultural Experiment Station Journal Series No. R-00000.

REFERENCES

Acock, B. and Allen, L.H. Jr. 1985. Crop responses to elevated carbon dioxide concentration. In: *Direct Effects of Increasing Carbon Dioxide on Vegetation*. DOE/ER-0238. B.R. Strain and J.D. Cure (eds.). US Dept. of Energy, Carbon Dioxide Res. Div., Washington DC. pp. 53-97.

Adams, R.M., Rosenzweig, C., Peart, R.M., Ritchie, J.T., McCarl, B.A., Glyer, J.D., Curry, R.B., Jones, J.W., Boote, K.J. and Allen, L.H. Jr. 1990. Global climate change and US agriculture. *Nature* **345**: 219-224.

Allen, L.H. Jr. 1990. Plant responses to rising carbon dioxide and potential interactions with air pollutants. *J. Environ. Qual.* **19**: 15-34.

Allen, L.H. Jr. 1991. Effects of increasing carbon dioxide levels and climate change on plant growth, evapotranspiration, and water resources. In: *Proceedings of a Colloquium on Managing Water Resources in the West Under Conditions of Climatic Uncertainty*. 14-16 Nov. 1990, Scottsdale, AZ. National Research Council, National Academy Press, Washington DC. pp. 101-147.

Allen, L.H. Jr. 1994. Carbon dioxide increase: Direct impacts on crops and indirect effects mediated through anticipated climatic changes. In: *Physiology and Determination of Crop Yield*. K.J. Boote, J.M. Bennett, T.R. Sinclair and G.M. Paulsen (eds.). American Society of Agronomy, Crop Science Society of America, and Soil Science Society of America, Madison, Wisconsin. pp. 425-459.

Allen, L.H. Jr. and Amthor, J.S. 1995. Plant physiological responses to elevated CO_2, temperature, air pollution, and UV-B radiation. In: *Biotic Feedbacks in the Global Climatic System: Will the Warming Increase the Warming?* G.M. Woodwell and F.T. Mackenzie (eds.). Oxford University Press, New York. pp. 51-84.

Allen, R.G. and Gichuki, F.N. 1989. Effects of projected CO_2-induced climatic changes on irrigation water requirements in the Great Plains States (Texas, Oklahoma, Kansas, and Nebraska). In: *The Potential Effects of Global Climate Change on the United States*. Appendix C, Agriculture, Vol. 1. EPA-230-05-89-053. J.B. Smith and D.A. Tirpak (eds.). US Environmental Protection Agency, Washington DC. pp. 6-1 to 6-42.

Allen, L.H. Jr., Jones, P. and Jones, J.W. 1985. Rising atmospheric CO_2 and evapotranspiration. In: *Advances in Evapotranspiration. ASAE Pub. 14-85*. American Society of Agricultural Engineers, St. Joseph, Michigan. pp. 13-27.

Allen, L.H. Jr., Boote, K.J., Jones, J.W., Jones, P.H., Valle, R.R., Acock, B., Rogers, H.H. and Dahlman, R.C. 1987. Response of vegetation to rising carbon dioxide: Photosynthesis, biomass, and seed yield of soybean. *Global Biogeochemical Cycles* **1**: 1-14.

Allen, L.H. Jr., Vu, J.C.V., Valle, R.R., Boote, K.J. and Jones, P.H. 1988. Nonstructural carbohydrates and nitrogen of soybean grown under carbon dioxide enrichment. *Crop Sci.* **28**: 84-94.

Allen, L.H. Jr., Valle, R.R., Mishoe, J.W., Jones, J.W. and Jones, P.H. 1990. Soybean leaf gas exchange responses to CO_2 enrichment. *Soil Crop Sci. Soc. Fla. Proc.* **49**: 192-198.

Allen, L.H. Jr., Bisbal, E.C., Boote, K.J. and Jones, P.H. 1991. Soybean dry matter allocation under subambient and superambient levels of carbon dioxide. *Agron. J.* **83**: 875-883.

Allen, L.H. Jr., Baker, J.T., Albrecht, S.L., Boote, K.J., Pan, D. and Vu, J.C.V. 1995. Carbon dioxide and temperature effects on rice. In: *Climate Change and Rice.* S. Peng, K.T. Ingram, H.-U. Neue, and L.H. Ziska (eds.). Springer-Verlag, Berlin, Heidelberg. pp. 256-277.

Amthor, J.S. 1991. Respiration in a future, higher-CO_2 world. *Plant Cell Environ.* **14**: 13-20.

Amthor, J.S. 1995. Plant respiration responses to elevated partial CO_2 pressures. In: *Advances in Carbon Dioxide Effects Research.* L.H. Allen, Jr., M.B. Kirkham, C. Whitman and D.M. Olzyk (eds.). Special Pub., American Society of Agronomy, Madison, Wisconsin.

Anklin, M., and thirty-nine other Greenland Ice-core Project (GRIP) members. 1993. Climate instability during the last interglacial period recorded in the GRIP ice core. *Nature* **364**: 203-207.

Arp, W.J. 1991. Effects of source-sink relations on photosynthetic acclimation to elevated CO_2. *Plant Cell Environ.* **14**: 869-875.

Baker, J.T. and Allen, L.H. Jr. 1993a. Contrasting crop species responses to CO_2 and temperature: Rice, soybean, and citrus. *Vegetatio* **104/105**: 239-260. Also: pp. 239-260. In: *CO_2 and Biosphere.* (Advances in Vegetation Science 14). J. Rozema, H. Lambers, S.C. van de Geijn and M.L. Cambridge (eds.). Kluwer Academic Publishers, Dordrecht.

Baker, J.T. and Allen, L.H. Jr. 1993b. Effects of CO_2 and temperature on rice: A summary of five growing seasons. *J. Agric. Meteorol. (Japan)* **48**: 575-582.

Baker, J.T. and Allen, L.H. Jr. 1994. Assessment of the impact of rising carbon dioxide and other potential climate changes on vegetation. *Environ. Pollut.* **83**: 223-235.

Baker, D.N. and Enoch, H.Z. 1983. Plant growth and development. In: *CO_2 and Plants.* E.R. Lemon (ed.). Westview Press, Boulder, Colorado. pp. 107-130.

Baker, D.N., Hesketh, J.D. and Duncan, W.G. 1972. Simulation of growth and yield in cotton: I. Gross photosynthesis, respiration, and growth. *Crop Sci.* **12**: 431-435.

Baker, J.T., Allen, L.H. Jr., Boote, K.J., Rowland-Bamford, A.J., Jones, J.W., Jones, P.H., Bowes, G. and Albrecht, S.L. 1988. Response of vegetation to carbon dioxide, Ser. No. 043. Response of rice to subambient and superambient carbon dioxide concentrations, 1986-1987. Report of the US Dept. of Agriculture, Agricultural Research Service (in cooperation with the University of Florida) for the US Dept. of Energy, Carbon Dioxide Research Division, Office of Energy Research, Washington DC.

Baker, J.T., Allen, L.H. Jr., Boote, K.J., Jones, P. and Jones, J.W. 1989. Response of soybean to air temperature and carbon dioxide concentration. *Crop Sci.* **29**: 98-105.

Baker, J.T., Allen, L.H. Jr., Boote, K.J., Jones, J.W. and Jones, P. 1990a. Developmental responses of rice to photoperiod and carbon dioxide concentration. *Agric.*

For. Meteorol. **50**: 201-210.

Baker, J.T., Allen, L.H. Jr., Boote, K.J., Jones, P. and Jones, J.W. 1990b. Rice photosynthesis and evapotranspiration in subambient, ambient, and superambient carbon dioxide concentrations. *Agron. J.* **82**: 834-840.

Baker, J.T., Allen, L.H. Jr. and Boote, K.J. 1992a. Response of rice to CO_2 and temperature. *Agric. For. Meteorol.* **60**: 153-166.

Baker, J.T., Allen, L.H. Jr. and Boote, K.J. 1992b. Temperature effects on rice at elevated CO_2 concentration. *J. Exp. Bot.* **43**: 959-964.

Baker, J.T., Laugel, F., Boote, K.J. and Allen, L.H. Jr. 1992c. Effects of daytime carbon dioxide concentration on dark respiration of rice. *Plant Cell Environ.* **15**: 231-239.

Baker, J.T., Allen, L.H. Jr. and Boote, K.J. 1995. Potential climate change effects on rice: Carbon dioxide and temperature. In: *Climate Change and Agriculture: Analysis of Potential International Impacts.* C. Rosenzweig, L.A. Harper, S.E. Hollinger, J.W. Jones and L.H. Allen, Jr. (eds.). ASA Special Pub. No. 59, American Society of Agronomy, Madison, Wisconsin. pp. 31-47.

Baker, J.T., Allen, L.H. Jr., Boote, K.J. and Pickering, N.B. 1996. Assessment of rice responses to global climate change: CO_2 and temperature. In: *Terrestrial Ecosystem Response to Elevated CO_2.* G.W. Koch and H.A. Mooney (eds.). Physiological Ecology Series, Academic Press, San Diego. pp. 265-282.

Bazzaz, F.A., Jasienski, M., Thomas, S.C. and Wayne, P. 1995. Microevolutionary responses in experimental populations of plants in CO_2-enriched environments. Parallel results from two model systems. *Proc. of the National Academy of Sciences* **S.92**: 8161-8165.

Boote, K.J., Pickering, N.B., Baker, J.T. and Allen, L.H. Jr. 1994. Modelling leaf and canopy photosynthesis of rice in response to carbon dioxide. *International Rice Research Notes* **19**: 47-48.

Broecker, W.S. 1987. Unpleasant surprises in the greenhouse? *Nature* **328**: 123-126.

Bunce, J.A. 1990. Short- and long-term inhibition of respiratory carbon dioxide efflux by elevated carbon dioxide. *Ann. Bot.* **65**: 637-642.

Campbell, W.J., Allen, L.H. Jr. and Bowes, G. 1988. Effects of CO_2 concentration on rubisco activity, amount, and photosynthesis in soybean leaves. *Plant Physiol.* **88**: 1310-1316.

Campbell, W.J., Allen, L.H. Jr. and Bowes, G. 1990. Response of soybean canopy photosynthesis to CO_2, light, and temperature. *J. Exp. Bot.* **41**: 427-433.

Crosson, P. 1993. Impacts of climate change on the agriculture and economy of the Missouri, Iowa, Nebraska, and Kansas (MINK) region. In: *Agricultural Dimensions of Global Climate Change*. H.M. Kaiser and T.E. Drennen (eds.). St. Lucie Press, Delray Beach, Florida. pp. 117-135.

Cure, J.D. and Acock, B. 1986. Crop responses to carbon dioxide doubling: A literature survey. *Agric. For. Meteorol.* **38**: 127-145.

Curry, R.B., Peart, R.M., Jones, J.W., Boote, K.J. and Allen, L.H. Jr. 1990a. Simulation as a tool for analyzing crop response to climate change. *Trans. ASAE* **33**: 981-990.

Curry, R.B., Peart, R.M., Jones, J.W., Boote, K.J. and Allen, L.H. Jr. 1990b. Response of crop yield to predicted changes in climate and atmospheric CO_2 using simulation. *Trans. ASAE* **33**: 1383-1390.

Curry. R.B., Jones, J.W., Boote, K.J., Peart, R.M., Allen, L.H. Jr. and Pickering, N.B. 1995. Response of soybean to predicted climate change in the USA. 1995. In: *Climate Change and Agriculture: Analysis of Potential International Impacts*. C. Rosenzweig, L.A. Harper, S.E. Hollinger, J.W. Jones, and L.H. Allen, Jr. (eds.). ASA Special Pub. No. 59, American Society of Agronomy, Madison, Wisconsin. pp. 163-182.

Delucia, E.H., Sasek, T.W. and Strain, B.R. 1985. Photosynthesis inhibition after long-term exposure to elevated levels of atmospheric carbon dioxide. *Photosyn. Res.* **7**: 175-184.

Dudek, D.J. 1989. Climate change impacts upon agriculture and resources: A case study of California. In: *The Potential Effects of Global Climate Change on the United States*. Appendix C, Agriculture, Vol. 1. EPA-230-05-89-053. J.B. Smith and D.A. Tirpak (eds.). US Environmental Protection Agency, Washington DC. pp. 5-1 to 5-38.

Hall, A.E. and Allen, L.H. Jr. 1993. Designing cultivars for the climatic conditions of the next century. In: *International Crop Science I*. D.R. Buxton, R. Shibles, R.A. Forsberg, B.L. Blad, K.H. Asay, G.M. Paulsen and R.F. Wilson (eds.). Crop Science Society of America, Madison, Wisconsin. pp. 291-297.

Hansen, J., Fung, I., Lacis, A., Lebedeff, S., Rind, D., Ruedy, R., Russell, G. and Stone, P. 1988. Global climate changes as forecast by the GISS 3-D model. *J. Geophys. Res.* **98**: 9341-9364.

Idso, S.B., Kimball, B.A., Anderson, M.G. and Mauney, J.R. 1987. Effects of atmospheric CO_2 enrichment on plant growth: The interactive role of air temperature. *Agric. Ecosystems Environ.* **20**: 1-10.

Jones, P., Allen, L.H. Jr., Jones, J.W. Boote, K.J. and Campbell, W.J. 1984. Soybean canopy growth, photosynthesis, and transpiration responses to whole-season carbon dioxide enrichment. *Agron. J.* **76**: 633-637.

Jones, P., Allen, L.H. Jr. and Jones, J.W. 1985a. Responses of soybean canopy photosynthesis and transpiration to whole-day temperature changes in different CO_2 environments. *Agron. J.* **77**: 242-249.

Jones, P., Allen, L.H. Jr., Jones, J.W. and Valle, R.R. 1985b. Photosynthesis and transpiration responses of soybean canopies to short- and long-term CO_2 treatments. *Agron. J.* **77**: 119-126.

Jones, P., Jones, J.W. and Allen, L.H. Jr. 1985c. Carbon dioxide effects on photosynthesis and transpiration during vegetative growth in soybeans. *Soil Crop Sci. Soc. Fla. Proc.* **44**: 129-134.

Keeling, C.D., Whorf, T.P., Wahlen, M. and van der Plicht, J. 1995. Interannual extremes in the rate of rise of atmospheric carbon dioxide since 1980. *Nature* **375**: 660-670.

Kimball, B.A. 1983. Carbon dioxide and agricultural yield: An assemblage and analysis of 430 prior observations. *Agron. J.* **75**: 779-788.

Kimball, B.A., Mauney, J.R., Nakayama, F.S. and Idso, S.B. 1993. Effects of increasing atmospheric CO_2 on vegetation. *Vegetatio* **104/105**: 65-75. Also: pp. 65-75. In: *CO_2 and Biosphere*. (Advances in Vegetation Science 14). J. Rozema, H. Lambers, S.C. van de Geijn and M.L. Cambridge (eds.). Kluwer Academic Publishers, Dordrecht.

Kramer, P.J. 1981. Carbon dioxide concentration, photosynthesis, and dry matter production. *Bioscience* **31**: 29-33.

Long, S.P. 1991. Modification of the response of photosynthetic productivity to rising temperature by atmospheric CO_2 concentrations: Has its importance been underestimated? *Plant Cell Environ.* **14**: 729-739.

Manabe, S. and Wetherald, R.T. 1987. Large scale changes of soil wetness induced by an increase in atmospheric carbon dioxide. *J. Atmos. Sci.* **44**: 1211-1235.

Parry, M.L., Carter, T.R. and Konijn, N.T. (eds.). 1988. *The Impact of Climatic Variations on Agriculture*. Vols. 1 and 2. Kluwer Academic Publishers, Dordrecht.

Peart, R.M., Jones, J.W., Curry, R.B., Boote, K.J. and Allen, L.H. Jr. 1989. Impact of climate change on crop yield in the Southeastern U.S.A.: A simulation study. In: *The Potential Effects of Global Climate Change on the United States*. Appendix C, Agriculture, Vol. 1, EPA-230-05-89-053. J.B. Smith and D.A. Tirpak (eds.). US Environmental Protection Agency, Washington DC. pp. 2-1 to 2-54.

Pickering, N.B., Jones, J.W. and Boote, K.J. 1995. Adapting SOYGRO V5.42 for prediction under climate change conditions. In: *Climate Change and Agriculture: Analysis of Potential International Impacts*. C. Rosenzweig, L.A. Harper, S.E. Hollinger, J.W. Jones and L.H. Allen, Jr. (eds.). ASA Special Pub. No. 59, American Society of Agronomy, Madison, Wisconsin. pp. 77-98.

Poorter, H. 1993. Interspecific variation in the response of plants to an elevated ambient CO_2 concentration. *Vegetatio* **104/105**: 77/97. Also: pp. 77-97. In: *CO_2 and Biosphere*. (Advances in Vegetation Science 14). J. Rozema, H. Lambers, S.C. van de Geijn and M.L. Cambridge (eds.). Kluwer Academic Publishers, Dordrecht.

Ritchie, J.T., Baer, B.D. and Chou, T.Y. 1989. Effect of global climate change on agriculture: Great Lakes Region. In: *The Potential Effects of Global Climate Change on the United States*. Appendix C, Agriculture, Vol. 1. EPA-230-05-89-053. J.B. Smith and D.A. Tirpak (eds.). US Environmental Protection Agency, Washington DC. pp. 1-1 to 1-42.

Rosenzweig, C. 1989. Potential effects of climate change on agricultural production in the Great Plains: A simulation study. In: *The Potential Effects of Global Climate Change on the United States*. Appendix C, Agriculture, Vol. 1. EPA-230-05-89-053. J.B. Smith and D.A. Tirpak (eds.). US Environmental Protection Agency, Washington DC. pp. 3-1 to 3-43.

Rosenzweig, C. and Iglesias, A. (eds.). 1994. *Implications of Climate Change for International Agriculture: Crop Modelling Study*. EPA 230-B-94-003. US Environmental Protection Agency, Washington DC.

Rosenzweig, C. and Parry, M.L. 1993. Potential impact of climate change on world food supply: A summary of a recent international study. In: *Agricultural Dimen-*

sions of Global Climate Change. H.M. Kaiser and T.E. Drennen (eds.). St. Lucie Press, Delray Beach, Florida. pp. 87-116.

Rosenzweig, C. and Parry, M.L. 1994. Potential impact of climate change on world food supply. *Nature* **367**: 133-138.

Rosenzweig, C., Harper, L.A., Hollinger, S.E., Jones, J.W. and Allen, L.H. Jr. (eds.). 1995. *Climate Change and Agriculture: Analysis of Potential International Impacts.* ASA Special Pub. No. 59, American Society of Agronomy, Madison, Wisconsin.

Rowland-Bamford, A.J., Allen, L.H. Jr., Baker, J.T. and Bowes, G. 1991. Acclimation of rice to changing atmospheric carbon dioxide concentration. *Plant Cell Environ.* **14**: 577-583.

Rozema, J., Lambers, H., van de Geijn, S.C. and Cambridge, M.L. (eds.). 1993. *CO_2 and Biosphere.* (Advances in Vegetation Science 14). Kluwer Academic Publishers, Dordrecht.

Satake, T. and Yoshida, S. 1978. High temperature induced sterility in indica rices at flowering. *Japan. J. Crop Sci.* **47**: 6-17.

Smith, J.B. and Tirpak, D.A. (eds.). 1989. *The Potential Effects of Global Climate Change on the United States.* EPA-230-05-89-050. US Environmental Protection Agency, Washington DC.

Thorne, J.M. and Koller, R.M. 1974. Influence of assimilate demand on photosynthesis, diffusive resistance, translocation and carbohydrate levels of soybean leaves. *Plant Physiol.* **54**: 201-207.

Valle, R., Mishoe, J.W., Campbell, W.J., Jones, J.W. and Allen, L.H. Jr. 1985. Photosynthetic responses of 'Bragg' soybean leaves adapted to different CO_2 environments. *Crop Sci.* **25**: 333-339.

Vergara, B.S. 1980. Rice plant growth and development. In: *Rice Production and Utilization.* B.S. Luh (ed.). Avi Publishing, Westport, CT. pp. 75-86.

Wilson, C.A. and Mitchell, J.F.B. 1987. A doubled CO_2 climate sensitivity experiment with a global model including a simple ocean. *J. Geophys. Res.* **92**: 13315-13343.

Wittwer, S.H. 1995. *Food, Climate, and Carbon Dioxide: The Global Environment and World Food Production.* CRC Press/Lewis Publishers, Boca Raton, Florida.

Woodwell, G.M. and Mackenzie, F.T. (eds.). 1995. *Biotic Feedbacks in the Global Climatic System: Will the Warming Increase the Warming?* Oxford University Press, New York.

Wullschleger, S.D., Post, W.M. and King, A.W. 1994. On the potential for a CO_2 fertilization effect in forests: Estimates of the Biotic Growth Factor based on 58 controlled-exposure studies. In: *Biotic Feedbacks in the Global Climatic System: Will the Warming Increase the Warming?* G.M. Woodwell and F.T. Mackenzie (eds.). Oxford University Press, New York. pp. 85-107.

Yoshida, S. 1977. Rice. In: *Ecophysiology of Tropical Crops.* P.T. Alvim and T.T. Kozlowski (eds.). Academic Press, London. pp. 57-87.

Yoshida, S. 1981. *Fundamentals of Rice Crop Science.* International Rice Research Institute, Los Baños, Philippines.

5. The Effects of Elevated CO_2 and Temperature Change on Transpiration and Crop Water Use

SIEBE C. VAN DE GEIJN
Research Institute for Agrobiology and Soil Fertility (AB-DLO), Wageningen, The Netherlands

JAN GOUDRIAAN
Wageningen Agricultural University, Dept. Theoretical Production Ecology, Wageningen, The Netherlands

The projected climatic effects of the continuously increasing concentrations of CO_2 and other radiatively active trace gasses in the atmosphere have caused concern over the last decades and increasingly attracted scientists' and policy makers' attention. The expected changes at a global level will be reflected in changed weather conditions in the growing season at regional and local levels that directly affect agriculture and natural vegetation. The exchange of water and energy is determined both by the climate and by the gas exchange properties of the vegetation. Feedback on land cover, changed vegetation properties and local and regional climate is therefore of prime importance.

So far, the General Circulation Models (GCMs) are not able to give accurate predictions of climate change with great geographical detail. Also, feedback processes by the biosphere are still poorly represented in the models. These shortcomings limit predictions at the regional level that are needed in impact studies for agricultural production. Qualitative changes in potential production may well be estimated on the basis of the scenarios, but quantitative changes in the actual production, including the limitations caused by the changed variability of weather conditions (daily and seasonal amplitudes of temperatures, seasonal patterns in precipitation, cloudiness) remain difficult to estimate.

In addition to changes in precipitation and total water availability for irrigation, that directly affect agricultural production, changes in the pattern of water use by crop plants throughout the season may affect the outcome. Of special concern is the change in the physiological functioning of the vegetation as a consequence of the changed atmospheric composition. Most plants react to the changed atmospheric CO_2 concentration with changed stomatal response, and not only is growth affected but also the transpiration. The complex nature of the physiological response in interaction with micrometeorological processes at the leaf and canopy level requires further attention.

This paper gives an indication of the state of the art regarding the effects of elevated CO_2 on stomatal behaviour, transpiration, water-use efficiency and total water-use. It briefly touches upon the interaction between vegetation and atmosphere at a larger scale. Aspects related to these effects at the level characteristic for a doubling of the present atmospheric CO_2 concentration [CO_2] are discussed. Changes in atmospheric concentration will not occur overnight but are expected to lag behind relative to changes in greenhouse gas concentration. Therefore uncertainties remain not only on the expected time course of modifications and adaptations, but also on the level at which new threats and opportunities arise in future conditions.

WATER USE AND WATER-USE EFFICIENCY

DEFINITIONS

It should be realized that various expressions for water-use efficiency are used (Figure 5.1), and that the physiological and agronomic definitions may differ. Time-averaged water use and water-use efficiency (seasonal or periodic whole-plant transpiration rate (WPTR), and water-use efficiency (WUE)) are often confounded with instantaneous transpiration rate and transpiration efficiency (ITE) (Morison, 1993; Eamus, 1991). The latter expression is best described as the ratio between the instantaneous CO_2 fixation (actual net photosynthesis) and water loss by transpiration (both measured with a leaf or plant chamber). Water-use efficiency is the ratio of the net gain in dry matter over a given period, divided by the water loss (from the vegetation alone or from soil and vegetation together) over the same period. Mostly the ITE, normally only measured during the light period, is higher than WUE, both because of variations in ITE over day and season and because of carbon losses by respiration in the dark. WUE seems to be the more relevant parameter to the question of impact on agricultural production.

INTERACTIONS WITH EXPERIMENTAL CONDITIONS

Most available data on water use and water-use efficiency have been collected from single plants grown in pots or hydroponically, and using porometer measurements supplemented with periodic weight determinations. Values for water-use efficiency and total water use in (semi-)field conditions over the full season, or an extended period, are scarce. Moreover, a complication for the correct estimation of the effect of elevated [CO_2] and temperature changes on transpiration is that almost all experiments in this field have been performed in environmentally controlled and generally well mixed and ventilated experimental set-ups (enclosures or Open-Top Chambers (OTCs)), where the indirect effects may not show up so prominently as in a real and

Figure 5.1. Different expressions for the relation water use – carbon assimilation and growth

Instantaneous transpiration efficiency (ITE)
calculated from photosynthesis/gas exchange

$$ITE = \frac{mmol\ CO_2}{mol\ H_2O}$$ instantaneous

(at leaf or canopy level)

Water-use efficiency (from gas exchange data)
integrated at leaf or canopy level, including respiration loss

$$WUE = \frac{(Net\ mg\ CO_2)}{(g\ H_2O)}$$ per day or extended period

(per m² leaf or soil surface)

Crop water-use efficiency / Seasonal water use efficiency
integrated over (part of) the season

$$(C)WUE = \frac{(g\ DM)}{(g\ H2O)}$$

(of canopy per m², with or without soil evaporation)

Crop / Plant water use
integrated over (part of) the season

$$WU = g\ H_2O\ per\ plant\ or\ per\ m^2$$

(incl. or excl. soil evaporation)

outside future climate (Unsworth *et al.*, 1984; Leuning and Foster, 1990). Similar interpretation problems have been met in the evaluation of air pollution research.

RESPONSE AT THE LEVEL OF STOMATA

STOMATAL DENSITY

The plant leaves lose water primarily by evaporation through the stomata. The stomatal density depends upon plant species, and can be related to the plant-ecotype (between 300 and 800 stomata/mm (Rowland-Bamford *et al.*, 1990; Woodward, 1987, 1993; Kimball *et al.*, 1986)). Woodward (1987) correlated the decrease in the stomatal density over time, observed in herbarium leaves collected over the last centuries, with the rising CO_2 concentrations and concluded from the shift in $\delta^{13}C$ (Woodward, 1993) that the water-use efficiency concomitantly has improved. The nitrogen content in the leaves had dropped, in line with most data from elevated CO_2 experiments (Penuelas and Matamala, 1990). Experimentally, an increase in [CO_2] up to about 310 µl/l decreased the stomatal density, but sometimes no effect is found above this [CO_2] (Woodward and Bazzaz, 1988). This matter is still under dispute (Körner, 1988; Woodward, 1993) although such a correlation has also been confirmed for palaeo-records (Van der Burgh et al., 1993). Among species large differences in response of stomatal density to elevated [CO_2] seem to exist. Experiments with a range of CO_2 concentrations (160-900 µl/l) (Rowland-Bamford et al., 1990; O'Leary and Knecht, 1981) showed an increase of stomatal density of rice and bean leaves, with a differential effect at abaxial (increasing) and adaxial sides. At subambient CO_2 concentrations stomatal density dropped. This is in contrast with the findings of Oberbauer et al. (1985) for tropical trees. The relative effect of changes in stomatal density and decreasing stomatal aperture at elevated [CO_2] for the water relations, however, have not been evaluated. A gradual change in [CO_2] over the next century may lead to a natural selection that favours cultivars having a lower stomatal density, especially for water-limited growth conditions. It should, however, be realized that other environmental factors like salt stress can also modify stomatal density (Rozema et al., 1991a). Whatever the net effect, the resulting stomatal conductance is primarily determined by stomatal functioning, and much less by density.

STOMATAL FUNCTIONING

In the pathway from the stomatal cavity to the leaf surface, and from ambient air to the photosynthetic machinery in the mesophyll, the stomata are a major resistance for gas transport between the leaf and the surrounding air. A change in gas exchange resistance of the stomatal pores therefore affects the entrance of CO_2 and even more the exit of water vapour (Figure 5.2). The opening status of the stomata is a compromise between water loss and uptake of CO_2 from ambient air (Farquhar et al., 1980; Mott, 1990; Wolfe, 1994; Stanghellini and Bunce, 1994; Leuning, 1995). In line with this proposition, stomatal response to elevated concentrations of CO_2 (C_a) in general is reflected in partial stomatal closure. The mechanism behind this stomatal closure is not yet clear (Mott, 1990; Wolfe, 1994). The observations are in line with the notion that plants tend to regulate the internal CO_2 concentration (C_i) in the substomatal cavity such that for a given water vapour deficit there is a constant ratio (C_i/C_a) with the atmospheric concentration (Mott, 1990; Goudriaan and Unsworth, 1990). Such a regulation would directly lead to the partial closure at elevated CO_2, as observed in many studies using porometers (Tyree and Alexander, 1993; Morison and Gifford, 1983; Morison, 1987). Jackson et al. (1994) measured photosynthesis and water relations of native grassland species and calculated C_i/C_a for C_3 and C_4 plants. They confirmed the conservation of the value, with only a small (not significant) tendency

Figure 5.2. Schematic cross-section of a stoma of a leaf showing the pathway of CO_2 and H_2O in the light. C_i, C_s, C_a: internal, surface and ambient CO_2 concentration; e_i, e_s, e_a: internal, surface and ambient air humidity. Bar indicates 100 µm

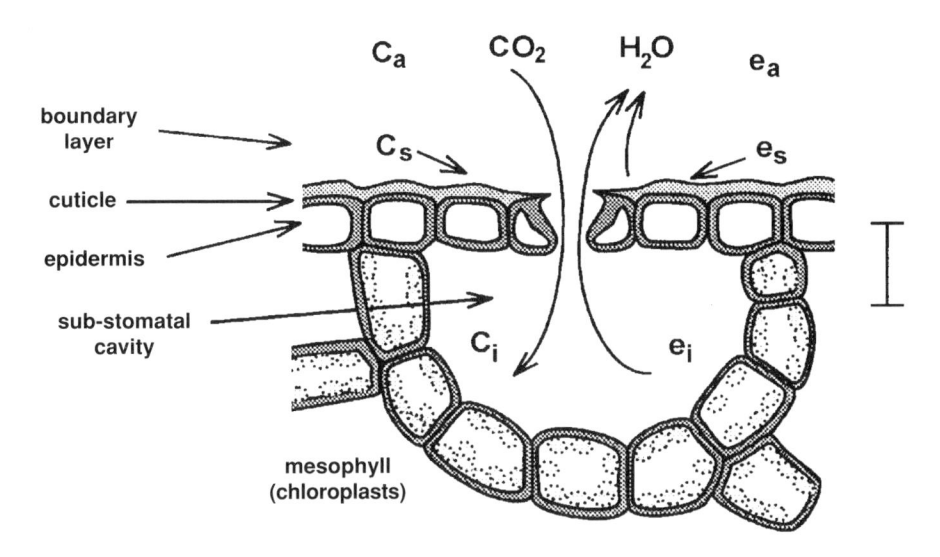

to rise with increasing [CO_2]. The rate of photosynthesis and therefore the required supply of carbon dioxide is directly coupled to light intensity. In line with this, the conductance of stomata is also highly correlated to light (Leuning, 1995). This relation can be modified by environmental conditions like drought or stress by air pollution. The ratio C_i/C_a in stationary conditions is about two-thirds for C_3 but about one-third for C_4 plants. The lower value for C_4 plants reflects the higher affinity for CO_2 of the C_4 photosynthetic pathway, and reflects the more efficient water use for these plants (Goudriaan and Unsworth, 1990; Kimball et al., 1993; Kimball et al., 1995).

ACCLIMATION OF STOMATAL MOVEMENT

Little is known about the acclimation of the stomatal movement to long-term exposure to elevated [CO_2]. Some studies suggest that the lower stomatal conductance not only persists over extended exposure periods to elevated [CO_2], but is also conserved after subsequent lowering of the [CO_2] (Gorissen, pers. comm.). For glasshouse horticulture it was reported that the stomatal conductance of high-[CO_2] tomato plants was less sensitive to short-term [CO_2] fluctuations, and higher than that of plants grown at ambient concentrations and measured at double present [CO_2] (Stanghellini and Bunce, 1994). This may indicate that the sensitivity of stomatal movement of high [CO_2] plants to changes in [CO_2] is reduced, but still a lower stomatal conductivity exists when compared to leaves growing and measured in present [CO$_2$] conditions.

EFFECTS AT THE LEVEL OF THE LEAF

The partial closure of stomata is reflected in the reduced conductance at the leaf level (Atkinson *et al.*, 1991; Sionit *et al.*, 1984). At a CO$_2$ concentration double the present, conductance is reduced by 30-40%, although large differences among species exist (Hendrey *et al.*, 1993; Morison, 1987).

LEAF TEMPERATURE AND VAPOUR PRESSURE DEFICIT

Water loss by transpiration is not only affected by the conductivity of the stomata, but also by the driving forces for exchange of the water vapour from the leaf surface to the surrounding atmosphere. Therefore the gradient in partial pressure of the water vapour at the leaf surface is also of importance (McNaughton and Jarvis, 1991). All other factors being equal, the existing vapour pressure deficit (VPD) between stomatal cavity and surrounding air (Figures 5.2 and 5.3), the boundary layer, will increase at a reduced transpiration rate, and feed back to stimulate transpiration. The reduced transpiration will cool the leaf less, and consequently a rise in temperature in the stomatal cavity and at the leaf surface may occur. Thus, in addition to the global greenhouse effect on air temperature, the temperature at the leaf surface may rise by 0.5 to 1.5 °C (Idso *et al.*, 1987; Morison, 1987; Kimball *et al.*, 1995; van de Geijn *et al.*, 1993). In their FACE experiments (Free Air Carbon dioxide Enrichment) in Arizona, Kimball *et al.* (1995) measured an average rise in canopy temperature of 0.56°C over the growing season. Such a higher leaf temperature may also have important consequences for the longevity and photosynthetic capacity of the individual leaves and at the canopy level, as ageing may be accelerated (Kimball *et al.*, 1995; Chaudhury *et al.*, 1989; Ellis *et al.*, 1990; Kuiper, 1993). It is presently not known how different crops respond to the increase in leaf and canopy temperature.

The mostly negative feedback on water vapour exchange will at least partly counteract the effect of the reduced stomatal conductance even without rise in ambient temperature at the global scale. A rise of ambient growing season temperatures due to the greenhouse effect will also tend to stimulate transpiration. Depending on changes in air humidity it may affect VPD at the leaf surface (Figure 5.3) determining the fluxes of sensible and latent heat (Massman and Ham, 1994) and thereby the energy balance. Increased water availability may, however, offset the relative warming at the leaf surface, especially if, through preceding water saving and lower instantaneous requirements, partial stomatal closure alleviates water stress during periods of high water requirements in the diurnal sequence (Tyree and Alexander, 1993). It requires detailed microclimatological data and complex calculations to integrate these various mechanisms in operation at the leaf level (Jacobs and de Bruin, 1992).

Figure 5.3. Stepwise transfer of water from the root environment to the atmosphere

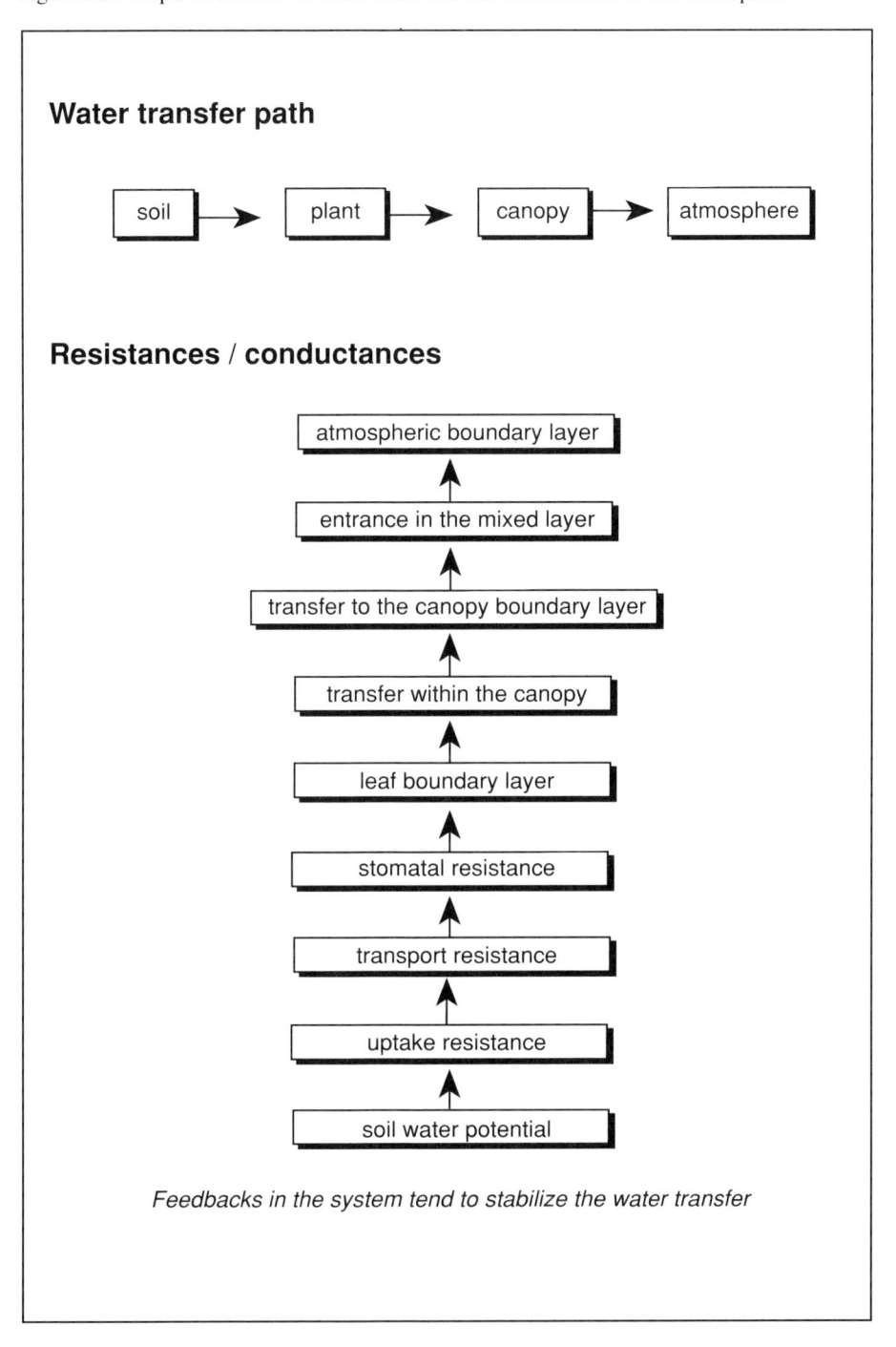

Water transfer path

soil → plant → canopy → atmosphere

Resistances / conductances

atmospheric boundary layer
↑
entrance in the mixed layer
↑
transfer to the canopy boundary layer
↑
transfer within the canopy
↑
leaf boundary layer
↑
stomatal resistance
↑
transport resistance
↑
uptake resistance
↑
soil water potential

Feedbacks in the system tend to stabilize the water transfer

LEAF AREA AND WEIGHT

Changes in area of individual leaves also have a direct effect on water loss at a per leaf basis. In general a substantial increase in individual leaf weight is found. This is primarily due to an increased leaf thickness and additional accumulation of non-structural carbohydrates. Leith *et al.* (1986) found for soybeans that a small but significant increase in leaf area with CO_2 concentration may be expected (about 5% increase with a doubling $[CO_2]$). This is consistent with trends reported in other papers (Rozema, 1993; Stanghellini and Bunce, 1994). In general, however, weight gain through increased leaf thickness is more important than that by the increase in leaf area. The relative gain in leaf area may also depend on growing conditions such as water shortage.

EFFECTS AT THE PLANT AND CANOPY LEVEL

Aspects of the water economy at the leaf level also apply at the canopy level. Moreover, light distribution in the canopy, leaf age, humidity gradients and coupling with the atmosphere for different canopy layers and canopy structures all have their effect. The direct effects on vegetation properties discussed here are linked to the change in CO_2 concentration and temperature, but indirect effects of a climate change related with changes in ambient temperature and air humidity, coupling to the lower atmosphere and albedo changes, may even dominate (Figure 5.3). Mismanagement or overexploitation of natural resources may lead to changes in the vegetation and even to desertification (Breman, 1992), and via a positive feedback aggravate changes in local and regional climate as studied in the EU-EFEDA-programme (European Field Experiment in Desertification-threatened Areas (Bolle *et al.*, 1993)). This field of research is still under development and the subject of international programmes like the IGBP-BAHC (International Geosphere-Biosphere Programme – Biospheric Aspects of the Hydrological Cycle (IGBP, 1993)).

LEAF AREA INDEX

The leaf area index (leaf surface area per unit soil surface area) of a crop with adequate water supply at elevated $[CO_2]$ increases, especially early in the season, as a result of earlier and more rapid leaf production in the vegetative growth phase (Ackerly *et al.*, 1992; Grashoff *et al.*, 1995; Morison and Gifford, 1984a,b). This applies especially for indeterminate growing species and under non-limiting supply of nutrients. Early development of the canopy will lead to an earlier full ground cover, and may thus limit water loss from direct soil evaporation. Depending on the local precipitation and available soil water reserves, such an early enhanced canopy development may also be favourable for the full utilization of water resources (Chaudhury *et al.*, 1990a,b). The higher transpiration early in the season may also lead to an earlier depletion of water reserves in the soil (Morison and Gifford, 1984a,b).

In their FACE experiments Hendrey et al. (1993) found that for cotton leaf area increased by 25% in dry plots, but by only 11% in irrigated plots as compared to the respective controls. Such enhancements in leaf area and biomass formation in dry conditions could improve the duration of the vegetative soil cover, and help counteract erosion and other forms of land degradation.

PLANT WATER STATUS

It has been shown that plant water status is generally improved at elevated CO_2. Part of the effect can be ascribed to a reduced transpiration demand (per unit leaf area), and therefore a partial alleviation of the water stress (Paez et al., 1984). Also for plants with a normal water supply, pre-dawn and midday water potentials have been found to be less negative in high [CO_2] (Clifford et al., 1993; Jackson et al., 1994). Even in saline conditions the effect of partial stomatal closure is reflected in a higher salt tolerance (Rozema et al., 1991b; Bowman and Strain, 1987). As a consequence, the restriction of photosynthesis under low water supply may be less severe, or be delayed.

CHANGES IN THE ROOTING PATTERN

Several studies have reported an increased input of carbon in below-ground processes (van de Geijn and van Veen, 1993; Rogers et al., 1994). An important aspect to be considered in reviewing the literature at this point is that most experiments concern plant growth over a short period and with a soil volume limited by pot or container size. In a study with container-grown grasses and clover Nijs et al. (1989) showed that in a rapidly developing terminal drought period water-use efficiency is about doubled, and stress is developing later under elevated [CO_2]. These conditions are thus improving the possibility to escape drought stress. Even though the trend is clear, elevated [CO_2] plants grown with restricted soil volume may not take full advantage of a larger root system. In (semi-)field experiments (Chaudhury et al., 1990b; Rogers et al., 1994) the larger root system indeed leads to a better exploration of the soil volume, and an earlier, or higher, rooting density at the larger depths. Experiments with potted plants can therefore not be expected to show the full extent to which elevated [CO_2] plants may profit in a (transient) drought period. The observation of Morison and Gifford (1984a,b) that the pattern of depletion of soil water reserves is rather similar may partly be an underestimation of the real moisture availability in field conditions because of a better access to deeper soil layers. It has also been observed that a faster recovery of the crop is possible when grown at elevated [CO_2] (Bhattacharya et al., 1990). The explanation of this observation has not yet been given, but could be related to deeper and more extensive rooting patterns.

TEMPERATURE EFFECTS

The importance of temperature at the canopy level is two-fold. Firstly higher temperatures increase transpiration by changing the VPD at the leaf surface, and secondly the higher canopy temperature may lead to an accelerated ageing of the foliage, and a shortening of the growing season, or, for example, grain-filling period. The latter type of effect is discussed extensively in crop growth modelling studies that describe changes in crop productivity due to climate change (Ellis *et al.*, 1990; Leuning *et al.*, 1993; Grashoff *et al.*, 1994; Van Keulen and Seligman, 1987; Acock and Acock, 1993; Kenny *et al.*, 1993).

On theoretical grounds it can be expected that transpiration losses will increase with higher air temperatures. It is likely that the evaporative demand as determined by the vapour pressure deficit would increase by about 5 to 6% per degree warming (McKenney and Rosenberg, 1993). The overall effect of a higher temperature alone is an intensification of the hydrological cycle. It should be realized that the combined result of a higher ambient temperature, leading to a higher evaporative demand, and partial stomatal closure, counteracting this, could be overridden by changes in other environmental and atmospheric conditions like soil water availability, precipitation patterns, cloudiness and air humidity.

EFFECTS OVER THE GROWING SEASON

The effects of elevated CO_2 alone on transpiration integrated over the season show that any of the cases (higher water use; no change; lower water use) can be found. Especially in the case of non-limiting water supply a higher water use can be expected even with a lower stomatal conductance as leaf area will increase with about the same factor as total biomass. This will compensate for the reduction in water exchange per leaf area. This applies especially in early season. In conditions with transient water shortage, profit from partial stomatal closure will be highest, as both the depletion of soil water reserves is delayed and plant sensitivity to midday water shortage (suppressed photosynthesis) is lessened.

In conditions with adequate water supply, water use over the whole season, and especially in periods in the season with a closed canopy (e.g., with leaf area index over 3), is little different if at all (Kimball *et al.*, 1995; Dijkstra *et al.*, 1993). Figure 5.4 shows some results of gas exchange measurements with wheat and faba bean grown for the whole season at two CO_2 concentrations (Dijkstra *et al.*, 1993). ITE and WUE (24h basis) are both higher (about 40-50%) at elevated [CO_2], but transpiration per unit soil surface is not changed significantly (CET-24h Table 5.1). Here treatment temperatures have been kept unchanged, tracking outside conditions.

Similarly, Baker and Allen (1993) presented results of canopy gas exchange for soybean and citrus at a range of temperatures (Table 5.2). Although daytime water-use efficiency (DWUE) fell with temperature, they measured at 800 µl/l [CO_2] a doubling (soybean) or more of the DWUE, independent of temperature (28-35°C). Seasonal water-use efficiency varies with crop and growth conditions (Table 5.3).

Figure 5.4. Effect of ambient (□) and elevated (■) [CO_2] on CET (Canopy Evapotranspiration rate, mmol/m^2/s) and CCER (Canopy Carbon Exchange Rate, μmol/m^2/s) on (A) spring wheat (1992) and (B) faba bean (1993) on Julian day 188

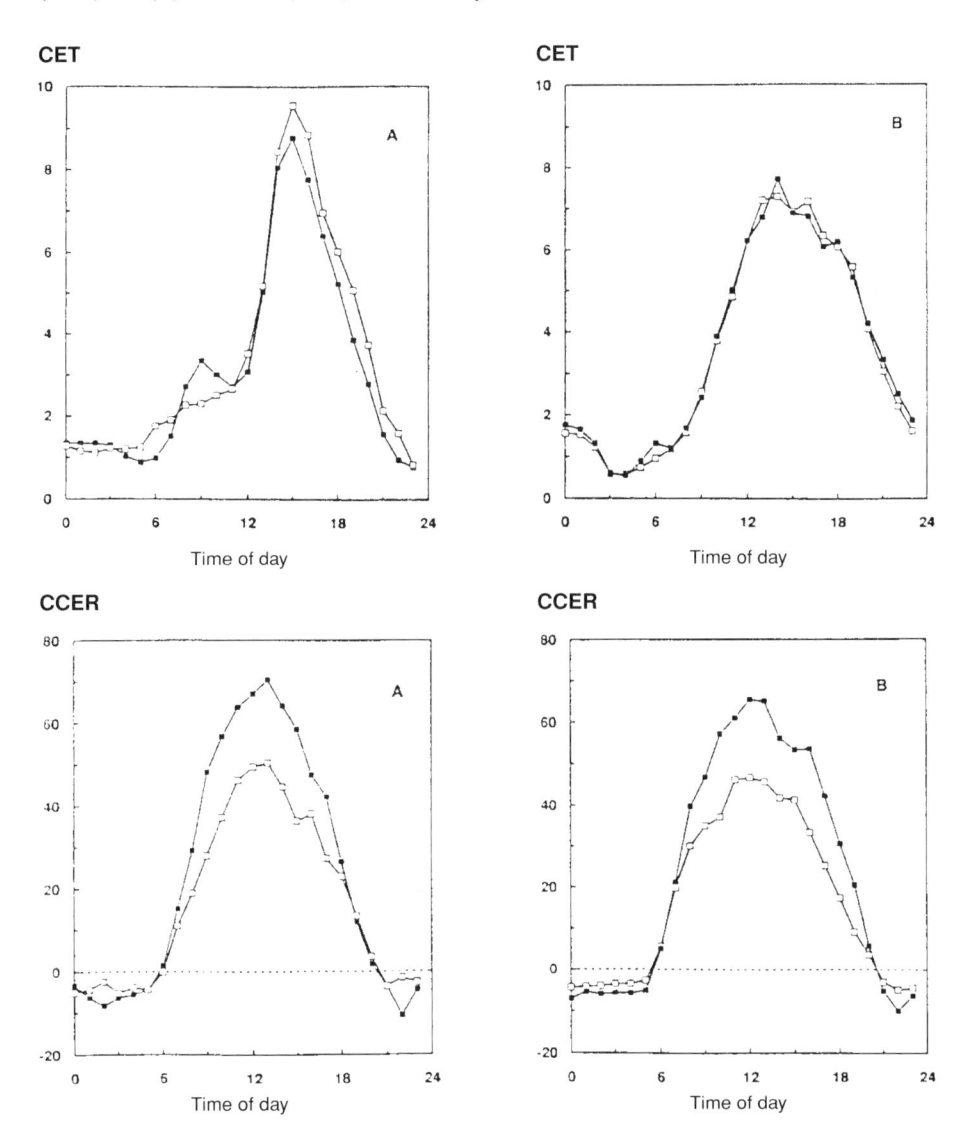

The improvement in seasonal water-use efficiency (SWUE) ranges from 23% (wheat) to 54% (faba bean), similar to the results reported by Dijkstra *et al.* (1993; Table 5.1).

Water-use efficiency rises especially in conditions with water limitations (Chaudhury *et al.*, 1990a). In their study of the potential changes of productivity of cool-season legumes, Grashoff *et al.* (1994) concluded that for rainfed faba bean crops in the

Table 5.1. Effect of [CO$_2$] on full canopy gas exchange parameters for spring wheat (1991) and faba bean (1992) both measured on Julian day 188. Data modified from Dijkstra et al. (1993)

	Spring wheat			Faba bean		
[CO$_2$] μmol/mol	350	700	ratio	350	700	ratio
CCERmax[1]	50.2	70.4	1.40	45.6	66.8	1.47
CDR[2]	−3.8	−6.0	1.59	−3.85	−6.50	1.69
CCER light period[7]	67.5	95.6	1.42	67.7	100.7	1.49
CDR dark period[7]	−5.4	−8.6	1.59	−5.5	−9.3	1.69
Net CCER shoot-24h[7]	62.1	87.0	1.40	62.2	91.4	1.47
CETmax[3]	9.55	8.77	0.92	7.29	7.71	1.06
CET-24h[4]	5.24	4.90	0.94	5.50	5.60	1.02
CCERmax/CETmax[5]	5.26	8.03	1.53	6.26	8.66	1.39
CCER-24h/CET-24h[6]	11.85	17.75	1.49	11.31	16.32	1.44

[1] CCERmax = maximum canopy CO$_2$ exchange rate on a day (μmol/m^2/s)
[2] CDR = mean canopy dark (night) respiration rate (μmol/m^2/s)
[3] CETmax = maximum canopy evapotranspiration rate on a day (mmol/m^2/s)
[4] CET-24h = integrated canopy evapotranspiration (mm/d)
[5] Or maximum ITE in mmol CO$_2$/mol H$_2$O
[6] Or daily WUE in g CO$_2$/kg H$_2$O
[7] g CO$_2$/m^2/d

Table 5.2. Daytime canopy water-use efficiency (D-WUE: photosynthesis per unit water transpired: mmol CO$_2$/mol H$_2$O) (data from Baker and Allen, 1993)

	[CO$_2$] μmol/mol	Temp. (°C)	DWUE	High/low ratio
Soybean	330	28	2.95	
		31	2.63	
		33	2.37	
		35	2.33	
	800	28	6.00	2.03
		31	5.11	1.94
		33	4.95	2.09
		35	4.71	2.02
Citrus	330	25	2.30	
		34	1.25	
	840	25	5.40	2.34
		34	3.52	2.81

Netherlands, Syria and Israel productivity would increase, especially under water limitation. Faba bean is a special case, because crop development rate is accelerated by drought. A small increase in temperature (1.7°C) would decrease yield, but the lower water requirement at higher [CO$_2$] (1.7°C and 460 μl/l for the climate in 2030) more than compensates for this (Table 5.4), raising yields from 2.8 to 4.7 t/ha (Syria),

Table 5.3. Seasonal crop water-use efficiency ((C)WUE in g DM/kg water) (data from Morison, 1993 and ref. therein)

	Ambient [CO_2]	Double [CO_2]	Ratio	
Sorghum	3.08	4.13	1.34	
Wheat well-watered	5.1	6.3	1.23	
Wheat water shortage	6.2	8.9	1.43	
Wheat	2.62	3.45	1.31	
Wheat well-watered	1.58	2.14	1.35	grain only
Wheat water shortage	1.27	1.86	1.46	grain only
Faba beans	4.91	7.82	1.54	
Water hyacinth	1.4	2.6	1.85	

Table 5.4. Average simulated seed yield in t/ha of rainfed and fully irrigated faba bean crops in Wageningen (Netherlands), Tel Hadya (Syria) and Migda (Israel) under present and changed climate conditions (year 2030: + 1.7°C and [CO_2] at 460 µmol/mol; year 2080: +3.0°C and [CO_2] at 700 µmol/mol; no precipitation change). Modified from Grashoff et al. (1994)

Location and Scenario	Rainfed			Fully irrigated		
	yield	change (%)	sd (%)	yield	change (%)	sd (%)
Wageningen						
Current climate	5.1	–	32	6.1	–	9
Year 2030	5.7	12	29	6.4	5	7
Year 2080	6.7	31	22	7.2	18	6
Tel Hadya						
Current climate	2.8	–	41	6.9	–	11
Year 2030	4.7	68	35	8.0	16	8
Year 2080	7.4	164	22	9.5	38	6
Migda						
Current climate	3.9	–	39	6.4	–	9
Year 2030	5.0	28	41	7.2	12	10
Year 2080	6.7	72	35	8.5	33	11

3.9 to 5.0 t/ha (Israel) and 5.1 to 5.7 t/ha (Netherlands). In fully irrigated conditions yields would be higher, but the relative CO_2 effect would be lower. Interestingly the standard deviation of the predicted yield (>10 year simulated average) shows a tendency to decrease in the changed climate scenario (Table 5.4).

EFFECTS ON THE REGIONAL VEGETATION-ATMOSPHERE WATER VAPOUR EXCHANGE

One of the more complex aspects of the estimation of changes in the water use of arable crops is the extrapolation from leaf and canopy to the field and region. This has long been recognized by the international scientific community, and has given rise to the establishment of several large-scale measuring campaigns (HAPEX:

Hydrologic Atmospheric Pilot Experiment (Sahel and Spain), GEWEX: Global Energy and Water Cycle Experiment (World Climate Research Programme)) and to the start of the IGBP-BAHC programme (IGBP, 1993). A major part of the efforts will go into the development of the SVAT-models (Soil-Vegetation-Atmosphere-Transfer models). The extrapolation from the stomatal and leaf level to the region have been the subject of experimental and theoretical studies (McNaughton, 1994; McNaughton and Jarvis, 1991; Jacobs and de Bruin, 1992; Jacobs, 1994; Hollinger *et al.*, 1994; Baldocchi, 1994a,b). In general the structure of the canopy determines to a large extent the transfer of heat and water vapour to the lower atmosphere. Most models to date neglect or oversimplify the feedback between the vegetated surface and the lower atmosphere, the planetary boundary layer (PBL). Jacobs (1994) showed in a study using coupled models (one-dimensional model for the PBL coupled to a modified Penman-Monteith big-leaf model) that in such a model the sensitivity of regional transpiration to changes in surface resistance is reduced (about halved) but that the sensitivity to changes in albedo is increased by between 25 and 250% relative to the values obtained without PBL feedback.

In addition to the sensitivity of stomata to photosynthesis and [CO_2], Jacobs (1994) introduced an additional correlative relationship between the humidity deficit at the leaf surface and the C_i/C_a ratio. The changes in transpiration caused by changes in surface resistance are damped. However, within the canopy the changes in the specific humidity lead to a positive feedback, e.g., the CO_2 effects on surface resistance are enhanced. The inclusion of PBL feedback and stomatal response to humidity deficit leads in his study to an estimated overall decrease of the regional transpiration by 10 to 30%. Modifying factors are surface roughness determined by the vegetation type, temperature and air humidity. In particular, information about the last factor, extrapolated from regionalized GCM output, is virtually absent or highly speculative.

Arable crops, especially in the temperate regions in general, have a low surface roughness, and in that case the transpiration is primarily determined by the radiation energy. This is also true for pastures. In rangelands with sparse higher vegetation the higher roughness leads to a stronger coupling, a situation comparable to forests and mixed vegetation (McNaughton and Jarvis, 1991; Hollinger *et al.*, 1994; McNaughton, 1994).

CONCLUDING REMARKS

Present knowledge does not allow a firm statement to be made concerning a future situation with respect to water-limited agricultural production. At a regional production level the available quantitative data from regionalized GCM runs are insufficient to be used as direct inputs for crop growth models. At a global level the simultaneous changes in rainfall pattern, air humidity and possible shifts in vegetation zones add to uncertainty, as many and often non-linear feedbacks are expected to operate.

The limited amount of available experimental data has as a trend that water use per unit soil surface area will change little (-10 to +10%). As, however, the general trend towards an improved water-use efficiency is clear, the productivity per unit of available water is expected to rise by 20-40%, probably much less than the value (100%) calculated from a reduced stomatal conductivity and an increased photosynthesis. Some studies show that in situations with marginal water availability the threshold for a successful crop may shift to lower values (Chaudhury *et al.*, 1990,b; Clifford *et al.*, 1993; Grashoff *et al.*, 1994). Whether at otherwise unchanged water availability this would open up possibilities to reverse existing trends towards desertification in certain areas is very questionable, as apart from precipitation falling short to maintain existing vegetation, other aspects like over-exploitation might dominate. It should, however, be emphasized that present knowledge on feedback among vegetation characteristics, gas exchange and albedo and the regional climate is insufficient to draw firm conclusions.

Some related problems have not been touched upon. For example, changes in precipitation distribution throughout the season may lead to shifts in the accessibility of fields for farm operations in early spring or late autumn. In addition they may cause changes in soil moisture and thereby modify the rate of mineralization of nutrients (nitrogen, phosphorus). Such interactions, combined with changes in (air and soil) temperatures, may markedly change soil fertility and thereby local farming systems and crop productivity. Depending on the aim of the study, such impacts have to be considered more or less relevant to the present subject.

In many studies the impact of climate change on crop growth and yield is analysed using crop simulation models (Van Keulen and Seligman, 1987; Kenny *et al.*, 1993; Acock and Acock, 1993; Grashoff *et al.*, 1994, 1995). A proper analysis of the performance of such models should be made to verify their reliability in the projected conditions of changed atmospheric composition and changed climate. The absolute values of the regionally and globally aggregated predicted crop yields, often calculated using a proportionality factor for the CO_2 response, are at present probably less reliable than the predicted sensitivity of the yield to various climate and management factors (see chapter of Tinker *et al.*).

The above considerations relate to a projected world, changed primarily in terms of climate and atmospheric composition. It should be emphasized that at many points agricultural practice is very dynamic, and will respond to changed conditions by adaptation (McKenney *et al.*, 1992). Crop and cultivar choice will, in most instances and in the most productive areas, change over time and gradually incorporate the traits necessary for adapted performance, or change to better adapted species. A rising temperature will for most current cultivars, for instance, accelerate crop development. This would in itself lead to a reduced water use over the shortened growth period, but also to a loss of potential yield. One may argue that farmers will repair such a loss of production potential by a proper choice of adapted cultivars or crop species, unless temperatures exceed the appropriate temperature window (Behl *et al.*, 1993).

These developments require, however, that new technologies and genetic resources be practically and economically accessible for all farmers, a situation that is not reached at present for farmers in arid and semi-arid regions in developing countries where the risks but probably also opportunities for agricultural production may be greatest.

ACKNOWLEDGEMENTS

Part of this work was done with support from the European Union, under contract EV5V-CT920169 (CROPCHANGE), and the Dutch National Programme on Global Air Pollution and Climate Change. The additional support from FAO is gratefully acknowledged.

REFERENCES

Ackerly, D.D., Coleman, J.S., Morse, S.R. and Bazzaz, F.A. 1992. CO_2 and temperature effects on leaf area production in two annual plant species. *Ecology* **73**: 1260-1269.

Acock, B. and Acock, M.C. 1993. Modelling approaches for predicting crop ecosystem responses to climate change. In: *International Crop Science I*. D.R. Buxton *et al.* (eds.). Crop Science Society of America, Madison, WI. pp. 299-306.

Atkinson, C.J., Wookey, P.A. and Mansfield, T.A. 1991. Atmospheric pollution and the sensitivity of stomata on barley leaves to absisic acid and carbon dioxide. *New Phytol.* **117**: 535-541.

Baker, J.T. and Allen, L.H. Jr. 1993. Contrasting crop species responses to CO_2 and temperature: rice, soybean and citrus. *Vegetatio* **104/105**: 239-260.

Baldocchi, D.D. 1994a. A comparative study of mass and energy exchange over a closed C3 (wheat) and an open C4 (corn) canopy. I: The partitioning of available energy into latent and sensible heat exchange. *Agric. For. Meteorol.* **67**: 191-220.

Baldocchi, D.D. 1994b. A comparative study of mass and energy exchange over a closed C3 (wheat) and an open C4 (corn) canopy. II: Canopy CO_2 exchange and water-use efficiency. *Agric. For. Meteorol.* **67**: 291-321.

Behl, R.K., Nainawatee H.S. and Singh, K.P. 1993. High temperature tolerance in wheat. In: *International Crop Science I*. D.R. Buxton *et al.* (eds.). Crop Science Society of America, Madison, WI. pp 349-355.

Bhattacharya, N.C., Hileman, D.R., Ghosh, P.P. and Musser, R.L. 1990. Interaction of enriched CO_2 and water stress on the physiology of and biomass production in sweet potato grown in open-top chambers. *Plant Cell Environ.* **13**: 933-940.

Bolle, H.-J., André, J.-C., Arrue, J.L., Barth, H.K., Bessemoulin, P., Brasa, A., De Bruin, H.A.R., Cruces, J., Dugdale, G., Engman, E.T., Evans, D.L., Fantechi, R., Fiedler, F., Van de Griend, A., Imeson, A.C., Jochum, A., Kabat, P., Kratzsch, T., Lagouarde, J.-P., Langer, I., Llamas, R., Lopez-Baeza, E., Melia Miralles, J., Muniosguren, L.S., Nerry, F., Noilhan, J., Oliver, H.R., Roth, R., Saatchi, S.S., Sanchez Diaz, J., De Santa Olalla, M., Shuttleworth, W.J., Søgaard, H., Stricker,

H., Thornes, J., Vauclin, M. and Wickland, D. 1993. EFEDA: European field experiment in a desertification-threatened area. *Ann. Geophysicae* **11**: 173-189.

Bowman, W.D. and Strain, B.R. 1987. Interaction between CO_2 enrichment and salinity stress in the C4 non-halophyte *Andropogon glomeratus* (Walter) BSP. *Plant Cell Environ.* **10**: 267-270.

Breman, H. 1992. Desertification control, the West African case; prevention is better than cure. *Biotropica* **24**: 328-334.

Chaudhury, U.N., Kanemasu, E.T. and Kirkham, M.B. 1989. Effect of elevated levels of CO_2 on winter wheat under two moisture regimes. *Report No. 50: Response of Vegetation to Carbon Dioxide.* US DOE, Washington DC. 49 p.

Chaudhury, U.N., Kirkham, M.B. and Kanemasu, E.T. 1990a. Carbon dioxide and water level effects on yield and water use of winter wheat. *Agron. J.* **82**: 637-641.

Chaudhury, U.N., Kirkham, M.B. and Kanemasu, E.T. 1990b. Root growth of winter wheat under elevated carbon dioxide and drought. *Crop Sci.* **30**: 853-857.

Clifford, S.C., Stronach, I.M., Mohamed, A.D., Azam-Ali, S.N. and Crout, N.M. 1993. The effects of elevated atmospheric carbon dioxide and water stress on light interception, dry matter production and yield in stands of groundnut (*Arachis hypogaea* L.). *J. Exp. Bot.* **44**: 1763-1770.

Dijkstra, P., Schapendonk, A.H.C.M. and Groenwold, J. 1993. Effects of CO_2 enrichment on canopy photosynthesis, carbon economy and productivity of wheat and faba bean under field conditions. In: *Climate Change; Crops and Terrestrial Ecosystems.* S.C. van de Geijn, J. Goudriaan and F. Berendse (eds.). Agrobiol. Themas 9. AB-DLO, Wageningen. pp. 23-41.

Eamus, D. 1991. The interaction of rising CO_2 and temperatures with water-use efficiency. *Plant Cell Environ.* **14**: 843-852.

Ellis, R.H., Hadley, P., Roberts, E.H. and Summerfield, R.J. 1990. Quantitative relations between temperature and crop development and growth. In: *Climate Change and Genetic Resources.* M. Jackson, B.V. Ford-Lloyd and M.L. Parry (eds.). Belhaven Press, London. pp. 85-115.

Farquhar, G.D., Schulze, E.-D. and Küppers, M. 1980. Responses to humidity by stomata of *Nicotiana glauca* L. and *Corylus avellana* L. are consistent with the optimization of carbon dioxide uptake with respect to water loss. *Austr. J. Plant Physiol.* **7**: 315-327.

Goudriaan, J. and Unsworth, M.H. 1990. Implications of increasing carbon dioxide and climate change for agricultural productivity and water resources. In: *Impact of Carbon Dioxide, Trace Gases, and Climate Change on Global Agriculture. ASA Spec. Pub No. 53.* pp. 111-130.

Grashoff, C., Rabbinge, R. and Nonhebel, S. 1994. Potential effects of global climate change on cool season food legume productivity. In: *Expanding the Production and Use of Cool Season Food Legumes.* F.J. Muehlbauer and W.J. Kaiser (eds.). Kluwer Academic, Dordrecht. pp. 159-174.

Grashoff, C., Dijkstra, P., Nonhebel, S., Schapendonk, A.H.C.M. and van de Geijn, S.C. 1995. Effects of climate change on productivity of cereals and legumes; model evaluation of observed year-to-year variability of the CO_2 response. *Global Change Biology* **1**(6): 417-428.

Hendrey, G.R., Lewin, K.F. and Nagy, J. 1993. Free air carbon dioxide enrichment: Development, progress, results. *Vegetatio* **104/105**: 17-31.

Hollinger, D.Y., Kelliher, F.M., Schulze, E.-D. and Köstner, B.M.M. 1994. Coupling of tree transpiration to atmospheric turbulence. *Nature* **371**: 60-62.

Idso, S.B., Kimball, B.A. and Mauney, J.R. 1987. Atmospheric carbon dioxide enrichment effects on cotton midday foliage temperature: Implications for plant water-use efficiency. *Agron. J.* **79**: 667-672.

IGBP [International Geosphere-Biosphere Programme]. 1993. Biospheric Aspects of the Hydrological Cycle (BAHC). The Operational Plan. *IGBP Report No. 27*. Stockholm, Sweden.

Jackson, R.B., Sala, O.E., Field, C.B. and Mooney, H.A. 1994. CO_2 alters water use, carbon gain, and yield for the dominant species in a natural grassland. *Oecologia* **98**: 257-262.

Jacobs, C.M.J. 1994. *Direct Impact of Atmospheric CO_2 Enrichment on Regional Transpiration*. Thesis. Wageningen Agricultural University. 179 p.

Jacobs, C.M.J. and de Bruin, H.A.R. 1992. The sensitivity of regional transpiration to land-surface characteristics: significance of feedback. *J. Climate* **5**: 683-698.

Kenny, G.J., Harrison, P.A., Olesen, J.E. and Parry, M.J. 1993. The effects of climate change on land suitability of grain maize, winter wheat and cauliflower in Europe. *Eur. J. of Agronomy* **2**: 325-338.

Kimball, B.A., Mauney, J.R., Radin, J.W., Nakayama, F.S., Idso, S.B., Hendrix, D.L., Akey, D.H., Allen, S.G., Anderson, M.G. and Hatung, W. 1986. Effects of increasing atmospheric CO_2 on the growth, water relations, and physiology of plants grown under optimal and limiting levels of water and nitrogen. In: *Response of Vegetation to Carbon Dioxide. Report No. 039*. US DOE, Carbon Dioxide Research Division, and USDA-ARS, Washington DC.

Kimball, B.A., Mauney, J.R., Nakayama, F.S. and Idso, S.B. 1993. Effects of increasing atmospheric CO_2 on vegetation. *Vegetatio* **104/105**: 65-75.

Kimball, B.A., Pinter, P.J. Jr., Garcia, R.L., LaMorte, R.L., Wall, G.W., Hunsaker, D.J., Wechsung, G., Wechsung, F. and Kartschall, Th. 1995. Productivity and water use of wheat under free-air CO_2 enrichment. *Global Change Biology* **1**(6): 429-442.

Körner, C. 1988. Does global increase of CO_2 alter stomatal density? *Flora* **181**: 253-257.

Kuiper, P.J.C. 1993. Diverse influences of small temperature increases on crop performance. In: *International Crop Science I*. D.R. Buxton *et al.* (eds.). Crop Science Society of America, Madison, WI. pp. 309-313.

Leith, J.H., Reynolds, J.F. and Rogers, H. 1986. Estimation of leaf area of soybeans grown under elevated carbon dioxide levels. *Field Crops Res.* **13**: 193-203.

Leuning, R. 1995. A critical appraisal of a combined stomatal-photosynthetic model for C3 plants. *Plant Cell Environ.* **18**: 339-355.

Leuning, R. and Foster, L.J. 1990. Estimation of transpiration by single trees:

comparison of a ventilated chamber, leaf energy budgets and a combination equation. *Agric. For. Meteorol.* **51**: 63-68.

Leuning, R., Wang, Y.P., de Pury, D., Denmead, O.T., Dunin, F.X., Condon, A.G., Nonhebel, S. and Goudriaan, J. 1993. Growth and water use of wheat under present and future levels of CO_2. *J. Agric. Meteorol.* **48**: 807-810.

Massman, W.J. and Ham, J.M. 1994. An evaluation of a surface energy balance method for partitioning ET data into plant and soil components for a surface with partial canopy cover. *Agric. For. Meteorol.* **67**: 253-267.

McKenney, M.S. and Rosenberg, N.J. 1993. Sensitivity of some potential evapotranspiration estimation methods to climate change. *Agric. For. Meteorol.* **64**: 81-110.

McKenney, M.S., Easterling, W.E. and Rosenberg, N.J. 1992. Simulation of crop productivity and responses to climate change in the year 2030: The role of future technologies, adjustments and adaptations. *Agric. For. Meteorol.* **59**: 103-127.

McNaughton, K.G. 1994. Effective stomatal and boundary-layer resistances of heterogeneous surfaces. *Plant Cell Environ.* **17**: 1061-1068.

McNaughton, K.G. and Jarvis, P.G. 1991. Effects of spatial scale on stomatal control of transpiration. *Agric. For. Meteorol.* **54**: 279-301.

Morison, J.I.L. 1987. Intercellular CO_2 concentration and stomatal response to CO_2. In: *Stomatal Function*. E. Zeiger, G.D. Farquhar and I.R. Cowan (eds.). Stanford Univ. Press, California. pp. 229-251.

Morison, J.I.L. 1993. Response of plants to CO_2 under water limited conditions. *Vegetatio* **104/105**: 193-209.

Morison, J.I.L. and Gifford, R.M. 1983. Stomatal sensitivity to carbon dioxide and humidity. *Plant Physiol.* **71**: 789-796.

Morison, J.I.L. and Gifford, R.M. 1984a. Plant growth and water use with limited water supply in high CO_2 concentrations. I. Leaf area, water use and transpiration. *Aust. J. Plant Physiol.* **11**: 361-374.

Morison, J.I.L. and Gifford, R.M. 1984b. Plant growth and water use with limited water supply in high CO_2 concentrations. II. Plant dry weight, partitioning and water-use efficiency. *Aust. J. Plant Physiol.* **11**: 375-384.

Mott, K.A. 1990. Sensing of atmospheric CO_2 by plants. *Plant, Cell Environ.* **13**: 731-737.

Nijs, I., Impens, I. and Behaeghe, T. 1989. Effects of long-term elevated atmospheric CO_2 concentration on *Lolium perenne* and *Trifolium repens* canopies in the course of a terminal drought stress period. *Can. J. Bot.* **67**: 2720-2725.

Oberbauer, S.O., Strain, B.R. and Fetcher, N. 1985. Effect of CO_2-enrichment on seedling physiology and growth of two tropical tree species. *Physiol. Plant.* **65**: 352-364.

O'Leary, J.W. and Knecht, G.N. 1981. Elevated CO_2 concentrations increase stomate numbers in *Phaseolus vulgaris* leaves. *Bot. Gaz.* **142**: 436-441.

Paez, A., Hellmers, H. and Strain, B.R. 1984. Carbon dioxide enrichment and water stress interaction on growth of two tomato cultivars. *J. Agric. Sci. (Camb.)* **102**: 687-693.

Penuelas, J. and Matamala, R. 1990. Changes in N and S leaf content, stomatal density and specific leaf area of 14 plant species during the last three centuries of CO_2 increase. *J. Exp. Bot.* **41**: 1119-1124.

Rogers, H.H., Runion, G.B. and Krupa, S.V. 1994. Plant responses to atmospheric CO_2 enrichment with emphasis on roots and the rhizosphere. *Environ. Pollution* **83**: 155-189.

Rowland-Bamford, A.J., Nordenbrock, C., Baker, J.T., Bowes, G. and Allen, L.H. Jr. 1990. Changes in stomatal density in rice grown under various CO_2 regimes with natural solar irradiance. *Envir. Exp. Bot.* **30**: 175-180.

Rozema, J. 1993. Plant responses to atmospheric carbon dioxide enrichment: interactions with some soil and atmospheric conditions. *Vegetatio* **104/105**: 173-190.

Rozema, J., Lenssen, G.M., Arp, W.J. and van de Staay, J.W.M. 1991a. Global change, the impact of the greenhouse effect (atmospheric CO_2 enrichment) and the increased UV-B radiation on terrestrial plants. In: *Ecological Responses to Environmental Stresses*. J. Rozema and J.A.C. Verkleij (eds.). Kluwer Academic, Dordrecht. pp. 220-231.

Rozema, J., Dorel, F., Janissen, R., Lenssen, G., Broekman, R., Arp, W. and Drake, B.G. 1991b. Effect of elevated atmospheric CO_2 on growth, photosynthesis and water relations of salt marsh grass species. *Aqu. Bot.* **39**: 45-55.

Sionit, N., Rogers, H.H., Bingham, G.E. and Strain, B.R. 1984. Photosynthesis and stomatal conductance with CO_2-enrichment of container- and field-grown soybeans. *Agron. J.* **76**: 447-451.

Stanghellini, C. and Bunce, J.A. 1994. Response of photosynthesis and conductance to light, CO_2, temperature and humidity in tomato plants acclimated to ambient and elevated CO_2. *Photosynthetica* **29**: 487-497.

Tyree, M.T. and Alexander, J.D. 1993. Plant water relations and the effects of elevated CO_2: a review and suggestions for future research. *Vegetatio* **104/105**: 47-62.

Unsworth, M.H., Heagle, A.S. and Heck, W.W. 1984. Gas exchange in open top field chambers. I. Measurement and analysis of atmospheric resistances. *Atmosph. Environ.* **18**: 373-380.

van de Geijn, S.C. and van Veen, J.A. 1993. Implications of increased carbon dioxide levels for carbon input and turnover in soils. *Vegetatio* **104/105**: 283-292.

van de Geijn, S.C., Goudriaan, J., van der Eerden, L.J. and Rozema, J. 1993. Problems and approaches to integrating the concurrent impacts of elevated CO_2, temperature, UV-B radiation and ozone on crop production. In: *International Crop Science I*. D.R. Buxton *et al.* (eds.). Crop Science Society of America, Madison, WI. pp. 333-338.

van der Burgh, J., Visscher, H., Dilcher, D.L. and Kurschner, W.M. 1993. Paleoatmospheric signatures in Neogene fossil leaves. *Science* **260**: 1788-1790.

van Keulen, H. and Seligman, N.G. 1987. *Simulation of Water Use, Nitrogen Nutrition and Growth of a Spring Wheat Crop*. Simulation Monographs, Pudoc, Wageningen, Netherlands.

Wolfe, D.W. 1994. Physiological and growth responses to atmospheric carbon dioxide concentration. In: *Handbook of Plant and Crop Physiology*. M. Pessarakli (ed.). Marcel Dekker, New York. pp. 223-242.

Woodward, F.I. 1987. Stomatal numbers are sensitive to increases in CO_2 from pre-industrial levels. *Nature* **327**: 617-618.

Woodward, F.I. 1993. Plant responses to past concentrations of CO_2. *Vegetatio* **104/105**: 145-155.

Woodward, F.I. and Bazzaz, F.A. 1988. The responses of stomatal density to CO_2 partial pressure. *J. Exp. Bot.* **39**: 1771-1781.

6. Effects of Higher Day and Night Temperatures on Growth and Yields of Some Crop Plants

YASH P. ABROL
Division of Plant Physiology, Indian Agricultural Research Institute, New Delhi, India

KEITH T. INGRAM
University of Georgia, College of Agricultural and Environmental Sciences, Georgia Agricultural Experiment Station, Griffin, Georgia, USA

Gaseous emissions from human activities are substantially increasing the concentrations of atmospheric greenhouse gases, particularly carbon dioxide, methane, chlorofluorocarbons and nitrous oxides. Global circulation models predict that these increased concentrations of greenhouse gases will increase average world temperature. Under the business-as-usual scenario of the Intergovernmental Panel on Climate Change (IPCC), global mean temperatures will rise 0.3°C per decade during the next century with an uncertainty of 0.2 to 0.5% (Houghton *et al.*, 1990). Thus global mean temperatures should be 1°C above the present values by 2025 and 3°C above the present value by 2100. Although global circulation models do not all agree as to the magnitude, most predict greenhouse warming. There is also general agreement that global warming will be greater at higher latitudes than in the tropics. Different global circulation models have predicted that global warming effects will vary diurnally, seasonally and with altitude.

It is also possible that there will be an autocatalytic component to global warming. Photosynthesis and respiration of plants and microbes increase with temperature, especially in temperate latitudes. As respiration increases more with increased temperature than does photosynthesis, global warming is likely to increase the flux of carbon dioxide to the atmosphere which would constitute a positive feedback to global warming.

This paper describes the effects of higher day and night temperatures on crop growth and yield. Temperature effects at different levels of organization – biochemical, physiological, morphological, agronomic and systems – are considered. This is followed by identification of options for germplasm improvement and crop management that may mitigate the adverse effects of higher day and night temperatures. The main focus is on wheat (*Triticum aestivum* L.) and rice (*Oryza sativa* L.).

MECHANISMS FOR HEAT TOLERANCE

Crop plants are immobile. They must adapt to prevalent soil and weather conditions. Except for transpirational cooling, plants are unable to adjust their tissue temperatures to any significant extent. On the other hand, plants have evolved several mechanisms that enable them to tolerate higher temperatures. These adaptive thermotolerant mechanisms reflect the environment in which a species has evolved and they largely dictate the environment where a crop may be grown.

Four major aspects of thermotolerance have been studied: (1) thermal dependence at the biochemical and metabolic levels; (2) thermal tolerance in relation to membrane stability; (3) induced thermotolerance through gradual temperature increase vis-à-vis production of heat shock proteins; and (4) photosynthesis and productivity during high temperature stress.

BIOCHEMICAL PROCESSES

Temperature effects on the rates of biochemical reactions may be modelled as the product of two functions, an exponentially increasing rate of the forward reaction and an exponential decay resulting from enzyme denaturation as temperatures increase (Figure 6.1a). The greatest concern is whether it is possible to increase the upper limit of enzyme stability to prevent denaturation.

Failure of only one critical enzyme system can cause death of an organism. This fact may explain why most crop species survive sustained high temperatures up to a relatively narrow range, 40 to 45°C. The relationship between the thermal environment for an organism and the thermal dependence of enzymes has been well established (Senioniti *et al.*, 1986).

The shape of this function also describes temperature effects on most biological functions, including plant growth and development. The function can be categorized by the three cardinal temperatures – minimum, optimum and maximum. Modellers frequently simplify the relationship into a stepwise linear function. The stepwise linear function has a plateau rather than an optimum temperature (Figure 6.1b).

The thermal dependence of the apparent reaction rate for selected enzymes may indicate the optimal thermal range for a plant. The range over which the apparent Michaelis-Menten constant for CO_2 (K_m) is minimal and stable is termed the thermal kinetic window (Mahan *et al.*, 1987). For crop plants, the thermal kinetic window (TKW) is generally established as a result of thermally induced lipid phase changes, rubisco activity and the starch synthesis pathway in leaves and reproductive organs (Burke, 1990).

In cotton and wheat, the time during which foliage temperature remained within the TKW was related to dry matter accumulation (Burke *et al.*, 1988). The cumulative time that rainfed crop foliage is outside the TKW provides an index of the degree of extreme temperature stress of the environment (Figure 6.2). Irrigation is one management option to reduce crop exposure to heat stress.

Temperatures that inhibit cellular metabolism and growth for a cool season C_3

EFFECTS OF HIGHER TEMPERATURES ON CROP GROWTH AND YIELD 125

Figure 6.1. (a) Exponential rate of reaction as a function of temperature. (b) Stepwise increase in reaction rate as a function of temperature

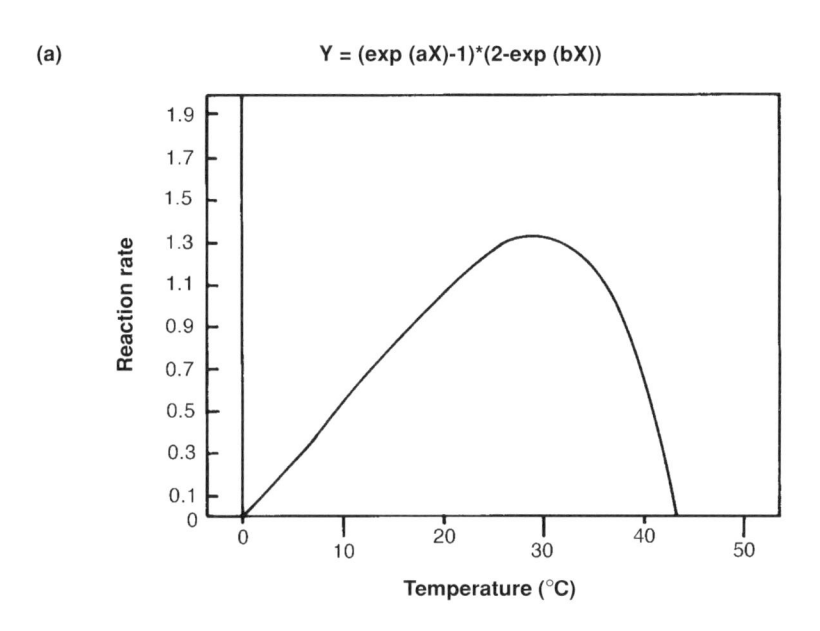

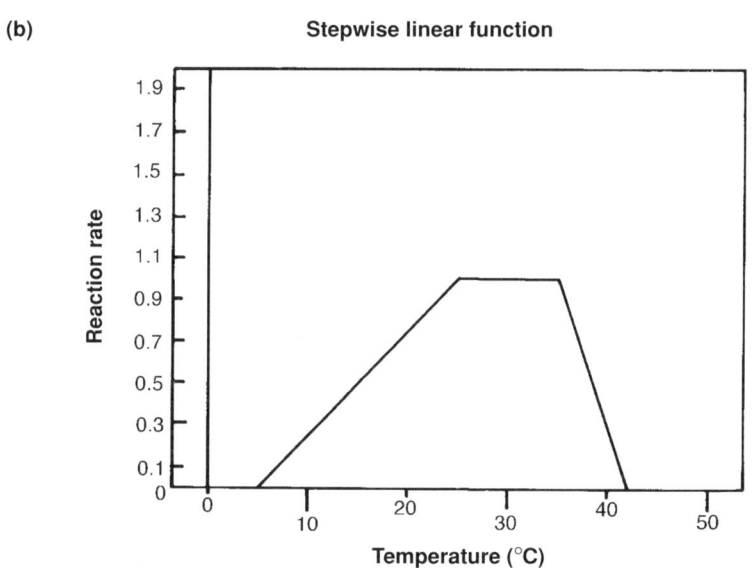

(a) $Y = (\exp(aX)-1)*(2-\exp(bX))$

(b) Stepwise linear function

species such as wheat may not inhibit warm-season C_3 species such as rice (*Oryza sativa* L.) and C_4 species such as sorghum, maize (*Zea mays* L.) and sugar cane (*Saccharum spontaneum* spp.). The identification of TKWs for different species can aid in the interpretation of the differential temperature stress responses for crop growth and development among species (Burke, 1990).

Figure 6.2. Seasonal foliage temperatures of wheat (cv. Kanking) and cotton (cv. Paymaster 145) grown at Lubbock, Texas. The vertical lines represent the temperature range that comprises the species-specific thermal kinetic window as determined from the changes in the apparent K_m of purified enzymes with temperature. Foliage temperatures were measured with a 50° field-of-view Teletemp Model 50 infrared thermometer (Teletemp Corp., Fullerton, CA) positioned at 1.5 m above the crop. Instruments were scanned at 1 min intervals with a 15 min average computed and stored. The infrared thermometer viewed an area of 0.75 m^2, with the same area continuously sampled (from Burke *et al.*, 1988; reproduced with permission)

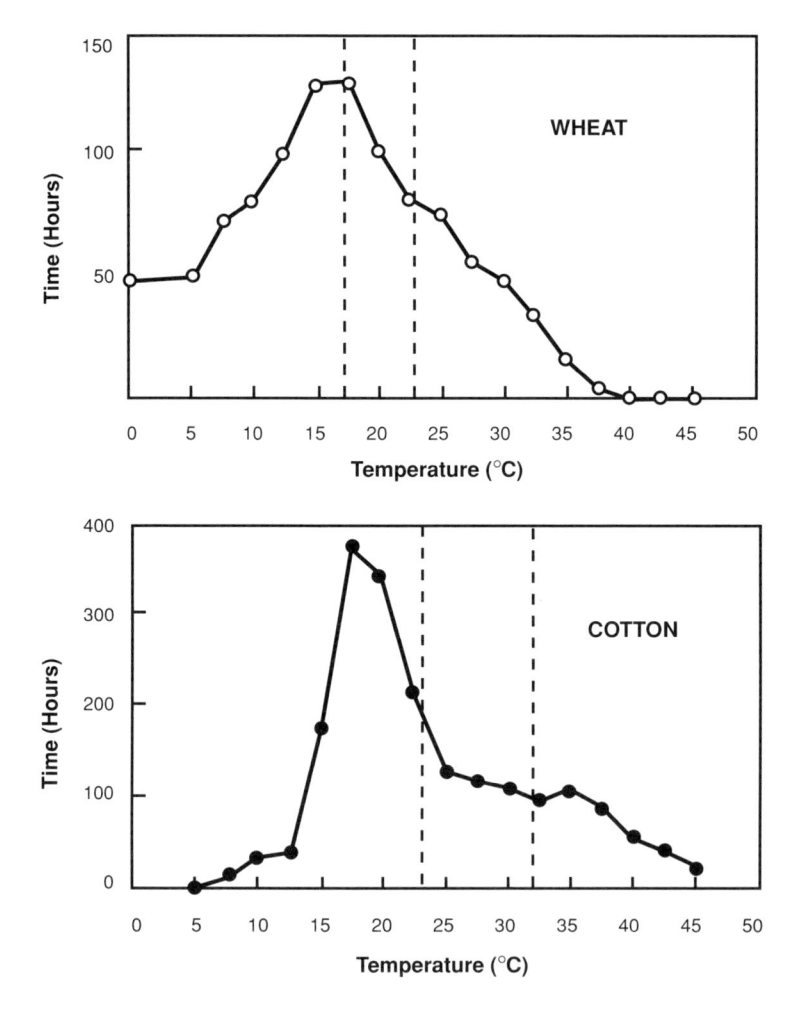

THERMAL STABILITY OF CELL MEMBRANES

The plasmalemma and membranes of cell organelles play a vital role in the functioning of cells. Any adverse effect of temperature stress on the membranes leads to disruption of cellular activity or death. Heat injury to the plasmalemma may be measured by ion leakage (Chaisompongpan et al., 1990; Hall, 1993). Injury to membranes from a sudden heat stress event may result from either denaturation of the membrane proteins or from melting of membrane lipids which leads to membrane rupture and loss of cellular contents (Ahrens and Ingram, 1988).

Heat stress may be an oxidative stress (Lee et al., 1983). Peroxidation of membrane lipids has been observed at high temperatures (Mishra and Singhal, 1992; Upadhyaya et al., 1990), which is a symptom of cellular injury. Enhanced synthesis of an anti-oxidant by plant tissues may increase cell tolerance to heat (Upadhyaya et al., 1990, 1991) but no such anti-oxidant has been positively identified.

A relationship between lipid composition and incubation temperature has been shown for algae, fungi and higher plants. In *Arabiodopsis*, exposed to high temperatures, total lipid content decreases to about one-half and the ratio of unsaturated to saturated fatty acids decreases to one-third of the levels at temperatures within the TKW (Somerville and Browse, 1991). Increase in saturated fatty acids of membranes increases their melting temperature and thus confers heat tolerance. An *Arabiodopsis* mutant, deficient in activity of chloroplast fatty acid W-9 desaturase, accumulates large amounts of 16:0 fatty acids, resulting in greater saturation of chloroplast lipids. This increases the optimum growth temperature (Kunst et al., 1989; Raison, 1986).

In cotton, however, heat tolerance does not correlate with degree of lipid saturation (Rikin et al., 1993) and similar differences in genotypic differences in heat tolerance have been unrelated to membrane lipid saturation in other species (Kee and Nobel, 1985). In such species, a factor other than membrane stability may be limiting growth at high temperature.

HEAT SHOCK PROTEINS

Synthesis and accumulation of proteins were ascertained during a rapid heat stress. These were designated as 'Heat Shock Proteins' (HSPs). Subsequently it was shown that increased production of these proteins also occurs when plants experience a gradual increase in temperature more typical of that experienced in a natural environment.

Three classes of proteins as distinguished by molecular weight account for most HSPs, namely HSP90, HSP70, and low molecular weight proteins of 15 to 30 kDa (LMW HSP). The proportions of the three classes differ among species. In general, heat shock proteins are induced by heat stress at any stage of development. Under maximum heat stress conditions, HSP70 and HSP90 mRNAs can increase ten-fold and LMW HSP increase as much as 200-fold. Three other proteins, though less important, are also considered to be heat shock proteins viz. 110 kDa polypeptides, ubiquitin, and GroEL proteins.

In arid and semi-arid regions, dryland crops may synthesize and accumulate substantial levels of heat shock proteins in response to elevated leaf temperatures. The induction temperature for synthesis and accumulation of heat shock proteins in laboratory-grown cotton ranged from 38 to 41°C (Burke et al., 1985). Soil water deficits resulting in midday canopy temperature of 40°C or greater for two to three weeks were used to study heat shock proteins in field-grown cotton (Figure 6.3). A comparison of polypeptide patterns of dryland and irrigated cotton leaves showed that at least eight new polypeptides accumulated in about half of the dryland leaves analysed. The polypeptides that accumulated in the dryland leaves but not irrigated cotton leaves had molecular weights of l00, 94, 89, 75, 60, 58 and 21 kDa. In a similar experiment with field-grown soybean (*Glycine max* (L.) Merr.), several heat shock proteins were observed in both irrigated and dryland treatments, although levels were greater in the non-irrigated treatments (Kimpel and Key, 1985).

Correlation between synthesis and accumulation of heat shock proteins and heat tolerance suggests, but does not prove, that the two are causally related. Further evidence for a causal relationship is that some cultivar differences in heat shock protein expression correlate with differences in thermotolerance. In genetic experiments, heat shock protein expression co-segregates with heat tolerance. Another evidence for the protective role of heat shock protein is that mutants unable to synthesize heat shock proteins, and cells in which HSP70 synthesis is blocked or inactivated, are more susceptible to heat injury.

Figure 6.3. Seasonal changes in the midday canopy temperatures of irrigated (○) and dryland (●) cotton. Arrow indicates the day on which dryland plots were irrigated with 10 to 12.5 cm of water (106 DAP). The cotton strain is T185. (From Burke et al., 1985: reproduced with permission)

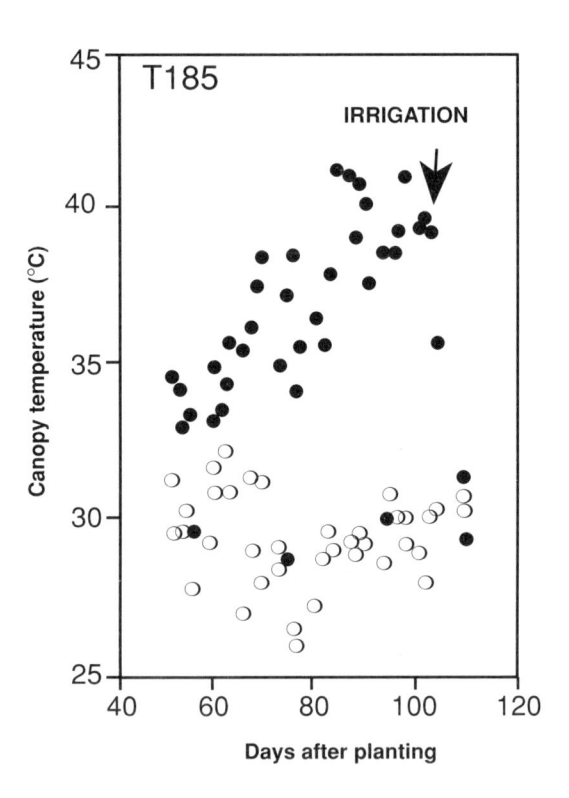

The mechanism by which heat shock proteins contribute to heat tolerance is still not certain. One hypothesis is that HSP70 participates in ATP-dependent protein unfolding or assembly/disassembly reactions and that they prevent protein denaturation during stress (Pelham, 1986). If this mechanism is true, then heat shock proteins may provide a significant basis for increasing heat tolerance of crop plants in a global warming situation. The LMW HSPs may play a structural role in maintaining cell membrane integrity during stress. Other heat shock proteins have been associated with particular organelles such as chloroplasts, ribosomes and mitochondria. In tomato (*Lycopersicon esculentum* L.), heat shock proteins aggregate into a granular structure in the cytoplasm, possibly protecting the machinery of protein synthesis.

HSPs provide a significant opportunity to increase heat tolerance of crops. To elucidate their mechanisms of action and to exploit their potential contribution to increasing heat tolerance, four lines of investigations are suggested:

1. Establish the biochemical activities of individual HSPs as a preliminary step.
2. Characterize the genetic variability of specific heat shock proteins across a wide range of germplasm. Develop iso-population and near isogenic lines selected for production of low and high levels of HSP synthesis.
3. As HSPs appear to participate in maintaining the conformation or assembly of other protein structures, analyse the molecular details of these processes and establish all participating protein substrates. Such biochemical studies are needed to understand how these processes protect or allow recovery from heat stress.
4. Identify specific HSP mutants or create transgenic mutant plants to complement molecular and biochemical understanding with genetic approaches.

PHOTOSYNTHESIS AND HIGH TEMPERATURE STRESS

Variability in leaf photosynthetic rates within or between species is often unrelated to differences in productivity. Similarly, high photosynthetic rates at high temperatures do not necessarily support high rates of crop dry matter accumulation. The temperature optimum for photosynthesis is broad, presumably because crop plants have adapted to a relatively wide range of thermal environments. A 1 to 2°C increase in average temperature is not likely to have a substantial impact on leaf photosynthetic rates. Further, there is a possibility that photosynthesis of crop plants can adapt to a slow increase in global average temperatures. Thus, global warming is not likely to affect photosynthetic rates per unit leaf area gradually or on a closed canopy basis over the next century.

While photosynthetic rates were found to be temperature-sensitive in other crops, wheat and rice appear to be different. In wheat, no measurable differences were found in photosynthetic rates per unit flag leaf area or on a whole-plant basis in the temperature range from 15 to 35°C (Bagga and Rawson, 1977). In rice, there is little temperature effect on leaf carbon dioxide assimilation from 20 to 40°C (Egeh *et al.*, 1994).

Recent research has shown significant variation among wheat cultivars with respect to reduction in photosynthesis at very high temperature. Photosynthesis of germplasm adapted to higher temperature environments was less sensitive to high temperature than was germplasm from cooler environments (Al-Khatib and Paulsen, 1990). When this germplasm was grown under moderate (22/17°C) and high (32/27°C) temperatures in the seedling stage or from anthesis to maturity, there was a highly significant correlation between photosynthesis rate and either seedling biomass (r=0.943***) or grain yield of mature plants (r=0.807**). Genotypes most tolerant to high temperatures had the most stable leaf photosynthetic rates across temperature regimes or they had the longest duration of leaf photosynthetic activity after anthesis and high grain weights. The above relationship was exemplified by 'Ventnor' from the high temperature area of Australia and 'Lancero' from the high altitude area of Chile (Table 6.1). See Al-Khatib and Paulsen (1990).

Despite observed negative effects of high temperature on leaf photosynthesis, the temperature optimum for net photosynthesis is likely to increase with elevated levels of atmospheric carbon dioxide. Several studies have concluded that CO_2-induced increases in crop yields are much more probable in warm than in cool environments (Idso, 1987; Gifford, 1989; Rawson, 1992, 1995). Thus, global warming may not greatly affect overall net photosynthesis.

NIGHT RESPIRATION RESPONSE TO CO_2 AND TEMPERATURE

In tomato (*Lycopersicon esculentum*) (Behboudian and Lai, 1994) and cotton (*Gossypium hirsutum*) (Thomas et al., 1993), elevated CO_2 increased dark respiration possibly because of increased carbohydrate accumulation in tissues. The latter has been shown to increase alternative pathway respiration as well (Amthor, 1991).

Apparent dark respiration may decline under elevated CO_2 if there is dark CO_2 fixation or if elevated CO_2 directly inhibits or inactivates respiratory enzymes as may occur through increased formation or carbamate (Wullschleger et al., 1994).

Table 6.1. Mean weekly photosynthetic rate (fmol $CO_2/m^2/s$) and duration of photosynthetic activity (weeks, in parentheses), and grain biomass of two wheat genotypes grown at two temperature regimes

GENOTYPE	Seedlings after two weeks of treatment		From anthesis to maturity		Grain biomass (g/tiller)	
	22/17°C	32/27°C	22/17°C	32/27°C	22/17°C	32/27°C
Ventor	8.6	7.0	6.4 (7)	5.3 (7)	0.93	0.8
Lancero	7.9	4.6	4.1 (10)	3.1 (7)	0.43	0.28

Figures in parentheses give duration of photosynthetic activity in weeks from anthesis to physiological maturity.
(Modified from Tables 2 and 3 of Al-Khatib and Paulsen (1990); reproduced with permission.)

Few studies have successfully partitioned the effects of elevated CO_2 on growth and maintenance respiration. Both components appear to decline, probably because of decrease in leaf protein levels which results in reduced construction and maintenance costs (Wullschleger *et al.*, 1994).

Elevated CO_2 reduced maintenance respiration of *Medicago sativa* and *Dactylis glomerata* at lower temperatures (15 to 20°C), whereas elevated CO_2 reduced growth respiration of *M. sativa* at 20 to 30°C and *D. glomerata* at 15 to 25°C (Ziska and Bunce, 1993).

CROP GROWTH AND DEVELOPMENT

Wheat development is customarily divided into vegetative and reproductive phases, with either ear emergence or anthesis as the event that separates the two phases. In the past 30 to 40 years, the sequence of pre-anthesis phenological events has been critically assessed with respect to grain yield potential and sensitivity to weather variables, particularly prevailing temperature and day length. Several systems are used to classify the sequence of phenological events. We identify five such developmental stages:

(i) germination – seeding to seedling emergence;
(ii) canopy development – emergence to first spikelet initiation, the double ridge stage;
(iii) spikelet production – first spikelet initiation to terminal spikelet formation;
(iv) spikelet development – terminal spikelet formation to anthesis;
(v) grain development – anthesis to maturity.

These stages are generally based on early recognized features of the apical meristem. They mark significant changes in morphology or physiology of different crop organs. Numbers of leaf and tiller primordia are determined before spikelet initiation but their subsequent growth and development are controlled by temperature and day length during the differentiation of spikes into spikelets. Similarly, floret number within each spikelet is established by anthesis, at which time the potential grain number per spike is established (Figure 6.4).

Productivity of wheat and other crop species falls markedly at high temperatures. Wheat in India is invariably exposed to extreme temperatures during some stages of development (Abrol *et al.*, 1991). In Australia, wheat is usually exposed to brief periods of heat stress during grain development.

All stages of development are sensitive to temperature. It is the main factor controlling the rate of crop development (Table 6.2). Development generally accelerates as temperature increases, a phenomenon that is often described as a linear function of daily average temperature. The growing degree day concept is a common example of a linear model of developmental response to temperature. While a linear model works well to describe wheat development as long as temperatures remain within 10 to 30°C, a non-linear model as in Figure 6.1 is needed to describe development when a crop is exposed to extreme temperature stress.

Figure 6.4. Schematic diagram of wheat growth and development, showing the stages of sowing (Sw), emergence (Em), first double ridge appearance (DR), terminal spikelet appearance (TS), heading (Hd), anthesis (At), beginning of the grain-filling period (BGF), physiological maturity (PM), and harvest (Hv). Patterned boxes indicate the period of differentiation or growth of specific organs. Bars represent the periods of development when different components of grain yield are produced (heavy lines refer to main shoots and light lines represent extension associated with tillers. (Adapted from Slafer and Rawson, 1994; reproduced with permission)

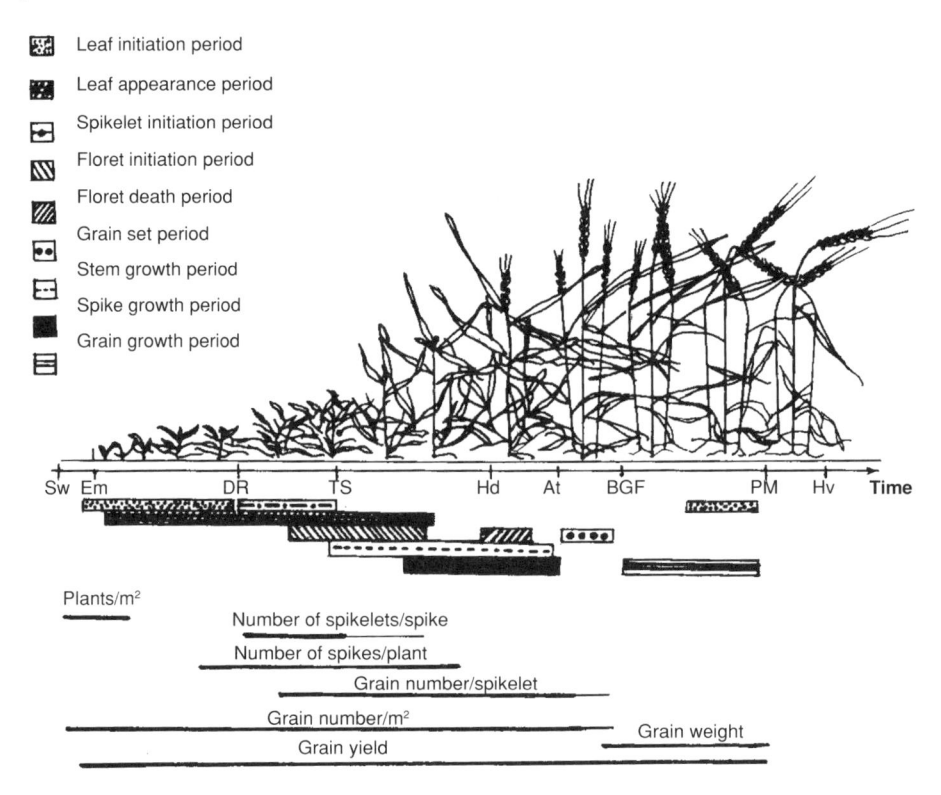

VEGETATIVE PHASE

Several experiments have observed the effects of temperature on the duration from sowing or emergence to heading under controlled environment and field conditions. Unfortunately, however, few experiments have been conducted with enough cultivars to assess the genetic variability in this trait. The major conclusions from these studies are:
  1. All genotypes are sensitive to temperature at one stage or another. Temperature sensitivity, however, varies greatly with genotype.
  2. Phenological stages differ in sensitivity to temperature.

Table 6.2. Response of phasic development to temperature photoperiod and vernalization

Developmental phase	Temperature	Photoperiod	Vernalization
Sowing-emergence	++++/+++++	0	0
Emergence-double ridges	+++/+++++	0/+++	0/+++++
Double ridges-terminal spikelet	+/+++	0/+++++	0/+++++
Terminal spikelet-heading	+++/+++++	0/+++	0/+(?)
Heading-anthesis	+++/+++++	0/+(?)	0
Anthesis-maturity	+++/+++++	0	0

For the estimation of sensitivity, the total life span was divided into the stages shown in Figure 6.4. An arbitrary scale was used to show when the effects are strong (+++++), moderate (+++) or slight (+). 0 denotes that the factor does not affect the process and question marks refer to uncertainties in the literature. For each factor, genetic variation in response was considered.

(From Slafer and Rawson (1994); reproduced with permission.)

3. The duration of phase from sowing to first spikelet initiation is less sensitive to change in temperature than are other phases, although genotypes do differ in thermotolerance during this phase.
4. The stages during which environment has the greatest impact on yield are from first spikelet initiation or terminal spikelet formation until anthesis. Spikelet number and floral number (potential grain number), both dominant yield contributing attributes, are established during these phases. Grain weight, on the other hand, appears to be much less sensitive to heat stress than is grain number.

GRAIN DEVELOPMENT PHASE

In experiments under controlled conditions from 25 to 35°C, mean grain weight declined 16% for each 5°C increase in temperature (Asana and Williams, 1965). In pot experiments, grain yield decreased by 17% for each 5°C rise (Wattal, 1965). For every 1°C rise in temperature, there is a depression in grain yield by 8 to 10%, mediated through 5 to 6% fewer grains and 3 to 4% smaller grain weight.

To elucidate the causal factor for reduced grain filling in wheat because of higher temperatures, Wardlaw (1974) studied the three main components of the plant system. The three components are: (a) source - flag leaf blade; (b) sink - ear; and (c) transport pathway - peduncle. He observed that photosynthesis had a broad temperature optimum from 20 to 30°C with photosynthesis declining rapidly at temperatures above 30°C. The rate of ^{14}C assimilate movement out of the flag leaf, phloem loading, was optimum around 30°C; the rate of ^{14}C assimilate movement through the stem was independent of temperature from 1 to 50°C. Thus, in wheat, temperature effects on translocation result indirectly from direct temperature effects on source and sink activities.

In a subsequent experiment with source-sink relationships altered through grain excision, defoliation and shading treatments, heat stress still reduced grain weight (Wardlaw *et al.*, 1980). This result supports the earlier findings that temperature effects on grain weight are direct effects rather than assimilate availability (Bremner and Rawson, 1978; Ford *et al.*, 1978; Spiertz, 1974). Furthermore, respiration effects do not appear to be the direct cause of decreased grain size in heat-stressed wheat (Wardlaw, 1974).

Reduction of grain weight by heat stress may be explained mostly by effects of temperature on rate and duration of grain growth. As temperature increased from 15/10°C to 21/16°C, duration of grain filling was reduced from 60 to 36 days and grain growth rate increased from 0.73 to 1.49 mg/grain/day with a result of minimal influence on grain weight at maturity. Further increase in temperature from 21/16°C to 30/25°C resulted in decline in grain filling during 36 to 22 days with a minimal increase in grain growth rate from 1.49 to 1.51 mg/grain/day. Thus, mature grain weight was significantly reduced at the highest temperature.

Research on the effects of brief periods of ear warming after anthesis on ear metabolism have identified differential responses of starch and nitrogen accumulation in grain of four wheat cultivars (Bhullar and Jenner, 1983, 1985, 1986; Hawker and Jenner, 1993, Jenner 1991a,b). Warming increased the rate of dry matter accumulation in all the cultivars but the increase was less in cv. Aus 22645 than in the other cultivars studied. Rate of increase in nitrogen accumulation was, however, higher than the increase in total dry matter accumulation (Table 6.3). Under long-term exposure to heat stress, increased grain nitrogen concentration is almost entirely as a result of decreased starch content rather than a change in total grain quality (Bhullar and Jenner, 1985). The conversion of sucrose to starch within the endosperm is decreased by elevated temperatures. Furthermore, heat stress effects on final grain weight were associated with reduced levels of soluble starch synthetase activity (Hawker and Jenner, 1993).

In summary, high temperature reduction of grain yield results from: (a) reduced numbers of grains formed; (b) shorter grain growth duration; and (c) inhibition of sucrose assimilation in grains.

EXTREME TEMPERATURE EFFECTS ON CROPS

There are two major forms of extreme temperature stress on crops – heat and cold. An increase in global temperatures may have either or both of these two acute effects: more frequent high temperature stress and less frequent cold temperature stress.

Increase in temperature will lengthen the effective growing season in areas where agricultural potential is currently limited by cold temperature stress. Thus, increased temperature will cause a poleward shift of the thermal limits to agriculture. This poleward shift will be especially important for crops such as rice that have tropical centres of origin and adaptation but are also grown in temperate latitudes during warm seasons. Global warming impact will be greater in the northern than southern

Table 6.3. Effect of whole-plant warming on the rate of total dry matter and nitrogen accumulation, between days 10 and 20 after anthesis, in the grains of four cultivars of wheat

Cultivar	Treatment	Rate of increase (mg/grain/day)	
		Total dry matter	N content
AUS 22645	C	1.94±09	0.03±002
	W	2.07±07(107)	0.48±003(160)
Kite	C	1.72±15	0.027±004
	W	2.28±17(133)	0.043±003(159)
Sonora	C	1.65±18	0.034±009
	W	2.06±19(125)	0.051±009(150)
WW15	C	1.89±20	0.037±005
	W	2.37±20(125)	0.053±005(143)

Plants were grown at 21/16°C and some (W) were warmed, between 10 and 20 days after anthesis, to 33/25°C and then returned to 21/16°C where they stayed until maturity. Control plants (C) were grown at 21/16°C throughout.
Values given are the means ±.e; values in parentheses are percentage of cotton values.
(From Bhullar and Jennar (1985); reproduced with permission.)

hemisphere because there is more high-latitude area cultivated in the northern hemisphere.

Increased temperature would also affect the crop calendar in tropical regions. In the tropics, however, global warming, though predicted to be of only small magnitude, is likely to reduce the length of the effective growing season, particularly where more than one crop per year is grown. In semi-arid regions and other agro-ecological zones where there is wide diurnal temperature variation, relatively small changes in mean annual temperatures could markedly increase the frequency of highest temperature injury. For example, canopy temperature is 10 to 15°C higher in dryland cotton (*Gossypium hirsutum* L.) than in irrigated cotton (Figure 6.2). Thus, global warming would reduce dry matter accumulation in dryland cotton because of increased respiration, and reduced photosynthesis and cellular energy.

In India, the growing season for wheat is limited by high temperatures at sowing and during maturation. As wheat is grown over a wide range of latitudes, it is frequently exposed to temperatures above the threshold for heat stress. For example, rainfed wheat depends on soil moisture remaining after the monsoon rains recede in September. High maximum and minimum temperatures in September (about 34/20°C), which adversely affect seedling establishment, accelerate early vegetative development, reduce canopy cover, tillering, spike size and yield. Hence, sowing is typically delayed until after mid-October when seedbeds have cooled, though much of the residual soil moisture may be lost. High temperatures in the second half of February (25/10°C), March (30/13°C) and April (30/20°C) reduce the numbers of viable florets and the grain-filling duration. High temperature stress particularly reduces yield of wheat sown in December/January which is necessitated in some regions because of the multiple cropping system.

The situation is similar for sorghum (*Sorghum bicolor* (L.) Moench) and pearl millet (*Pennisetum glaucum* (L.) R.Br.) which are exposed to extreme high temperatures in Rajasthan, India. After sowing, air and soil temperatures often exceed 40°C and midday soil surface temperatures above 50°C are common (Figure 6.5).

Acute effects of high temperature are most striking when heat stress occurs during anthesis. In rice, heat stress at anthesis prevents anther dehiscence and pollen shed, to reduce pollination and grain numbers (Mackill *et al.*, 1982; Zheng and Mackill, 1982).

Clearly, many crops in tropical areas are already subjected to heat stress. If temperatures increase further, crop failure in some traditional areas would become more commonplace.

LONG-TERM EFFECTS OF HIGH TEMPERATURES ON CROPS

More important than acute effects of extreme temperature stress are the chronic effects of continuously warmer temperatures on crop growth and development. Chronic effects of high temperature include effects on grain growth discussed above. Record crop yields clearly reflect the importance of season-long effects on crop yields: crops generally yield the most where temperatures are cool during growth of the harvested component.

Crop growth simulations show that rice yields decrease 9% for each 1°C increase in seasonal average temperature (Kropff *et al.*, 1993). This chronic effect of high temperature differs significantly from the acute effect of short-term temperature events,

Figure 6.5. Diurnal temperature data recorded in Fatehpur, Rajasthan, India. (Latitude 27°C 37'N) in June 1989). Each measurement is the mean value from three thermocouples placed at either 5 cm depth of soil (▲): 0.5 cm depth of soil (●): or 150 cm above the soil surface (■). (From Howarth (1991); reproduced with permission.)

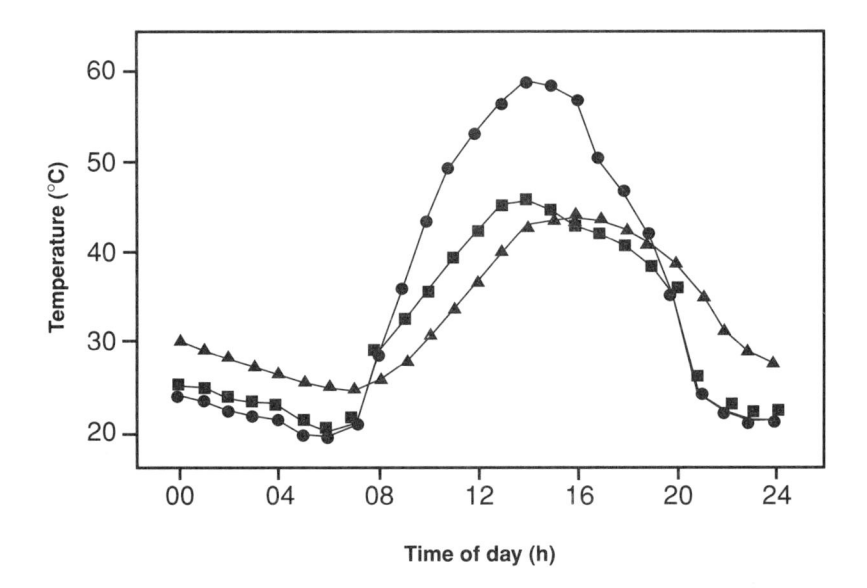

because seasonal temperature effects are mostly a result of effects on crop development. For most grain crops, there is much greater genotypic variation in thermal requirements for vegetative than for reproductive development. As long-term temperatures increase, grain-filling periods decrease, and there appears to be little scope to manipulate this effect through existing genetic variation within species.

REFERENCES

Abrol, Y.P., Bagga, A.K., Chakravorty, N.V.K. and Wattal, P.N. 1991. Impact of rise in temperature on productivity of wheat in India. In: *Impact of Global Climatic Change on Photosynthesis and Plant Productivity*. Y.P. Abrol *et al.* (eds.). Oxford & IBH Publishers, New Delhi. pp. 787-798.

Ahrens, M.J. and Ingram, D.L. 1988. Heat tolerance of citrus leaves. *Hort Sci.* **23**: 747-748.

Al-Khatib, K. and Paulsen, G.M. 1990. Photosynthesis and productivity during high temperature stress of wheat genotypes from major world regions. *Crop Sci.* **30**: 1127-1132.

Amthor, J.S. 1991. Respiration in a future higher CO_2 world. *Plant Cell Environ.* **14**: 13-20.

Asana, R.D. and Williams, R.F. 1965. The effect of temperature stress on grain development in wheat. *Aust. J. Agric. Res.* **16**: 1-13.

Bagga, A.K. and Rawson, H.M. 1977. Contrasting responses of morphologically similar wheat cultivars to temperatures appropriate to warm temperate climates with hot summers: a study in controlled environment. *Aust. J. Plant Physiol.* **4**: 877-887.

Behboudian, M.H. and Lai, R. 1994. Carbon dioxide in 'Virosa' tomato plants; responses to enrichment duration and to temperature. *Hort. Sci.* **29**: 1456-1459.

Bhullar, S.S. and Jenner, C.F. 1983. Responses to brief periods of elevated temperatures in ears and grain of wheat. *Aust. J. Plant Physiol.* **10**: 549-560.

Bhullar, S.S. and Jenner, C.F. 1985. Differential responses to high temperature of starch and nitrogen accumulation in the grain of four cultivars of wheat. *Aust. J. Plant Physiol.* **12**: 313-325.

Bhullar, S.S. and Jenner, C.F. 1986. Effect of a brief episode of elevated temperature on grain filling in wheat ears cultured on solution of sucrose. *Aust. J. Plant Physiol.* **13**: 617-626.

Bremner, P.M. and Rawson, H.M. 1978. Weights of individual grains of the wheat ear in relation to their growth potential, the supply of assimilate and interaction between grains. *Aust. J. Plant Physiol.* **5**: 51-72.

Bunce, J.A. 1994. Responses of respiration to increasing atmospheric carbon dioxide concentration. *Physiol. Plant* **90**: 427-430.

Burke, J.J. 1990. High temperature stress and adaptation in crops. In: *Stress Response in Plants: Adaptation and Acclimation Mechanisms*. R.G. Alscher and J.R. Cummings (eds.). Wiley-Liss, New York. pp. 295-309.

Burke, J.J., Hatfield, J.L., Klein, R.R. and Mullet, J.E. 1985. Accumulation of heat shock proteins in field grown soybean. *Plant Physiol.* **78**: 394-398.

Burke, J.J., Mahan, J.R. and Hatfield, J.L. 1988. Crop specific thermal kinetic windows in relation to wheat and cotton biomass production. *Agron. J.* **80**: 553-556.

Chaisompongopan, N., Li, P.H., Davis, D.W. and Mackhart, A.H. 1990. Photosynthetic responses to heat stress in common bean genotypes differing in heat acclimation potential. *Crop Sci.* **30**: 100-104.

Egeh, A.O., Ingram, K.T. and Zamora, O.B. 1994. High temperature effects on leaf exchange. *Phil. J. Crop Sci.* **17**: 21-26.

Ford, M.A., Pearman, I. and Thorne, G.N. 1978. Effects of variation in ear temperature on growth and yield of spring wheat. *Aust. J. Plant Physiol.* **3**: 337-347.

Gifford, R.M. 1989. The effect of the build-up of carbon dioxide in the atmosphere on crop productivity. Proceedings of the Fifth Australian Agronomy Conference, Perth, WA. pp. 312-322.

Hall, A.E. 1993. Breeding for heat tolerance. *Plant Breed Res.* **10**: 129-168.

Hawker, J.S. and Jenner, C.F. 1993. High temperature effects on the activity of enzymes in the committed pathway of starch synthesis in developing wheat endosperm. *Aust. J. Plant Physiol.* **20**: 197-200.

Houghton, J.T., Collander, B.A. and Ephraums, J.J. (eds.). 1990. *Climate Change – The IPCC Scientific Assessment.* Cambridge University Press, Cambridge. 135 p.

Howarth, C.J. 1991. Molecular response of plants to an increased incidence of heat shock. *Plant Cell Environ.* **14**: 831-841.

Idso, S.B., Kimball, B.A., Anderson, M.G. and Mouney, J.R. 1987. Effects of atmospheric CO_2 enrichment on plant growth; the interactive role of air temperature. *Agric. Ecosystems & Environment* **20**: 1-10.

Jenner, C.F. 1991a. Effects of exposure of wheat ears to high temperature on dry matter accumulation and carbohydrate metabolism in the grain of two cultivars. I. Immediate response. *Aust. J. Plant Physiol.* **18**: 165-177.

Jenner, C.F. 1991b. Effect of exposure of wheat ears to high temperature on dry matter accumulation and carbohydrate metabolism in the grain of two wheat cultivars. II. Carry over effects. *Aust. J. Plant Physiol.* **18**: 179-190.

Kee, S.C. and Nobel, P.S. 1985. Fatty acid composition of chlorenchyma membrane fractions from three desert succulents grown at moderate and high temperature. *Biochim. Biophys. Acta* **820**: 100-106.

Kimpel, J.A. and Key, J.L. 1985. Presence of heat shock mRNAs in field-grown soybean. *Plant Physiol.* **79**: 622-678.

Kropff, M.J., Centeno, G., Bachelet, D., Lee, M.H., Mohan Dass, S., Horie, T., De feng, S., Singh, S. and Penning de Vries, F.W.T. 1993. Predicting the impact of CO_2 and temperature on rice production. *IRRI Seminar Series on Climate Change and Rice.* International Rice Research Institute, Los Baños, Philippines (unpublished).

Kunst, L., Browse, J. and Somerville. 1989. Enhanced thermal tolerance in a mutant of *Arabidopsis* deficient in palmitic acid unsaturation. *Plant Physiol.* **91**: 401-408.

Lee, P.C., Bochner, B.R. and Ames, B.N. 1983. A heat shock stress and cell oxidation. *Proc. Natl. Acad. Sci., USA* **80**: 7496-7500.

Mackill, D.J., Coffman, W.R. and Rutger L.J. 1982. Pollen shedding and combining ability for high temperature tolerance in rice. *Crop Sci.* **20**: 730-733.

Mahan, J.R., Burke, J.J. and Orzech, K.A. 1987. The 'thermal kinetic window' as an indicator of optimum plant temperature. *Plant Physiol.* **82**: 518-522.

Mishra, R.K. and Singhal, G.S. 1992. Function of photosynthetic apparatus of intact wheat leaves under high light and heat stress and its relationship with thylakoid lipids. *Plant Physiol.* **98**: 1-6.

Pelham, H. 1986. Speculations on the major heat shock and glucose regulated proteins. *Cell* **46**: 959-961.

Raison, J.K. 1986. Alterations in the physical properties and thermal response of membrane lipids: correlations with acclimation to chilly and high temperature. In: *Frontiers of Membrane Research in Agriculture*. J.B. St. John, E. Berlin and P.C. Jackson (eds.). Rowman and Allanheld, Totoma, NJ. pp. 383-401.

Rawson, H.M. 1992. Plant responses to temperature under conditions of elevated CO_2. *Aust. J. Plant Physiol.* **40**: 473-490.

Rawson, H.M. 1995. Yield response of two wheat genotypes to carbon dioxide and temperature in field studies using temperature gradient tunnels. *Aust. J. Plant Physiol.* **22**: 23-32.

Rikin, A., Dillworth, J.W. and Bergman, D.K. 1993. Correlation between circadian rythym of resistance to extreme temperature and changes in fatty acid composition in cotton seedlings. *Plant Physiol.* **101**: 31-36.

Senioniti, E., Manetos, Y. and Gavales, N.A. 1986. Co-operative effects of light and temperature on the activity of phosphoenolpyruvate carboxylase from *Amaranthus paniculatus*. *Plant Physiol.* **82**: 518-522.

Slafer, G.A. and Rawson, R.M. 1994. Sensitivity of wheat phasic development to major environmental factors: a re-examination of some assumptions made by physiologists and modellers. *Aust. J. Plant Physiol.* **22**: 393-426.

Somerville, C. and Browse, J. 1991. Plant lipids, metabolism and membranes. *Science* **252**: 80-87.

Spiertz, J.H.J. 1974. Grain growth and distribution of dry matter in wheat plants as influenced by temperature, light energy and ear size. *Neth. J. Agric. Sci.* **22**: 207-220.

Thomas, R.B., Reid, C.D., Ybema, R. and Strain, B.R. 1993. Growth and maintenance components of leaf respiration of cotton grown in elevated carbon dioxide partial pressure. *Plant Cell Environ.* **16**: 533-546.

Upadhyaya, A., Davis, T.D., Larsen, M.H., Walsen, R.H. and Sankhla, M. 1990. Uniconazole-induced thermotolerance in soybean seedling root tissue. *Physiol. Plant.* **79**: 78-84.

Upadhyaya, A., Davis, T.D. and Sankhla, M. 1991. Heat shock tolerance and antioxidant activity in moth bean seedlings treated with tetayclasis. *Plant Growth Regulation* **10**: 215-222.

Wardlaw, I.F. 1974. Temperature control of translocation. In: *Mechanism of Regula-*

tion of Plant Growth. R.L. Bielske, A.R. Ferguson and M.M. Cresswell (eds.). Bull. Royal Soc. New Zealand, Wellington. pp. 533-538.

Wardlaw, I.F., Sofield, I. and Cartwright, P.M. 1980. Factors limiting the rate of dry matter in the grain of wheat grown at high temperature. *Aust. J. Plant Physiol.* **7**: 387-400.

Wattal, P.N. 1965. Effect of temperature on the development of the wheat grain. *Indian J. Plant Physiol.* **8**: 145-159.

Wullschleger, S.D., Ziska, L.H. and Bunce, J.A. 1994. Respiratory responses of higher plants to atmospheric CO_2 enrichment. *Physiol. Plant* **90**: 221-229.

Zheng, K.L. and Mackill, D.T. 1982. Effect of high temperature on anther dehiscence and pollination in rice. *Sabrao J.* **14**: 61-66.

Ziska, L.M. and Bunce, J.A. 1993. Inhibition of whole plant respiration by elevated CO_2 as modified by growth temperature. *Physiol. Plant.* **87**: 459-466.

7. Adverse Effects of Elevated Levels of Ultraviolet (UV)-B Radiation and Ozone (O₃) on Grop Growth and Productivity

SAGAR V. KRUPA
Department of Plant Pathology, University of Minnesota, St. Paul, USA

HANS-JURG JÄGER
Institut für Pflanzenökologie der Justus-Liebig-Universität, Giessen, Germany

Surface-level ultraviolet (UV)-B radiation (280-320 nm) and ozone (O_3) are components of the global climate and any increases in their levels can lead to adverse effects on crop growth and productivity on a broad geographic scale (Krupa and Kickert, 1993). Possible increases in surface UV-B radiation are attributed to the depletion of the beneficial stratospheric O_3 layer (Cicerone, 1987). On the other hand, increases in surface-level O_3, that in many regions are largely the result of photochemical oxidant pollution, are also part of the general increase in the concentrations of the so-called 'greenhouse' gases (e.g., carbon dioxide, CO_2; methane, CH_4; nitrous oxide, N_2O; chlorofluorocarbons, CFCs) that may lead to global warming. In the context of climate change, it is therefore important to maintain a holistic view and recognize that UV-B and O_3 levels at the surface are only parts of the overall system of atmospheric processes and their products (Runeckles and Krupa, 1994).

EFFECTS OF ELEVATED SURFACE-LEVEL UV-B RADIATION OR O_3 ON CROPS

In recent years there have been a number of technical reviews or assessments of the direct effects of elevated surface-level UV-B or O_3 on crops and other terrestrial vegetation (UV-B: Caldwell, 1981; Worrest and Caldwell, 1986; Tevini and Teramura, 1989; Krupa and Kickert, 1993; O_3: Guderian, 1985; Heck *et al.*, 1988; Kickert and Krupa, 1991; Lefohn, 1992; Runeckles and Chevone, 1992; Runeckles and Krupa, 1994). Table 7.1 provides a comparative summary of the general effects of UV-B radiation and O_3 on crops. This summary is primarily based on artificial exposure studies.

In their analysis of the UV-B exposure studies, Krupa and Kickert (1989) noted that very different crop responses had been observed for the same crop species by different investigators at different times and locations. In many cases, these differences reflect cultivar and varietal differences in sensitivity within a given species (Teramura et al., 1990, 1991). Another reason for the differences is probably from the use of different UV-B lamps, exposure systems and action spectra for computing biologically effective UV-B flux densities (Runeckles and Krupa, 1994). Action spectra play a key role in the understanding of biological impacts of UV-B because of: (1) the differential sensitivities of various responses across the range of wavelengths in the UV-B region, and (2) the magnitudes of the differences in surface flux densities at these wavelengths resulting from the shape of the O_3 absorption spectrum (Caldwell et al., 1986).

The results obtained with a given species are frequently contradictory when comparing the effects of UV-B exposures in growth chambers or greenhouses to those under field conditions (e.g., Dumpert and Knacker, 1985). Such differences may well arise not only from differences in the microclimatic radiant and heat energy budgets extant in the different exposure systems at the times of exposure, but also because of differences induced by the environmental conditions under which the plants were

Table 7.1. Effects of elevated surface-level UV-B radiation or O_3 on crops[1]

Plant characteristic	Effect	
	UV-B	O_3
Photosynthesis	Reduced in many C_3 and C_4 species (at low light intensities)	Decreased in most species
Leaf conductance	Reduced (at low light intensities)	Decreased in sensitive species and cultivars
Water-use efficiency	Reduced in most species	Decreased in sensitive species
Leaf area	Reduced in many species	Decreased in sensitive species
Specific leaf weight	Increased in many species	Increased in sensitive species
Crop maturation rate	Not affected	Decreased
Flowering	Inhibited or stimulated	Decreased floral yield, fruit set and yield, delayed fruit set
Dry matter production and yield	Reduced in many species	Decreased in most species
Sensitivity between cultivars (within species)	Response differs between cultivars	Frequently large variability
Drought stress sensitivity	Plants become less sensitive to UV-B, but sensitive to lack of water	Plants become less sensitive to O_3 but sensitive to drought
Mineral stress sensitivity	Some species become less while others more sensitive to UV-B	Plants become more susceptible to O_3 injury

[1] Summary conclusions from artificial exposure studies. However, there can be exceptions. Modified from: Krupa and Kickert (1989) by Runeckles and Krupa (1994).

grown beforehand. For example, although the cuticle may act as a barrier to UV-B (Steinmüller and Tevini, 1985), greenhouse-grown plants are known to have a much thinner and less well-developed cuticle than field-grown plants (Martin and Juniper, 1970) and thus might exhibit greater sensitivity.

However, probably the factor of greatest importance in determining the relevance of many growth chamber and greenhouse studies is the intensity of Photosynthetically Active Radiation (PAR) to which the plants were exposed (Runeckles and Krupa, 1994). For example, Biggs et al. (1981) acknowledged that the sensitivities to UV-B of the soybean cultivars they studied were enhanced by the low (one-eighth of full sunlight) PAR levels used, because of the minimal activity of photorepair processes.

The results of field studies using selective filters to remove UV-B or filter/UV-lamp combinations are themselves far from conclusive (Runeckles and Krupa, 1994). For example, Becwar et al. (1982) found no significant effects on dry matter accumulation of pea, potato, radish and wheat plants grown for 50 days at a high elevation site (3000 m) in Colorado with high PAR and UV-B fluxes, even when filtered lamps increased the effective UV-B radiation by 52%. The only significant effect observed was an early slight decrease in wheat plant height growth that had disappeared by the time of final harvest. In contrast, Teramura et al. (1990) reported net adverse effects of enhanced UV-B on the yield of the sensitive soybean cultivar, Essex, based on a six-year study, both for individual years and when averaged over the years. However, this effect was only observed at the higher UV-B enhancement used (computed to be equivalent to a 25% stratospheric O_3 depletion). There was no adverse effect on the cultivar, Williams, when averaged over the six years; indeed a lower UV-B enhancement (equivalent to 16% O_3 depletion) resulted in a significant average increase in yield. The authors attributed the wide range of responses observed for either cultivar over the years to a strong influence of seasonal microclimate.

Recent studies at the University of Lancaster have used growth chambers in which UV-B enhancement was provided at high overall light intensities approximating two-thirds full sunlight (N. Paul and A.R. Wellburn, pers. comm.). Under these artificial conditions that approach those typical of the field with respect to light intensities, the vegetative growth of pea plants and their rates of photosynthesis were found to be unaffected by increased UV-B levels. However, yields were found to be somewhat depressed presumably because of adverse effects of UV-B at various stages in the process of sexual reproduction. The vegetative growth of a range of barley cultivars was unaffected by increased UV-B flux, although the cultivar, Scout, that contains little if any flavonoids, suffered visible injury. Such observations tend to confirm the conclusions of others (Runeckles and Krupa, 1994). For example, Beyschlag et al. (1988) found no adverse effect of UV-B on the photosynthetic rates of competing wheat and wild oat plants, although the competitiveness of wheat was increased because of UV-B-induced inhibition of the height growth of the wild oat plants. Increased flavonoid production has long been hypothesized as a protective feature of many species and cultivars (Beggs et al., 1986). Nevertheless, Table 7.2 provides a summary of cases where exposure to elevated UV-B radiation resulted in a decrease in biomass accumulation.

As noted in the case of UV-B effects, there are also appreciable differences among species and cultivars with regard to their response to O_3 (Runeckles and Krupa, 1994). Such differences in sensitivity exist with respect to both growth responses and the induction of the visible symptoms of acute injury. Seasonal and locational differences also contribute to the variability of response, as illustrated in many of the extensive field studies undertaken as part of the US National Crop Loss Assessment Network Program (NCLAN; Heck et al., 1988) and the US National Acid Precipitation Assessment Program (US NAPAP, 1991).

Much of the information that is available on the effects of O_3 on growth and productivity under field conditions has been obtained using open-top exposure chambers (Manning and Krupa, 1992). Although widely adopted for gaseous pollutant exposure studies, Runeckles and Wright (1988) and Manning and Krupa (1992) have pointed out that the use of such chambers and the associated experimental protocols for providing a range of O_3 exposure treatments have serious limitations in their abilities to reflect true ambient exposures. Because of significant differences in the microclimate and in the O_3 exposure potential between the chamber and open-field environments (Heagle et al., 1988a; Sanders et al., 1991; Krupa et al., 1994), the relevance of the effects observed using such chamber systems to exposures in free air is in question. Nevertheless, the bulk of evidence obtained using a variety of experimental exposure systems clearly indicates the phytotoxic effects of O_3 on crop growth and/or productivity (Table 7.3).

Table 7.2. Adverse response of crops to elevated UV-B radiation. Based on decreases in biomass accumulation

Crop[1]	Response variable	Exposure environment[2]
Alfalfa (*Medicago sativa*)	Tot dry wt	gh, gc
Barley (*Hordeum vulgare*)	Tot dry wt	gh, gc, field
Bean (*Phaseolus* sp.)	Tot dry wt	gh, gc
	Prim leaf dry wt	gc
Broccoli (*Brassica oleracea, Botrytis*)	Tot dry wt	gh, gc, field
	Crop yield	field
Brussels sprouts (*Brassica oleracea, Gemmifera*)	Tot dry wt	gh, gc
Cabbage (*Brassica oleracea, Capitata*)	Tot dry wt	gh, gc
Cantaloupe (*Cucumis melo* var. *Cantalupensis*)	Tot dry wt	gh, gc
Carrot (*Daucus carota*)	Tot dry wt	gc
Cauliflower (*Brassica oleracea, Botrytis*)	Tot dry wt	gh, gc
Chard (*Beta vulgaris, Cicla*)	Tot dry wt	gh, gc
Collards (*Brassica oleracea, Acephala*)	*Tot dry wt*	*gh, gc*
Corn (*Zea mays*)	Tot dry wt	gh, gc, field
	Crop yield	field
Cotton (*Gossypium hirsutum*)	Tot dry wt	gh
	Cotyledon dry wt	gh
Cowpea (*Vigna sinensis*)	Tot dry wt	gc, field
	Crop yield	field

(Cont.)

Table 7.2. (continued)

Crop[1]	Response variable	Exposure environment[2]
Cucumber (*Cucumis sativus*)	Crop yield	gc
	Leaf dry wt	gh
	Tot dry wt	gh, gc
	Cotyledon dry wt	gc
Eggplant (*Solanum melongena*)	Cotyledon dry wt	gc
Kale (*Brassica oleracea*, Acephala)	Tot dry wt	gh, gc
Kohlrabi (*Brassica oleracea*, Gongylodes)	Tot dry wt	gh, gc
Lettuce (*Lactuca sativa*)	Tot dry wt	gh, gc
Mustard (*Brassica* sp.)	Tot dry wt	gh, gc, field
	Crop yield	field
Oats (*Avena sativa*)	Tot dry wt	gh, gc
Okra (*Hibiscus esculentus*)	Tot dry wt	gh, gc
Onion (*Allium cepa*)	Tot dry wt	gc, field
Pea (*Pisum sativum*)	Tot dry wt	gh, gc, field, solarium
	Crop yield	field
Peanut (*Arachis hypogaea*)	Tot dry wt	gc, field
	Crop yield	field
Pepper (*Capsicum frutescens*)	Tot dry wt	field
	Crop yield	field
Potato (*Solanum tuberosum*)	Tot dry wt	field
	Crop yield	field
Pumpkin (*Cucurbita pepo*)	Tot dry wt	gh, gc
Radish (*Raphanus sativus*)	Tot dry wt	gh, gc
	Cotyledon dry wt	gc
	Cotyledon fresh wt	gc
Rice (*Oryza sativa*)	Tot dry wt	gh, gc, field, solarium
	Crop yield	field
Rye (*Secale cereale*)	Tot dry wt	gh, gc
Sorghum (*Sorghum vulgare*)	Tot dry wt	gh, gc, field
Soybean (*Glycine max*)	Root dry wt	field
	Crop yield	field
	Tot dry wt	gh, gc, field, solarium
Spinach (*Spinacia oleracea*)	Tot dry wt	gh
Squash (*Cucurbita* sp.)	Crop yield	field
	Tot dry wt	gh, gc, field
Sugar beet (*Beta vulgaris*)	Tot dry wt	gc, field
Sugar cane (*Saccharum officinarum*)	Tot dry wt	gh
	Crop yield	gh
Sweet corn (*Zea mays* var. *Saccharata*)	Tot dry wt	gh
	Crop yield	field
Tomato (*Lycopersicon esculentum*)	Tot dry wt	gh, gc, field
	Crop yield	field
Turnip (*Brassica rapa*)	Tot dry wt	field, solarium
Watermelon (*Citrullus vulgaris*)	Tot dry wt	gh, gc
Wheat (*Triticum aestivum*)	Tot dry wt	gh, gc, field
	Crop yield	field
White mustard (*Sinapis alba*)	Tot dry wt	field

[1] Modified from: Krupa and Kickert (1989).
[2] gh = greenhouse; gc = growth chamber.

As previously noted, any changes in UV-B radiation and surface-level O_3 are only a part of the overall global climate change. In the context of the direct effects of climate change on crops, a key consideration is the increasing concentrations of atmospheric CO_2. While any increases in surface-level UV-B and O_3 can lead to adverse effects on crops over a broad geographic scale, elevated concentrations of CO_2 are considered to provide a fertilization effect (Table 7.4, also Rozema *et al.*, 1993; Rogers *et al.*, 1994). According to Krupa and Kickert (1989), an analysis of the available voluminous literature suggests that sorghum, oats, rice, pea, bean, potato, lettuce, cucumber and tomato are among the crop species that appear to exhibit a high degree of responsiveness to the joint effects of all three environmental variables (CO_2, O_3 and UV-B). A critical limitation in this type of assessment is the fact that almost all of our knowledge has been primarily derived from studies on crop response to single rather than to all three variables (CO_2, UV-B or O_3). Compounding this limitation is the lack of consideration of the impacts of possible changes in temperature and moisture regimes (Krupa and Kickert, 1989). Independent of these and other concerns, Table 7.5 provides a summary of world production statistics during 1989 for crops considered to be sensitive to elevated surface-level UV-B radiation and/or O_3. As an example, the two most populated regions, People's Republic of China and South Asia, are also the two largest producers of rice and cotton (Table 7.6). While rice is considered to be sensitive to elevated levels of UV-B, cotton is known to be sensitive to O_3 (Table 7.4). Both China and India are regions of high photochemical smog (O_3) at the present time and most likely will remain so into the future. Similarly the most productive (kg/ha) regions for most crops listed in Table 7.5 (North America and Western Europe) are also within the photochemical smog regions. In comparison, because of the virtual lack of data, it is not possible to provide a similar geographic analysis for UV-B at the present time. However, one could expect significant temporal and spatial variability at the local scale for both UV-B and O_3. What is not known with any degree of certainty is how various crops sensitive to elevated levels of UV-B and/or O_3 will respond on a consistent basis to these two variables in the presence of elevated levels of CO_2. The limited amount of information that is presently available on this subject has recently been reviewed by Krupa and Kickert (1993) and the reader is also referred to other chapters in this volume.

Table 7.3. Adverse effects of ozone on crop growth and/or productivity: A select summary

Species	O_3 concentration	Exposure duration	Variable	Effect	Reference
Alfalfa (Medicago sativa)	14-98 ppb, 12 h mean	32 days	Dry weight	2.4% reduction at 40 ppb, 18.3% reduction at 66 ppb	Temple et al. (1987)
Alfalfa	20-53 ppb, 12 h mean	11 weeks	Dry weight	22% reduction at 53 ppb	Takemoto et al. (1988a)
Alfalfa	10-109 ppb, 12 h mean	208 and 200 days during 2 growing seasons	Dry weight	0-25% reduction at levels of 38 ppb and above	Temple et al. (1988a)
Alfalfa	60-80 ppb, 6 h/day	5 days/week, for 8 weeks	Relative growth rate	Reduced up to 40% in the variety Saranac	Cooley and Manning (1988)
Alfalfa	18-66 ppb, 12 h mean	11 weeks	Shoot dry weight	22% reduction at 36 ppb	Takemoto et al. (1988b)
Barley (spring) (Hordeum vulgare)	0.8-83 ppb, 8 h mean	97, 198 and 98 days during 3 growing seasons	Seed weight	0-13% reduction	Adaros et al. (1991b)
Bean (fresh) (Phaseolus vulgaris)	35-132 ppb, 7 h mean	42 days	Green pod weight	Significant yield reductions of >10% in 8 lines at 63 ppb, 7 h mean	Eason and Reinert (1991)
Bean (fresh)	11-40 ppb, 12 h mean, 7-42 ppm-h	69 days	Pod weight	15.5% reduction at 45 ppb (39 ppm-h)	Schenone et al. (1992)
Bean (fresh)	26-126 ppb, 7 h mean	26 and 44 days, early and late in season	Pod weight	3.5-26% reduction in resistant and sensitive cultivars at 55-60 ppb	Heck et al. (1988)
Bean (fresh)	24-109 ppb, 8 h mean	43 and 34 days, 2 growing seasons	Pod weight	20% reduction at 80 ppb	Bender et al. (1990)
Bean (dry)	15-116 ppb, 12 h mean, 339 ppb highest hour	54 days	Seed yield	55-75% reduction at 72 ppb, 12 h mean, 198 ppb highest hourly	Temple (1991)
Bean (dry)	10-50 ppb, 7 h mean	86 days	Seed weight	26-42% reduction at 38-50 ppb	Sanders et al. (1992)
Celery (Apium graveolens)	18-66 ppb, 12 h mean	11 weeks	Shoot dry weight	12% reduction at 66 ppb	Takemoto et al. (1988b)
Cotton (Gossypium hirsutum)	15-111 ppb, 12 h mean	123 days	Leaf, stem and root dry weight	Up to 42% reduction in leaf and stem, and 61% reduction in root dry weights	Temple et al. (1988c)

Table 7.3. (continued)

Species	O_3 concentration	Exposure duration	Variable	Effect	Reference
Cotton	10-90 ppb, 12 h mean	102 days	Lint weight	40-71% reduction at highest concentration, determinate cultivars more sensitive	Temple (1990a)
Cotton	25-74 ppb, 12 h mean	123 days	Lint weight	Predicted loss of 26.2% at 74 ppb	Temple et al. (1988b)
Cotton	22-44 ppb, 12 h mean	124 days	Lint weight	Predicted loss of 19% at 44 ppb	Heagle et al. (1988b)
Cotton	26-104 ppb, 7 h mean	119 days	Lint weight	Predicted loss of 11% at 53 ppb	Heagle et al. (1986a)
Green pepper (Capsicum annuum)	19-66 ppb, 12 h mean	77 days	Fresh fruit weight	12% reduction at 66 ppb	Takemoto et al. (1988b)
Green pepper	18-66 ppb, 12 h mean	11 weeks	Fresh fruit weight	13% reduction in fruit weight at 66 ppb	Takemoto et al. (1988b)
Lettuce (Lactuca sativa)	21-128 ppb, 7 h mean	52 days	Head weight	Significant reduction at 83 ppb, 35% at 128 ppb	Temple et al. (1986)
Radish (Raphanus sativus)	20 or 70 ppb, 24 h mean	27 days	Shoot and root growth	36 and 45% reduction at 70 ppb	Barnes and Pfirrman (1992)
Rape (spring)/Brassica napus)	0.8-83 ppb, 8 h mean	89, 113 and 84 days during 3 growing seasons	Seed weight	9.4-16% reduction at 30 or 51 ppb	Adaros et al. (1991b)
Rape (spring)	43-60 ppb, 8 h mean	89, 113 and 84 days during 3 growing seasons	Seed weight	12-27% reduction	Adaros et al. (1991c)
Rice (Oryza sativa)	0-200 ppb, 5 h/day	5 days/week, 15 weeks	Seed weight	12-21% reduction at 200 ppb	Kats et al. (1985)
Soybean (Glycine max)	17-122 ppb, 7 h mean	69 days	Seed yield	From 8% at 35 ppb to 41% at 122 ppb	Kohut et al. (1986)
Soybean	18 or 24 ppb vs. 59 or 72 ppb, 9 h mean	13 weeks, 2 growing seasons	Seed yield	12.5% reduction vs. charcoal-filtered air, averaged over cultivars. Intercultivar differences as great as the ozone effect.	Mulchi et al. (1988)
Soybean	23, 40 and 66 ppb, 7 h mean	84 days	Seed yield	15.8 and 29% reduction vs. 23 ppb control	Mulchi et al. (1992)

Table 7.3. (continued)

Species	O_3 concentration	Exposure duration	Variable	Effect	Reference
Soybean	97 ppb vs. 38, 23, 16 ppb, 7 h mean	Four 31 day periods, 1 growing season	Seed yield	30-56% reduction vs. charcoal-filtered air (control), most loss in mid to late growth stage	Heagle et al. (1991)
Soybean	25 and 50 ppb, 7 h mean	About 90 days	Seed yield	Predicted loss of 10%	Heagle et al. (1986b)
Soybean	20 and 50 ppb, 12 h mean	107 days	Seed yield	Predicted loss of 13%	Miller et al. (1989)
Soybean	25 and 55 ppb, 7 h mean	64, 70 and 62 days, 3 growing seasons	Seed yield	Predicted loss of 15%	Heggestad and Lesser (1990)
Soybean	27 and 54 ppb, 7 h mean	About 109 and 103 days, 2 growing seasons	Seed yield	Predicted loss of 12 and 14%	Heagle et al. (1987)
Soybean	10-130 ppb	8 weeks, 6.8 h/day	Biomass	Predicted reduction of 16 or 33% at 60 and 100 ppb vs. 25 ppb control	Amundson et al. (1986)
Tomato (Lycopersicon esculentum)	13-109 ppb, 12 h mean, 79.5 ppm-h	75 days	Fresh weight	17-54% reduction at 109, no reduction at ambient	Temple (1990b)
Tomato	10-85 ppb, 6 h/day	12-21 days	Shoot dry weight	35-62% reduction	Mortensen (1992b)
Watermelon (Citrullus lanatus)	15-27 ppb, 7 h mean	81 days	Marketable fresh weight and number	20.8 and 21.5% reduction at 27 ppb	Snyder et al. (1991)
Wheat (spring) (Triticum aestivum)	14-46 ppb, 24 h mean	79, 92 and 79 days during 3 growing seasons	Seed weight	13% reduction at 40 ppb	Fuhrer et al. (1989)
Wheat (spring)	21.6-80 and 24.6-93.5 ppm-h	82 and 88 days during 2 growing seasons	Seed weight	48-54% reduction at 80 and 93.5 ppm-h	Grandjean and Fuhrer (1989)
Wheat (spring)	3-56 ppb, 7 h mean	61 and 55 days during 2 growing seasons	Seed weight	7% reduction at 15 and 22 ppb	Pleijel et al. (1991)

Table 7.3. (continued)

Species	O$_3$ concentration	Exposure duration	Variable	Effect	Reference
Wheat (spring)	8-101 and 20-221 ppb, 8 h mean	118 and 98 days during 2 growing seasons	Seed weight	10% reduction at 17-23 ppb	Adaros et al. 1991a
Wheat (spring)	0-38 ppb	Entire growing season	Seed weight	5% reduction at 38 ppb	De Temmerman et al. (1992)
Wheat (spring)	17-77 ppb, 7 h mean	90 and 87 days during 2 growing seasons	Seed weight	9.5-11.6 reduction at 37 and 45 ppb	Fuhrer et al. (1992)
Wheat (spring)	6-10 ppb, 6 h/day	21 days	Shoot dry weight	Decreased 35-60% at 101 ppb, in low and high light	Mortensen (1990a)
Wheat (spring)	10-125 ppb, 6 h/day	21 and 17 days	Top dry weight	Reduced by up to 35%	Mortensen (1990b)
Wheat (winter) (Triticum aestivum)	11-42 ppb, 14 week mean	109 days	Seed weight	No effect	Olszyk et al. (1986)
Wheat (winter)	30-93 ppb, 4 h mean	39 and 40 days during 2 growing seasons, 5 days/week, 4 h/day	Seed weight	Exposures >60 ppb during anthesis reduced yield	Slaughter et al. (1989)
Wheat (winter)	27-96 ppb, 7 h mean	36 days	Seed weigh/head	50% reduction at 96 ppb	Amundson et al. (1987)
Wheat (winter)	22-96 ppb, 7 h mean	65 and 36 days during 2 growing seasons	Seed weight	33 and 22% reduction at 42 and 54 ppb	Kohut et al. (1987)
Wheat (winter)	23-123 ppb, 4 h/day	5 days at anthesis	Seed weight	Up to 28% reduction	Mulchi (1986)
Ladino clover (Trifolium repens f. lodigense)	28-46 ppb, 12 h mean	180 and 191 days during 2 growing seasons	Dry weight	Predicted yield of mixture reduced 10%, with 19% decrease in clover and 19% increase in fescue at 46 ppb	Heagle et al. (1989)
Ladino clover-tall fescue (Fescue sp.) pasture	22-114 ppb, 12 h mean	Five 3-4 week exposure periods. Six 3-4 week exposures during 2 years	Shoot dry weight (SDW). Root dry weight (RDW)	18-50% reduction SDW at 40-47 ppb in clover, 25% reduction RDW. SDW increased by up to 50% in fescue	Rebbeck et al. (1988)

Table 7.3. (continued)

Species	O$_3$ concentration	Exposure duration	Variable	Effect	Reference
Red clover (*Trifolium pratense*)	6-59 ppb, 7 h mean	5 weeks	Shoot dry weight	30% reduction at 59 ppb	Mortensen (1992a)
Red clover	19-62 ppb, 12 h mean	83 and 91 days during 2 growing seasons	Dry weight	11% reduction at 62 ppb	Kohut *et al.* (1988)
Meadow grass (*Poa pratensis*)	10-55 ppb, 7 h mean	5 weeks	Shoot dry weight	28% reduction at 55 ppb	Mortensen (1992a)
Pasture grass (*Dactylis glomerata*)	10-55 ppb, 7 h mean	5 weeks	Shoot dry weight	28% reduction at 55 ppb	Mortensen (1992a)
Pasture grass (*Festuca pratensis*)	10-55 ppb, 7 h mean	5 weeks	Shoot dry weight	16% reduction at 55 ppb	Mortensen (1992a)
Red fescue (*Festuca rubra*)	10-55 ppb, 7 h mean	5 weeks	Shoot dry weight	23% reduction at 55 ppb	Mortensen (1992a)
Timothy (*Phleum pratense*)	10-55 ppb, 7 h mean	5 weeks	Shoot dry weight	45% reduction at 55 ppb	Mortensen (1992a)

Table 7.4. Comparison of sensitivities of agricultural crops to enhanced CO_2 (mean relative yield increases of CO_2-enriched to control) (after Kimball, 1983a,b, 1986; Cure, 1985; Cure and Acock, 1986) for CO_2 concentrations of 1200 ppm or less (Kimball, 1983a,b), or 680 ppm (Cure and Acock, 1986); to enhanced surface UV-B radiation; and to ground-level O_3. Species considered to be sensitive to all three factors are indicated in bold print. Plant species are listed by the order of their response to elevated CO_2 (column no. 3).

Crop type	Crop[1]	Enhanced CO_2, mean relative yield increase[2]	Sensitivity to enhanced UV-B radiation	Sensitivity to O_3
Fibre	Cotton[a]	3.09	Tolerant	Sensitive
C_4 grain	**Sorghum**	**2.98**	**Sensitive**	**Intermediate**
Fibre	Cotton[a]	2.59-1.95		
Fruit	Eggplant	2.54-1.88	Tolerant	Unknown
Legume	**Peas**	**1.89-1.84**	**Sensitive**	**Sensitive**
Root and tuber	Sweet potato	1.83	Unknown	Unknown
Legume	**Beans**	**1.82-1.61**	**Sensitive**	**Sens./Intermed.**
C_3 grain	Barley[b]	1.70	Sensitive	Tolerant
Leaf	Swiss chard	1.67	Sensitive	Unknown
Root and tuber	**Potato**[c]	**1.64-1.44**	**Sens./Toler.**	**Sensitive**
Legume	Alfalfa	1.57[3-4]	Tolerant	Sensitive
Legume	**Soybean**[d]	**1.55**[5]	**Sensitive**	**Sens./Toler.**
C_4 grain	Corn[e]	1.55	Tolerant	Sensitive
Root and tuber	Potato[c]	1.51		
C_3 grain	**Oats**	**1.42**	**Sensitive**	**Sensitive**
C_4 grain	Corn[e]	1.40[5]		
C_3 grain	Wheat[f]	1.37-1.26	Tolerant	Sens./Intermed.
Leaf	**Lettuce**	**1.35**	**Sensitive**	**Sensitive**
C_3 grain	Wheat[f]	1.35		
Fruit	**Cucumber**	**1.30-1.43**	**Sensitive**	**Intermediate**
Legume	Soybean[d]	1.29		
C_4 grain	Corn[e]	1.29		
Root and tuber	Radish	1.28	Tolerant	Intermediate
Legume	Soybean[d]	1.27-1.20		
C_3 grain	Barley[b]	1.25		
C_3 grain	**Rice**[g]	**1.25**	**Sensitive**	**Intermediate**
Fruit	Strawberry	1.22-1.17	Unknown	Tolerant
Fruit	Sweet pepper	1.20-1.60	Sens./Toler.	Unknown
Fruit	**Tomato**	**1.20-1.17**	**Sensitive**	**Sens./Intermed.**
C_3 grain	Rice[g]	1.15		
Leaf	Endive	1.15	Unknown	Intermediate
Fruit	Muskmelon	1.13	Sensitive	Unknown
Leaf	Clover	1.12	Tolerant	Sensitive
Leaf	Cabbage	1.05	Tolerant	Intermediate

[1] Crops with superscript have more than one ranking.
[2] From Kimball (1983a,b) and, if shown, the second value is from Kimball (1986).
[3] Mean relative yield increase of CO_2-enriched (680 ppm) to control crop (300-350 ppm), after Cure and Acock (1986).
[4] Based on biomass accumulation; yield not available.
[5] Field-based result from Rogers et al. (1983a,b).
From: Krupa and Kickert (1989).

Table 7.5. Summary statistics of world crop production for 1989. Only those major crops considered to be relatively sensitive to elevated levels of surface UV-B radiation and/or ozone are listed (refer to Table 7.4)

Parameter	North America	Central America	The Caribbean	South America	Western Europe	EC-12	Eastern Europe	Former USSR	Sub-Saharan Africa	North Africa & Near East	South Asia	Southeast Asia & Pacific Islands	China, People's Republic	East Asia	Australia & New Zealand
(1) Population (× million)															
	275.1	32.8	32.7	293.1	359.0	326.2	139.5	288.7	523.5	316.1	1097.3	447.6	1 102.4	215.4	26.0
(2) % Population growth/year															
	1.0	1.4	2.4	2.0	0.4	0.3			3.2	2.6	3.1	2.2	2.0	1.3	1.5
(3) Crop production (× 1000 Mt)															
(a) Barley	20 456	430	1 199	52 321	46 481	19 226	50 000	1 444	13 221	1 885	—	3 200	1 135	4 417	340
(b) Corn	197 597	14 055	390	36 790	28 292	26 605	27 089	17 000	32 547	7 261	10 441	16 563	78 930	5 032	286
(c) Cotton lint	2 663	229	5	1 290	311	311	10	2 656	1 060	1 297	3 645	75	3 925	9	449
(d) Cottonseed	4 323	366	9	2 287	590	590	19	5 100	1 787	2 152	7 271	149	7 838	17	1 746
(e) Oats	8 974	105	—	1 138	8 074	4 443	3 697	17 000	87	292	—	—	600	115	805
(f) Rice, paddy	7 007	1 192	1 116	17 088	1 941	1 941	307	2 525	7 921	5 009	42 056	112 321	180 130	29 029	1 165
(g) Sorghum	15 694	5 657	200	3 066	523	523	109	180	12 986	1 292	12 651	290	5450	117	113
(h) Soybeans	53 659	1061	—	32 635	2052	2040	798	920	341	389	1914	2 063	10 230	998	14 455
(i) Wheat	79 789	4408	—	18 672	82 818	78 460	44 370	90 500	4145	37 861	70 282	230	90 800	1 900	
(4) Crop yield (kg/ha)															
(a) Barley	2 538	1 655	—	1 639	3 951	3 970	3 999	1 883	1411	1 160	1 447	1 000	3 333	2 350	1 567
(b) Corn	6 998	1 674	1 185	2 096	6 772	6 722	4 346	3 552	1 625	3 231	1 332	1 847	3 719	6 076	5 336
(c) Oats	2 001	1 000	—	1 487	3 057	2 765	2 697	1 563	120	967	—	—	1 500	1 757	1 434
(d) Rice, paddy	6 444	2 719	3 568	2 507	5 571	5 571	3 524	3 861	1 518	4 205	2 563	2 950	5 549	6 367	7 682
(e) Sorghum	3 478	2 561	1 004	2 057	4 625	4 625	1 765	1 192	800	1 361	738	1 281	3 191	1 964	1 990
(f) Soybeans	2 184	1 930	—	1 880	2 935	2 946	1 684	1 179	1 059	1 999	712	1 118	1 301	1 543	1 833
(g) Wheat	2 057	4 005	—	1 844	4 852	4 835	3 909	1 900	1 520	1 496	2 128	1 854	3 054	2 580	1 617

Modified from: US Department of Agriculture, Economic Research Service, Agriculture and Trade Analysis Division (1990).

Table 7.6. Three largest producers during 1989 of crops considered to be sensitive to elevated surface levels of UV-B radiation and/or O_3. See Table 7.5 for the actual production statistics

Crop	World rank		
	No. 1	No. 2	No. 3
Barley[1]	South America	Eastern Europe	Western Europe
Corn[2]	North America	China	South America
Cotton (lint)[2]	China	South Asia	North America
Cotton (seed)[2]	China	South Asia	Former USSR
Oats[1,2]	Former USSR	North America	Western Europe
Rice[1]	China	Southeast Asia	South Asia
Sorghum[1]	North America	Sub-Saharan Africa	South Asia
Soybeans[2]	North America	South America	China
Wheat[2]	China	Former USSR	Western Europe

[1] Sensitive to elevated levels of surface UV-B radiation.
[2] Sensitive to elevated levels of surface O_3.

EFFECTS OF ELEVATED UV-B RADIATION OR O_3 ON THE INCIDENCE OF CROP PESTS

According to Runeckles and Krupa (1994) and Manning and v. Tiedemann (1995) the available evidence clearly shows that the effects of UV-B on the incidence and development of pathogen-induced diseases on crops is dependent upon the crop cultivar and age, pathogen inoculum level and the timing and duration of the UV-B exposure. In their studies, Orth *et al.* (1990) exposed three cultivars of cucumber (*Cucumis sativus*) to a daily dose of 11.6 k/m biologically effective UV-B radiation in an unshaded greenhouse before and/or after inoculation with *Colletotrichum laginarium* (anthracnose) or *Cladosporium cucumerinum* (scab). Pre-inoculation exposure of one to seven days to UV-B resulted in greater disease severity from both pathogens on the susceptible cultivar, Straight-8. Post-inoculation UV-B exposure was much less effective. Although the resistant cultivars Poinsette and Calypso Hybrid showed increased severity of anthracnose under a heavy pathogen inoculum load when exposed to UV-B (both pre- and post-inoculation), this effect was observed only on the cotyledons and not on the leaves. From their UV-B exposure and crop cultivar studies, Biggs *et al.* (1984) suggested that rust disease on wheat is more likely to exhibit increased severity when a susceptible rather than a resistant cultivar is exposed to UV-B radiation. In contrast, severity of *Cercospora* leaf spot on clonally propagated sugar beet *(Beta vulgaris)* increased in combined exposures to UV-B (Panagopoulos *et al.*, 1992).

Carns *et al.* (1978) found that when the anthracnose-resistant cucumber cultivar Poinsette was exposed to UV-B doses injurious to the crop, the mycelial growth of *Colletotrichum laginarium* was partially inhibited and spore germination was severely decreased. Owens and Krizek (1980) showed that *Cladosporium cucumerinum* spore germination was also significantly inhibited by UV-B radiation. Similarly, conidial germ tubes of *Diplocarpon rosae* appear to be sensitive to UV-B prior to the penetration of rose leaves (Semeniuk and Stewart, 1981). These types of direct adverse effects of UV-B on micro-organisms are well known (Sussman and Halvorsen, 1966; Leach, 1971). In contrast, Orth *et al.* (1990) in their studies on cucumber and *Colletotrichum* concluded that UV-B action on the host was apparently more important than on the fungus *per se*, since there was no difference in the disease severity between plants that received only pre-inoculation UV-B treatment and those that received both pre- and post-inoculation UV-B treatment.

A similar range of types of response is true of the interactions of pathogens and O_3 (Manning and v. Tiedemann, 1995). Effects on the host plant, on the pathogen, or on both may lead to stimulations or inhibitions of disease incidence or severity (Heagle, 1982). Dowding (1988) has stressed the critical importance of the coincidence of the timing of exposure and the infective period to any effect of O_3 on the establishment of disease.

As with UV-B, in the case of many fungal pathogens, potential effects of exposure to O_3 at the spore stage appear to be minimal, but the organisms are vulnerable following deposition, since their carbohydrate energy reserves are rapidly depleted on germination. The diverse interactions with O_3 on the leaf surface have been discussed by Dowding (1988) and include effects on cuticular chemistry and surface properties, exuded materials and stomatal response. The growth and development of the pathogen on the host may be inhibited directly by O_3, since toxicity has been observed in axenic culture (Krause and Weidensaul, 1978). Alternatively, important O_3-induced changes in the host may have profound effects on the successful growth and development of the pathogen. Similar interactions may be involved with bacterial pathogens, although infection is usually dependent upon successful entry into host tissues via wounds or insect vectors.

The development of the pathogen may affect the susceptibility of the host plant. Since the early report of 'protection' against 'smog injury' afforded to bean or sunflower leaves by infection with the rust fungi, *Uromyces phaseoli* and *Puccinia helianthi*, respectively (Yarwood and Middleton, 1954), there have been numerous reports of infection with viruses, bacteria and fungi leading to reduced host susceptibility to O_3. The converse enhancement of the impact of O_3 on host plant growth has been observed with nematode infection (Bisessar and Palmer, 1984). While there is considerable evidence indicating that exposure to O_3 can reduce infection, invasion and sporulation of fungal pathogens, including 'obligate' pathogens such as the rust fungi (Heagle, 1973), examples also exist of increased infection of O_3-injured plants (Manning *et al.*, 1969). There is no clear understanding of the mechanisms involved in O_3-host-pathogen-environment interactions, but one generalization that can be made is that pathogens which can benefit from injured host cells and disordered transport mechanisms will be enhanced by earlier exposure of the host to O_3, while those that depend on 'healthy' host tissue will be disadvantaged (Dowding, 1988).

Herbivorous insects and spider mites are major causes of crop loss but little is known of the effects of UV-B on plant-insect interactions and, although there is a sizeable body of information about air pollutant effects, relatively few of these concern O_3 (Manning and Keane, 1988). As with pathogens and disease, the topic can be subdivided into the influence of O_3 on insect attack and population dynamics (whether direct or mediated by changes induced in the plant), and the converse effects of insect attack on plant response.

Host plant resistance to insect attack may be modified through metabolic changes which affect feeding preference and insect behaviour, development and fecundity. O_3-induced changes in both major and secondary metabolites may be qualitative and quantitative, and while there is abundant evidence that such changes can influence insect growth and development, there have been few experimental investigations of the specific effects of O_3. Trumble *et al.* (1987) reported that the tomato pinworm (*Keiferia lycopersicella*) developed faster on O_3-injured tomato plants, although fecundity and female longevity were unaffected. The Mexican bean beetle (*Epilachna varivestis*) was found to show a preference for O_3-treated soybean foliage that increased with increased exposure (Endress and Post, 1985). Such preferences can lead to increased larval growth rates, as shown by the work of Chappelka *et al.* (1988). Other examples are reviewed in Runeckles and Chevone (1992).

Only one report appears to exist with regard to insect attack modifying the effects of O_3 on a herbaceous host. Rosen and Runeckles (1976) showed that the combination of extremely low-level O_3 (0.02 ppmv) and infestation with the greenhouse whitefly (*Trialeurodes vaporariorum*) acted synergistically in inducing accelerated chlorosis and senescence of bean leaves. They speculated that the effect might be the result of reaction of O_3 with enhanced ethylene production resulting from whitefly injury.

In summary, the information available on these biotic interactions involving UV-B or O_3 is fragmentary and precludes a clear unravelling of the complexities of the relationships, a situation that will only be remedied by further systematic investigation.

Similarly, as a comparison, the interaction of high CO_2 and plant insect pests has been shown (Fajer *et al.*, 1989; Osbrink *et al.*, 1987). Lincoln *et al.* (1984) found that insect (butterfly larvae) feeding rates rose as CO_2 in the plant growth atmosphere was increased. This was related to the nitrogen and water content of soybean leaves. In contrast, more recent studies have suggested that leaf-feeding caterpillars do not do as well on plants grown at high CO_2, presumably due to increased carbon:nitrogen ratio (lower nutritive value) (Akey and Kimball, 1989). These types of studies need to be expanded to determine not only the increases or decreases in the populations of a given insect species, but also in population shifts among multiple species under elevated CO_2 concentrations.

EFFECTS OF ELEVATED UV-B RADIATION OR O_3 ON CROP-WEED COMPETITION

The dynamics of changes in plant populations and communities is a product of the intra- and inter-specific competition for resources needed for growth and productivity of the competing species, as influenced by abiotic and biotic environmental factors. In view of the ranges in sensitivities of individual species and varieties to both UV-B and O_3 (Table 7.4), it is to be expected that mixed plantings will show differential responses to either stress. In the case of O_3 stress, a widely held view is that changes in community composition favouring tolerant species will occur (Treshow, 1968).

There have been few investigations of the effects of either UV-B or O_3 on interspecific competition, and even fewer on the density-dependent intra-specific competition typical of monocultures (Runeckles and Krupa, 1994). With both stresses, plant spacing of monocultures and mixtures will influence the exposures received. In the case of UV-B, this will result from canopy density and shading; with O_3, canopy structure will determine the flux of gas into the foliage as well as dictating the microclimatic conditions (Runeckles, 1992). Of the few published studies, some have merely reported differential effects on biomass responses observed with established mixed plantings, e.g., O_3 on grass/clover (Blum *et al.*, 1983), while others have attempted to analyse the nature of the competition by the use of replacement series experiments and the calculation of relative crowding coefficients (Wit, 1960). Although such coefficients provide a measure of competitiveness, Jolliffe *et al.* (1984) have drawn attention to their limitations in interpreting the results of replacement series experiments.

Fox and Caldwell (1978) examined the effects of an artificial increase in UV-B radiation on the competitive interactions of several pairs of species, using replacement series. Three types of associations were included: crop-weed, montane forage species, and weeds of disturbed habitats. The data showed statistically significant shifts in the competitive balance of two of the pairs: *Amaranthus-Medicago* (alfalfa) and *Poa* (bluegrass)-*Geum*. In both cases, UV-B caused a shift in favour of the crop (alfalfa or bluegrass) over the weed species. Only in the case of the *Alyssum-Pisum* (pea) mixture did enhanced UV-B irradiation result in a significant reduction in total mixture biomass, although changes occurred in the proportions of biomass contributed by the individual species of the other pairs. In further studies of the competition between wheat (*Triticum aestivum* cv. Bannock) and wild oat (*Avena fatua*) and between wheat and goatgrass (*Aegilops cylindrica*), W.G. Gold (as cited in Gold and Caldwell, 1983) and most recently, Barnes *et al.* (1994) also found a competitive advantage for the crop species (wheat) and increased UV-B enhancement in the former or in both mixtures.

Most of the limited number of studies of the effects of O_3 on competition have utilized crop species. Bennett and Runeckles (1977) used replacement series to study effects on the competition between Italian ryegrass (*Lolium multiflorum*) and crimson clover (*Trifolium incarnatum*). The relative crowding coefficients for ryegrass increased with the O_3 exposure compared to clover on both a dry weight and leaf area basis, indicating a shift in favour of ryegrass.

Other grass-legume studies of the yields of individual competing species have confirmed the effect of O_3 in favouring the grass species (Blum *et al.*, 1983; Kohut *et al.*, 1988; Rebbeck *et al.*, 1988). However, recent work by Evans and Ashmore (1992) on a semi-natural grassland dominated by the grasses *Agrostis capillaris, Festuca rubra* and *Poa pratensis* showed that O_3 adversely affected their growth relative to the growth of forb species including clover (*Trifolium repens*) and the major weed species, *Veronica chamaedris*, *Plantago major* and *Stellaria gramineae*. Since tests of the individual species indicated that, in isolation, the grasses were less sensitive than the forbs, the authors concluded that 'the response of plant communities cannot be predicted from the responses of individual component species'. This contrasts with the conventional wisdom that O_3 will induce changes in community composition favouring tolerant species. While this may well be true in the case of prolonged exposure to high O_3 levels, it appears that the nature of any changes will be highly dependent upon the severity of the O_3 stress imposed on a community and its composition (Runeckles and Krupa, 1994).

CONSIDERATIONS RELEVANT TO THE STUDY OF CROP RESPONSES TO ELEVATED LEVELS OF UV-B RADIATION AND O_3

Krupa and Kickert (1989) have described a range of possible joint exposure scenarios for UV-B and O_3. Table 7.7 presents modified descriptions of these possibilities (Runeckles and Krupa, 1994). Case 1 describes geographic locations where there is no predicted or observed stratospheric O_3 depletion (i.e., no increase in UV-B radiation) and no marked increase in the tropospheric O_3 concentrations. Here we should only expect 'normal' UV-B effects and 'background' O_3 effects on plants.

Case 2 defines the situation at locations where there is no predicted or observed stratospheric O_3 depletion and hence no increase in UV-B, but continued upward trends in tropospheric O_3 concentrations. This case is subdivided, since in some situations (Case 2a), the surface ambient O_3 levels may be high enough to cause local adverse effects but insufficient to increase the total tropospheric column and thereby reduce the surface UV-B flux significantly. In Case 2b, the tropospheric O_3 column is sufficient to attenuate the surface UV-B levels significantly. This would lead to subnormal UV-B levels coinciding with elevated O_3 and result in an alternation between exposures to UV-B and O_3. These cases define situations such as southern California and many other low to mid-latitude locations that are subjected to photochemical oxidant pollution, in which ambient O_3 is the dominant factor. It is the impact of this type of situation that air pollution - plant effects scientists have addressed for several years. However, we are unaware of any information to indicate whether the numerous published studies of the effects of O_3 would fall into Case 2a or 2b, since in no cases was UV-B flux density reported.

In those geographic areas where stratospheric O_3 depletion might occur, one might expect an increase in UV-B if spring-summer cloud conditions are not significantly increased. At locations not subjected to boundary layer O_3 concentrations significantly above the background, we might expect some plants to respond to increased UV-B (case 3). Any interactive effect of enhanced UV-B with an atmospheric variable would probably involve increased ambient CO_2. Although we have not specifi-

Table 7.7. Possible patterns of environmental stress for higher plants with respect to O_3 and UV-B, depending upon stratospheric O_3 status and surface boundary layer

Surface boundary layer O_3 status	Stratospheric O_3 status	
	No O_3 depletion	O_3 depletion
Background O_3 only	(1) 'Normal' UV-B plant effects only, with no O_3 effects	(3) Enhanced UV-B plant effects only, with no O_3 effects
Elevated O_3	(2a) 'Normal' UV-B plant effects, with O_3 effects	(4a) Enhanced incoming UV-B may be attenuated in boundary layer leading to 'normal' UV-B effects, with O_3 effects (similar to Case 2a)
	(2b) 'Subnormal' UV-B effects due to attenuation in boundary layer, co-occurring or intermittent with O_3 effects	(4b) Enhanced UV-B effects on plants co-occurring or intermittent with O_3 effects

Modified from: Krupa and Kickert (1989) by Runeckles and Krupa (1994).

cally examined the interactive effects of increased CO_2, most of the photobiology research cited in this paper, and especially in the reports of CIAP (Climatic Impact Assessment Program, USA) and BACER (Biological and Climatic Effects Research, USA) projects in the early and mid-1970s, used this type of situation as a frame of reference.

The most complex situations depicted as Cases 4a and 4b are possible scenarios for some geographic regions, and involve: (1) stratospheric O_3 depletion with increased UV-B, and (2) continued increases in O_3 within the boundary layer. In one case (4a), the timing and geography could lead to high boundary layer O_3 concentrations along with enhanced UV-B, but with the high O_3 concentrations offsetting the UV-B enhancement. The net result would simply be due to the effects of O_3 on the vegetation at 'normal' UV-B levels, and hence would resemble the impact of Case 2a.

Case 4b envisions situations where there is a net enhancement in UV-B during the growing season, occurring intermittently and inversely with O_3 episodes in the boundary layer. When ground-level O_3 concentrations are not high enough to absorb all of the enhanced UV-B, both factors would impact the vegetation. In this situation, O_3 and UV-B would compete with or enhance each other in affecting a particular plant response process.

As stated previously, to all of these scenarios can be added the effects of increased CO_2 levels. However, we are unaware of any vegetation response studies that have focused on the effects of changes in all three factors. In addressing this question,

Krupa and Kickert (1989) suggest three alternative possibilities for the joint effects of elevated CO_2, UV-B and O_3.
1. There might be no interaction between the three factors. The 'Law of Limiting Factors' might prevail in which the most important factor overrides plant response.
2. There might be a cumulative effect in which the net plant response is simply the sum of stress effects from O_3 and increased UV-B regardless of the temporal patterns of exposure.
3. There might be a more than additive effect where the plant response is more severe than would be found from either stress singly. There is also the possibility of a less than additive interaction in the sense that high ambient CO_2 might allow sufficient repair processes to proceed in some plants so that sensitivity to increased UV-B and/or ambient O_3 may be reduced.

For more details on these issues, the reader should consult other appropriate chapters in this volume.

CONCLUSIONS AND FUTURE RESEARCH DIRECTIONS

There has been no global network for monitoring surface-level UV-B radiation. Long-term UV-B data are sparse and not very reliable. Nevertheless, numerous investigators have examined the effects of UV-B radiation on crops in artificial exposures, but large uncertainties in the relevance to climate change of much of the information obtained remain. According to Runeckles and Krupa (1994), the transfer of results from growth chamber or greenhouse experiments to the ambient environment has been particularly difficult. This appears to be due to the differences in the characteristics of plants grown under these environments and to photorepair under the high photosynthetic photon flux densities encountered in the ambient environment. Studies of the effects of UV-B (or of O_3) on physiological processes such as photosynthesis and on modes of action are appropriately examined under controlled environment conditions. However, the integration of their effects on the processes affected within the whole organism that ultimately lead to growth can only reliably be investigated using plants growing under true field conditions.

Our knowledge of the effects of O_3 is also beset by uncertainties related largely to the lack of information about plant responses under such field conditions. Here the problem is not one of interrelating growth and field observations, but concerns the relevance of the results from the most frequently used open-top exposure chamber method. There is no question of the phytotoxicity of O_3. However, the results obtained in many studies of its effects are primarily supported by controversial statistical techniques (Kickert and Krupa, 1991) and, therefore, the fact remains that the results cannot be validated and show considerable variability from season to season and from location to location most likely because of the types of experimental designs used.

Perhaps the most pressing need at the moment is to obtain field information about the effects of UV-B and O_3 that are clearly identified with one or more of the different scenarios outlined in Table 7.7. To such studies should be added elevated levels of CO_2, in view of the preliminary observations that indicate significant interaction with the effects of O_3. While such information is needed for direct effects on crop species, the studies must also include information about the possible long-term effects on growth, joint effects with other pollutants, incidence of pathogens and insect pests, intra-species competition, and crop-weed relationships (Krupa and Kickert, 1993; Runeckles and Krupa, 1994).

Such studies should also permit the acquisition of information about the processes involved, such as the partitioning of assimilates and the induction of morphological changes. In contrast to these gross mechanisms affecting growth and development, ongoing studies at the biochemical and metabolic level are needed in order to provide a sound understanding of the fine mechanisms involved.

In view of the evidence that suggests that UV-B has little adverse effect on photosynthesis or growth under field conditions, it appears that concern over increases in UV-B irradiation resulting from stratospheric O_3 depletion should focus on longer-term effects probably involving the consequences of damage to nucleic acids. In contrast, increased tropospheric O_3 levels will undoubtedly have immediate adverse effects on most species, independent of any longer-term effects brought about by either adaptation or genetic selection.

In view of the urgency of acquiring information on the potential impacts of the various components of climatic change, including UV-B and tropospheric O_3, that can be realistically envisioned, every effort should therefore be made to avoid wasting research effort and resources on studies that will do nothing to reduce the uncertainties associated with our present information, and which fail to recognize the potential importance of the interactions among the various components (Runeckles and Krupa, 1994).

ACKNOWLEDGEMENTS

The senior author would like to acknowledge the invaluable help of Leslie Johnson (word processing) in the preparation of this manuscript. This effort was supported in kind to the senior author by the University of Minnesota Agriculture Experiment Station, St. Paul, Minnesota.

REFERENCES

Adaros, G., Weigel, H.-J. and Jäger, H.-J. 1991a. Impact of ozone on growth and yield parameters of two spring wheat cultivars (*Triticum aestivum* L.). *Z. Pflanzenkr. Pflanzenschutz*. **98**: 113-124.

Adaros, G., Weigel, H.-J. and Jäger, H.-J. 1991b. Concurrent exposure to SO_2 and/or

NO$_2$ alters growth and yield responses of wheat and barley to low concentrations of O$_3$. *New Phytol.* **118**: 581-591.

Adaros, G., Weigel, H.-J. and Jäger, H.-J. 1991c. Single and interactive effects of low levels of O$_3$, SO$_2$ and NO$_2$ on the growth and yield of spring rape. *Environ. Pollut.* **72**: 269-286.

Akey, D.H. and Kimball, B.A. 1989. Growth and development of the beet armyworm on cotton grown in an enriched carbon dioxide atmosphere. *Southwestern Entomol.* **14**: 255-260.

Amundson, R.G., Raba, R.M., Schoettle, A.W. and Reich, P.B. 1986. Response of soybean to low concentrations of ozone: II. Effects on growth, biomass allocation, and flowering. *J. Environ. Qual.* **15**: 161-167.

Amundson, R.G., Kohut, R.J., Schoettle, A.W., Raba, R.M. and Reich, P.B. 1987. Correlative reductions in whole-plant photosynthesis and yield of winter wheat caused by ozone. *Phytopathology* **77**: 75-79.

Barnes, J.D. and Pfirrman, T. 1992. The influence of CO$_2$ and O$_3$ singly and in combination on gas exchange, growth and nutrient status of radish *Raphanus sativus* L. *New Phytol.* **121**: 403-412.

Barnes, P.W., Flint, S.D. and Caldwell, M.M. 1994. Early-season effects of solar UV-B enhancement on plant canopy structure, simulated photosynthesis and competition. (Abstr.). *Bull. Ecol. Soc. Amer. (Suppl.)* **75**(2): 9-10.

Becwar, M., Moore III, F.D. and Burke, M.J. 1982. Effects of deletion and enhancement of ultraviolet-B (280-315 nm) radiation on plants grown at 3000 m elevation. *J. Amer. Soc. Hort. Sci.* **107**(5): 771-774.

Beggs, C.J., Schneider-Ziebert, U. and Wellmann, E. 1986. UV-B radiation and adaptive mechanisms in plants. In: *Stratospheric Ozone Reduction, Solar Ultraviolet Radiation and Plant Life; NATO Advanced Research Workshop on the Impact of Solar Ultraviolet Radiation Upon Terrestrial Ecosystems: 1. Agricultural Crops, 27-30 September 1983, Bad Windsheim, West Germany.* R.C. Worrest and M.M. Caldwell (eds.). NATO ASI Series, Series G, Ecological Sciences, Vol. 8, Springer-Verlag, Berlin. pp. 235-520.

Bender, J., Weigel, H.-J. and Jäger, H.-J. 1990. Regression analysis to describe yield and metabolic responses of beans (*Phaseolus vulgaris*) to chronic ozone stress. *Angew. Bot.* **64**: 329-344.

Bennett, J.P. and Runeckles, V.C. 1977. Effects of low levels of ozone on growth of crimson clover and annual ryegrass. *Crop Sci.* **17**: 443-445.

Beyschlag, W., Barnes, P.W., Flint, S.D. and Caldwell, M.M. 1988. Enhanced UV-B radiation has no effect on photosynthetic characteristics of wheat (*Triticum aestivum* L.) and wild oat (*Avena fatua* L.) under greenhouse and field conditions. *Photosynthetica* **22**: 516-525.

Biggs, R.H., Kossuth, S.V. and Teramura, A.H. 1981. Response of 19 cultivars of soybeans to ultraviolet-B irradiance. *Physiol. Plant.* **53**(1): 19-26.

Biggs, R.H., Webb, P.G., Garrard, L.A., Sinclair, T.R. and West, S.H. 1984. *The Effects of Enhanced Ultraviolet-B Radiation on Rice, Wheat, Corn, Citrus and Duck-*

weed. Interim Report 80:8075-03, US Environmental Protection Agency, Washington DC.

Bisessar, S. and Palmer, K.T. 1984. Ozone, antioxidant spray and *Meloidogyne hapla* effects on tobacco. *Atmos. Environ.* **18**: 1025-1027.

Blum, U., Heagle, A.S., Burns, J.C. and Linthurst, R.A. 1983. The effects of ozone on fescue-clover forage regrowth, yield and quality. *Environ. Exp. Bot.* **23**: 121-132.

Caldwell, M.M. 1981. Plant responses to solar ultraviolet radiation. In: *Encyclopedia of Plant Physiology, New Series, Vol. 12A, Physiological Plant Ecology*. O.L. Lange, P.S. Nobel, C.B. Osmond and B.H. Ziegler (eds.). Springer-Verlag, Berlin. pp. 169-197.

Caldwell, M.M., Camp, L.B., Warner, C.W. and Flint, S.D. 1986. Action spectra and their key role in assessing biological consequences of solar UV-B radiation change. In: *Stratospheric Ozone Reduction, Solar Ultraviolet Radiation and Plant Life; NATO Advanced Research Workshop on the Impact of Solar Ultraviolet Radiation Upon Terrestrial Ecosystems: 1. Agricultural Crops, 27-30 September 1983, Bad Windsheim, West Germany*. R.C. Worrest and M.M. Caldwell (eds.). NATO ASI Series, Series G, Ecological Sciences, Vol. 8, Springer-Verlag, Berlin. pp. 87-111.

Carns, H.R., Graham, J.H. and Ravitz, S.J. 1978. Effects of UV-B radiation on selected leaf pathogenic fungi and on disease severity. In: *UV-B Biological and Climatic Effects Research (BACER), FY 77-78 Research Report on Impacts of Ultraviolet-B Radiation on Biological Systems: A Study Related to Stratospheric Ozone Depletion, Final Report, Vol. I*. EPA-IAG-6-0168, USDA EPA, Stratospheric Impact Research and Assessment Program (SIRA), US Environmental Protection Agency, Washington DC. 43 p.

Chappelka, A.H., Kraemer, M.E., Mebrahtu, T., Rangappa, M. and Benepal, P.S. 1988. Effects of ozone on soybean resistance to the Mexican bean beetle (*Epilachna varivestis* Mulsant). *Environ. Exp. Bot.* **28**: 53-60.

Cicerone, R.J. 1987. Changes in stratospheric ozone. *Science* **237**: 35-42.

Cooley, D.R. and Manning, W.J. 1988. Ozone effects on growth and assimilate partitioning in alfalfa, *Medicago sativa* L. *Environ. Pollut.* **49**: 19-36.

Cure, J.D. 1985. Carbon dioxide doubling response: A crop survey. In: *Direct Effects of Increasing Carbon Dioxide on Vegetation*. B.R. Strain and J.D. Cure (eds.). DOE/ER-0238, US Dept. of Energy, Washington DC. pp. 99-116.

Cure, J.D. and Acock, B. 1986. Crop responses to carbon dioxide doubling: A literature survey. *Agr. Forest Meteorol.* **38**: 127-145.

De Temmerman, L., Vandermeiren, K. and Guns, M. 1992. Effects of air filtration on spring wheat grown in open-top field chambers at a rural site. 1. Effect on growth, yield and dry matter partitioning. *Environ. Pollut.* **77**: 1-5.

Dowding, P. 1988. Air pollutant effects on plant pathogens. In: *Air Pollution and Plant Metabolism*. S. Schulte-Hostede, N.M. Darrall, L.W. Blank and A.R. Wellburn (eds.). Elsevier, London. pp. 329-355.

Dumpert, K. and Knacker, T. 1985. A comparison of the effects of enhanced UV-B radiation on some crop plants exposed to greenhouse and field conditions. *Biochem. Physiol. Pflanz.* **180**(8): 599-612.

Eason, G. and Reinert, R.A. 1991. Responses of closely related Bush Blue Lake snap bean cultivars to increasing concentrations of ozone. *J. Amer. Soc. Hort. Sci.* **116**: 520-524.

Endress, A.G. and Post, S.L. 1985. Altered feeding preference of Mexican bean beetle *Epilachna varivestis* for ozonated soybean foliage. *Environ. Pollut.* **39**: 9-16.

Evans, P.A. and Ashmore, M.R. 1992. The effects of ambient air on a semi-natural grassland community. *Agr. Ecosyst. Environ.* **38**: 91-97.

Fajer, E.D., Bowers, M.D. and Bazzaz, F.A. 1989. The effects of enriched carbon dioxide atmospheres on plant-insect herbivore interactions. *Science* **243**: 1198-1200.

Fox, F.M. and Caldwell, M.M. 1978. Competitive interaction in plant populations exposed to supplementary ultraviolet-B radiation. *Oecologia* **36**: 173-90.

Fuhrer, J., Egger, A., Lehnherr, B., Grandjean, A. and Tschannen, W. 1989. Effects of ozone on the yield of spring wheat (*Triticum aestivum* L., cv. Albis) grown in open-top field chambers. *Environ. Pollut.* **60**: 273-289.

Fuhrer, J., Grimm, A.G., Tschannen, W. and Shariatmadari, H. 1992. The response of spring wheat (*Triticum aestivum* L.) to ozone at higher elevations. 2. Changes in yield, yield components and grain quality in response to ozone flux. *New Phytol.* **121**: 211-219.

Gold, W.G. and Caldwell, M.M. 1983. The effects of ultraviolet-B radiation on plant competition in terrestrial ecosystems. *Physiol. Plant.* **58**: 435-444.

Grandjean, A. and Fuhrer, J. 1989. Growth and leaf senescence in spring wheat (*Triticum aestivum*) grown at different ozone concentrations in open-top field chambers. *Physiol. Plant* **77**: 389-394.

Guderian, R. (ed.). 1985. *Air Pollution by Photochemical Oxidants: Formation, Transport, Control and Effects on Plants.* Springer-Verlag, New York. 346 p.

Heagle, A.S. 1973. Interactions between air pollutants and plant parasites. *Annu. Rev. Phytopathol.* **11**: 365-388.

Heagle, A.S. 1982. Interactions between air pollutants and parasitic plant diseases. In: *Effects of Gaseous Air Pollution on Agriculture and Horticulture.* M.H. Unsworth and D.P. Ormrod (eds.). Butterworths, London. pp. 333-348.

Heagle, A.S., Heck, W.W., Lesser, V.M., Rawlings, J.O. and Mowry, F.L. 1986a. Injury and yield response of cotton to chronic doses of ozone and sulfur dioxide. *J. Environ. Qual.* **15**: 375-382.

Heagle, A.S., Lesser, V.M., Rawlings, J.O., Heck, W.W. and Philbeck, R.B. 1986b. Responses of soybeans to chronic doses of ozone applied as constant or proportional additions to ambient air. *Phytopathology* **76**: 51-56.

Heagle, A.S., Heck, W.W., Lesser, V.M. and Rawlings, J.O. 1987. Effects of daily ozone exposure duration and concentration fluctuation on yield of tobacco. *Phytopathology* **77**: 856-862.

Heagle, A.S., Kress, L.W., Temple, P.J., Kohut, R.J., Miller, J.E. and Heggestad, H.E.

1988a. Factors influencing ozone dose-yield response relationships in open-top field chamber studies. In: *Assessment of Crop Loss From Air Pollutants*. W.W. Heck, O.C. Taylor and D.T. Tingey (eds.). Elsevier Applied Science, London. pp. 141-179.

Heagle, A.S., Miller, J.E., Heck, W.W. and Patterson, R.P. 1988b. Injury and yield response of cotton to chronic doses of ozone and soil moisture deficit. *J. Environ. Qual.* **17**(4): 627-635.

Heagle, A.S., Rebbeck, J., Shafer, S.R., Blum, U. and Heck, W.W. 1989. Effects of long-term ozone exposure and soil moisture deficit on growth of a ladino clover-tall fescue pasture. *Phytopathology* **79**: 128-136.

Heagle, A.S., McLaughlin, M.R., Miller, J.E., Joyner, R.L. and Spruill, S.E. 1991. Adaptation of a white clover population to ozone stress. *New Phytol.* **119**(1): 61-68.

Heck, W.W., Taylor, O.C. and Tingey, D.T. (eds.). 1988. *Assessment of Crop Loss from Air Pollutants*. Elsevier Applied Science, London. 552 p.

Heggestad, H.E. and Lesser, V.M. 1990. Effects of ozone, sulfur dioxide, soil water deficit, and cultivar on yields of soybean. *Environ. Qual.* **19**(3): 488-94.

Hidy, G.M., Mahoney J.R. and Goldsmith, B.J. 1978. *International Aspects of the Long Range Transport of Air Pollutants, Final Report*. US Department of State, Washington, DC.

Jolliffe, P.A., Minjas, A.N. and Runeckles, V.C. 1984. A reinterpretation of yield relationships in replacement series experiments. *J. Appl. Ecol.* **21**: 227-243.

Kats, G., Dawson, P.J., Bytnerowicz, A., Wolf, J.W., Thompson, C.R. and Olszyk, D.M. 1985. Effects of ozone or sulfur dioxide on growth and yield of rice. *Agr. Ecosyst. Environ.* **14**: 103-117.

Kickert, R.N. and Krupa, S.V. 1990. Forest responses to tropospheric ozone and global climate change: An analysis. *Environ. Pollut.* **68**: 29-65.

Kickert, R.N. and Krupa, S.V. 1991. Modeling plant response to tropospheric ozone: A critical review. *Environ. Pollut.* **70**: 271-383.

Kimball, B.A. 1983a. *Carbon Dioxide and Agricultural Yield: An Assemblage and Analysis of 770 Prior Observations*. Report 14, USDA, US Water Conservation Laboratory, Phoenix, AZ. 71 p.

Kimball, B.A. 1983b. Carbon dioxide and agricultural yield: an assemblage and analysis of 430 prior observations. *Agron. J.* **75**: 779-788.

Kimball, B.A. 1986. Influence of elevated CO_2 on crop yield. In: *Carbon Dioxide Enrichment of Greenhouse Crops, Vol. II – Physiology, Yield, and Economics*. H.Z. Enoch and B.A. Kimball (eds.). CRC Press, Boca Raton, FL. pp. 105-115.

Kohut, R.J., Amundson, R.G. and Laurence, J.A. 1986. Evaluation of growth and yield of soybean exposed to ozone in the field. *Environ. Pollut.* **41**: 219-234.

Kohut, R.J., Amundson, R.G., Laurence, J.A., Colavito, L., Van Leuken, P. and King, P. 1987. Effects of ozone and sulfur dioxide on yield of winter wheat. *Phytopathology* **77**: 71-74.

Kohut, R.J., Laurence, J.A. and Amundson, R.G. 1988. Effects of ozone and sulfur dioxide on yield of red clover and timothy. *J. Environ. Qual.* **17**(4): 580-585.

Krause, C.R. and Weidensaul, T.C. 1978. Effects of ozone on the sporulation, germination and pathogenicity of *Botrytis cinerea*. *Phytopathology* **68**: 196-198.

Krupa, S.V. and Kickert, R.N. 1989. The greenhouse effect: impacts of ultraviolet-B (UV-B) radiation, carbon dioxide (CO_2), and ozone (O_3) on vegetation. *Environ. Pollut.* **61**(4): 263-393.

Krupa, S.V. and Kickert, R.N. 1993. *The Effects of Elevated Ultraviolet (UV)-B Radiation on Agricultural Production*. Report submitted to the Formal Commission on 'Protecting the Earth's Atmosphere' of the German Parliament, Bonn, Germany. 432 p.

Krupa, S.V., Grünhage, L., Jäger, H.-J., Nosal, M., Manning, W.J., Legge, A.H. and Hanewald, K. 1994.. Ambient ozone (O_3) and adverse crop response: A unified view of cause and effect. *Environ. Pollut.* **87**: 119-126.

Leach, C.M. 1971. A practical guide to the effects of visible and ultraviolet light on fungi. *Methods Microbiol.* **4**: 609-664.

Lefohn, A.S. 1992. *Surface Level Ozone Exposures and Their Effects on Vegetation*. Lewis Publishers, Chelsea, MI. 366 p.

Lincoln, D.E., Sionit, N. and Strain, B.R. 1984. Growth and feeding response of *Pseudoplusia includens* (Lepidoptera: Noctuidae) to host plants grown in controlled carbon dioxide atmospheres. *Environ. Entomol.* **13**: 1527-1530.

Manning, W.J. and Keane, K.D. 1988. Effects of air pollutants on interactions between plants, insects and pathogens. In: *Assessment of Crop Loss From Air Pollutants*. W.W. Heck, O.C. Taylor and D.T. Tingey (eds.). Elsevier Applied Science, London. pp. 365-386.

Manning, W.J. and Krupa, S.V. 1992. Experimental methodology for studying the effects of ozone on crops and trees. In: *Surface Level Ozone Exposures and Their Effects on Vegetation*. A.S. Lefohn (ed.). Lewis Publishers, Chelsea, MI. pp. 93-156.

Manning, W.J. and v. Tiedemann, A. 1995.. Climate change: Potential effects of increased atmospheric carbon dioxide (CO_2), ozone (O_3), and ultraviolet-B (UV-B) radiation on plant diseases. *Environ. Pollut.* **88**: 219-246.

Manning, W.J., Feder, W.A., Perkins, I. and Glickman, M. 1969. Ozone injury and infection of potato leaves by *Botrytis cinerea*. *Plant Dis. Rept.* **53**: 691-693.

Martin, J.T. and Juniper, B.E. 1970. *The Cuticles of Plants*. St. Martins, New York. 347 p.

Miller, J.E., Heagle, A.S., Vozzo, S.F., Philbeck, R.B. and Heck, W.W. 1989. Effects of ozone and water stress, separately and in combination, on soybean yield. *J. Environ. Qual.* **18**: 330-336.

Mortensen, L.M. 1990a. The effects of low O_3 concentrations on growth of *Triticum aestivum* L. at different light and air humidity levels. *Norw. J. Agr. Sci.* **4**: 337-342.

Mortensen, L.M. 1990b. Effects of ozone on growth of *Triticum aestivum* L. at different light, air humidity, and CO_2 levels. *Norw. J. Agr. Sci.* **4**: 343-348.

Mortensen, L.M. 1992a. Effects of ozone on growth of seven grass and one clover species. *Acta Agr. Scand. Sect. B Soil Plant Sci.* **42**: 235-239.

Mortensen, L.M. 1992b. Effects of ozone concentration on growth of tomato at various light, air humidity and carbon dioxide levels. *Sci. Hort. (Amsterdam)* **49**: 17-24.

Mulchi, C.L., Sammons, D.J. and Baenziger, P.S. 1986. Yield and grain quality responses of soft red winter wheat exposed to ozone during anthesis. *Agron. J.* **78**: 593-600.

Mulchi, C.L., Lee, E.H., Tuthill, K. and Olinick, E.V. 1988. Influence of ozone stress on growth processes, yields and grain quality characteristics among soybean cultivars. *Environ. Pollut.* **53**: 151-169.

Mulchi, C.L., Slaughter, L., Saleem, M., Lee, E.H., Pausch, R. and Rowland, R. 1992. Growth and physiological characteristics of soybean in open-top chambers in response to ozone and increased atmospheric CO_2. *Agr. Ecosyst. Environ.* **38**: 107-118.

Olszyk, D.M., Bytnerowicz, A., Kats, G., Dawson, P.J., Wolf, J. and Thompson, C.R. 1986. Effects of sulfur dioxide and ambient ozone on winter wheat and lettuce. *J. Environ. Qual.* **15**: 363-369.

Orth, A.B., Teramura, A.H. and Sisler, H.D. 1990. Effects of ultraviolet-B radiation on fungal disease development in *Cucumis sativus*. *Amer. J. Bot.* **77**: 1188-1192.

Osbrink, W.L.A., Trumble, J.T. and Wagner, R.E. 1987. Host suitability of *Phaseolus lunata* for *Trichoplusia ni* (Lepidoptera: Noctuidae) in controlled carbon dioxide atmospheres. *Environ. Entomol.* **16**: 639-644.

Owens, O.V.H. and Krizek. D.T. 1980. Multiple effects of UV radiation (265-330 nm) on fungal spore emergence. *Photochem. Photobiol.* **32**: 41-49.

Panagopoulos, I., Bornman, J.F. and Björn, L.O. 1992. Response of sugar beet plants to ultraviolet-B (280-320 nm) radiation and Cercospora leaf spot disease. *Physiol. Plant.* **84**: 140-145.

Pleijel, H., Skärby, L., Wallin, G. and Selldén, G. 1991. Yield and grain quality of spring wheat (*Triticum aestivum* L., cv. Drabant) exposed to different concentrations of ozone in open-top chambers. *Environ. Pollut.* **69**: 151-168.

Rebbeck, J., Blum, U. and Heagle, A.S. 1988. Effects of ozone on the regrowth and energy reserves of a ladino clover-tall fescue pasture. *J. Appl. Ecol.* **25**: 659-681.

Rogers, H.H., Bingham, G.E., Cure, J.D., Smith, J.M. and Surano, K.A. 1983a. Responses of selected plant species to elevated carbon dioxide in the field. *J. Environ. Qual.* **12**(4): 569-574.

Rogers, H.H., Thomas, J.F. and Bingham, G.E. 1983b. Response of agronomic and forest species to elevated atmospheric carbon dioxide. *Science* **220**: 428-429.

Rogers, H.H., Runion, G.B. and Krupa, S.V. 1994. Plant responses to atmospheric CO_2 enrichment, with emphasis on roots and the rhizosphere. *Environ. Pollut.* **83**: 155-189.

Rosen, P.M. and Runeckles, V.C. 1976. Interaction of ozone and greenhouse whitefly in plant injury. *Environ. Conserv.* **3**: 70-71.

Rozema, J., Lambers, H., van de Geijn and Cambridge, M.L. (eds.). 1993. *CO_2 and Biosphere*. Kluwer Academic Publishers, Dordrecht, The Netherlands. 484 p.

Runeckles, V.C. 1992. Uptake of ozone by vegetation. In: *Surface Level Ozone Expo-*

sures and Their Effects on Vegetation. A.S. Lefohn (ed.). Lewis Publishers, Chelsea, MI. pp. 157-188.

Runeckles, V.C. and Chevone, B.I. 1992. Crop responses to ozone. In: *Surface Level Ozone Exposures and Their Effects on Vegetation.* A.S. Lefohn (ed.). Lewis Publishers, Chelsea, MI.

Runeckles, V.C. and Krupa, S.V. 1994. The impact of UV-B radiation and ozone on terrestrial vegetation. *Environ. Pollut.* **83**: 191-213.

Runeckles, V.C. and Wright, E.F. 1988. Crop response to pollutant exposure – The 'ZAPS' approach. In: *Proceedings of the 81st Annual Meeting of Air Pollution Control Association, Dallas, TX.* Paper #88-69.4, Air Pollution Control Association, Pittsburgh, PA.

Sanders, G.E., Clark, A.G. and Colls, J.J. 1991. The influence of open-top chambers on the growth and development of field bean. *New Phytol.* **117**: 439-447.

Sanders, G.E., Robinson, A.D., Geissler, P.A. and Colls, J.J. 1992. Yield stimulation of a commonly grown cultivar of *Phaseolus vulgaris* L. at near-ambient ozone concentrations. *New Phytol.* **122**: 63-70.

Schenone, G., Botteschi, G., Fumagalli, I. and Montinaro, F. 1992. Effects of ambient air pollution in open-top chambers on bean (*Phaseolus vulgaris* L.). I. Effects on growth and yield. *New Phytol.* **122**: 689-697.

Semeniuk, P. and Stewart, R.N. 1981. Effect of ultraviolet (UV-B) irradiation on infection of roses by *Diplocarpon rosae* Wolf. *Environ. Exp. Bot.* **21**: 45-50.

Slaughter, L.H., Mulchi, C.L., Lee, E.H. and Tuthill, K. 1989. Chronic ozone stress effects on yield and grain quality of soft red winter wheat. *Crop Sci.* **29**: 1251-1255.

Snyder, R.G., Simon, J.E., Reinert, R.A., Simini, M. and Wilcox, G.E. 1991. Effects of air quality on growth, yield and quality of watermelon. *HortScience* **26**: 1045-1047.

Steinmüller, D. and Tevini, M. 1985. Action of ultraviolet radiation (UV-B) upon cuticular waxes in some crop plants. *Planta (Berl)* **164**(4): 557-564.

Sussman, A.S. and Halvorsen, H.O. 1966. *Spores.* Harper and Row, New York. 354 p.

Takemoto, B.K., Hutton, W.J. and Olszyk, D.M. 1988a. Responses of field-grown *Medicago sativa* L. to acidic fog and ambient ozone. *Environ. Pollut.* **54**: 97-107.

Takemoto, B.K., Bytnerowicz, A. and Olszyk, D.M. 1988b. Depression of photosynthesis, growth, and yield in field-grown green pepper (*Capsicum annuum* L.) exposed to acidic fog and ambient ozone. *Plant Physiol.* **88**: 477-482.

Temple, P.J. 1990a. Growth form and yield responses of four cotton cultivars to ozone. *Agron. J.* **82**: 1045-1050.

Temple, P.J. 1990b. Growth and yield responses of processing tomato (*Lycopersicon esculentum* Mill.) cultivars to ozone. *Environ. Exp. Bot.* **30**: 283-291.

Temple, P.J. 1991. Variations in responses of dry bean (*Phaseolus vulgaris*) cultivars to ozone. *Agr. Ecosyst. Environ.* **36**: 1-11.

Temple, P.J., Taylor, O.C. and Benoit, L.F. 1986. Yield response of head lettuce (*Lactuca sativa* L.) to ozone. *Environ. Expt. Bot.* **26**(1): 53-58.

Temple, P.J., Lennox, R.W., Bytnerowicz, A. and Taylor, O.C. 1987. Interactive effects of simulated acidic fog and ozone on field-grown alfalfa. *Environ. Expt. Bot.* **27**(4): 409-417.

Temple, P.J., Benoit, L.F., Lennox, R.W., Reagan, C.A. and Taylor, O.C. 1988a. Combined effects of ozone and water stress on alfalfa growth and yield. *J. Environ. Qual.* **17**(1): 108-113.

Temple, P.J., Kupper, R.S., Lennox, R.L. and Rohr, K. 1988b. Injury and yield responses of differentially irrigated cotton to ozone. *Agron. J.* **80**: 751-755.

Temple, P.J., Kupper, R.S., Lennox, R.L. and Rohr, K. 1988c. Physiological and growth responses of differentially irrigated cotton to ozone. *Environ. Pollut.* **53**: 255-263.

Teramura, A.H., Sullivan, J.H. and Lyden, J. 1990. Effects of UV-B radiation on soybean yield and seed quality: A 6-year field study. *Physiol. Plant.* **80**: 5-11.

Teramura, A.H., Ziska, L.H. and Sztein, A.E. 1991. Changes in growth and photosynthetic capacity of rice with increased UV-B radiation. *Physiol. Plant.* **83**: 373-380.

Tevini, M. and Teramura, A.H. 1989. UV-B effects on terrestrial plants. *Photochem. Photobiol.* **50**(4): 479-487.

Treshow, M. 1968. The impact of air pollutants on plant populations. *Phytopathology* **58**: 1108-1113.

Trumble, J.T., Hare, J.D., Musselman, R.C. and McCool, P.M. 1987. Ozone-induced changes in host-plant suitability: Interactions of *Keiferia lycopersicella* and *Lycopersicon esculentum*. *J. Chem. Ecol.* **13**: 203-218.

US Department of Agriculture, Economic Research Service, Agriculture and Trade Analysis Division. 1990. World Agriculture: Trends and Indicators, 1970-89. Statistical Bulletin No. 815, The Service, Washington DC. pp. 19-77.

US NAPAP. 1991. *Acidic Deposition: State of Science and Technology, Summary Report*. US National Acid Precipitation Assessment Program, Washington DC.

Wit, C.T. de 1960. *On Competition*. Versl. Landbouwk. Onderzoek. 66.8. pp. 82.

Worrest, R.C. and Caldwell, M.M. (eds.). 1986. *Stratospheric Ozone Reduction, Solar Ultraviolet Radiation and Plant Life; NATO Advanced Research Workshop on the Impact of Solar Ultraviolet Radiation Upon Terrestrial Ecosystems: 1. Agricultural Crops; 27-30 September 1983, Bad Windsheim, West Germany*. NATO ASI Series, Series G, Ecological Sciences, Vol. 8, Springer-Verlag, Berlin. 374 p.

Yarwood, C.E. and Middleton, J.T. 1954. Smog injury and rust infection. *Plant Physiol.* **29**: 393-395.

8. Combined Effects of Changing CO_2, Temperature, UV-B Radiation and O_3 on Crop Growth

MICHAEL H. UNSWORTH
Center for Analysis of Environmental Change, Oregon State University, Corvallis, Oregon, USA

WILLIAM E. HOGSETT
US Environmental Protection Agency, Environmental Research Laboratory, Corvallis, Oregon, USA

Our understanding of the relationships between crop growth and the atmospheric environment has developed substantially in the past few decades. Improvements in technology have enabled crop physiologists to study processes such as transpiration and carbon dioxide exchange in the field. Mechanistic understanding of the factors controlling photosynthesis and respiration has improved greatly, but we still have no mechanistic understanding of the ways in which many abiotic stresses affect these processes. Our knowledge of factors that control crop development and the partitioning of assimilate between organs is still at the descriptive stage, and poses one of the main limitations on our ability to develop fully mechanistic mathematical models of crop growth. Nevertheless, crop science has reached a stage where reliable semi-empirical simulation models of some of the major world crops have been constructed and used to explore the potential for crop growth and yield in different environments. For the reasons discussed above, these models are restricted in their application to situations where the empirical relationships in them are well defined, but they are often useful in the design of particular experiments to investigate new environments, for example, to apply them to scenarios of climatic change. It should be emphasized that, although we have a much improved understanding of the growth and yield potential of major agricultural crops, this knowledge needs to be extended to include other crops, especially those important for developing countries.

However, as we steadily improve our understanding of how crop yield is influenced by the atmospheric environment, that environment is constantly changing as a result of human activities, and the challenge today is to predict how crops will respond to the changed environments of tomorrow. Traditionally, crop scientists have used controlled environments and field-based facilities to investigate crop responses to individual factors, such as radiation, temperature, humidity and water availability. More recently, manipulation of air quality has been used to study crop responses to air pollutants, such as SO_2, O_3, NO_x, and some of the technology developed for these experiments has been successfully applied to study responses to increased CO_2. Although this 'single component' approach is a highly effective way of discovering how one atmospheric factor influences crops, in practice the atmosphere in the future will alter in several ways simultaneously.

Some of the impending changes are more certain than others. Without doubt, CO_2 concentration is increasing worldwide at about 1.8 µmol/mol per year and under all likely scenarios will continue to do so. This increase will enhance the global greenhouse effect, and the consensus of scientific opinion is that it will result in global warming and in changed global distributions of rainfall and other weather components. The amount of climatic change is, however, very uncertain, because even the best computer simulation models do not adequately describe features such as ocean-atmosphere linkages, and feedbacks associated with changes in atmospheric aerosols or cloud amount and distribution. At present, expert opinion is that, when the CO_2 concentration reaches double its present value of about 350 µmol/mol, global temperatures will be about 2.5°C warmer, but this figure and the associated global variability are very uncertain; even more uncertain are any changes in rainfall amount and distribution (Houghton et al., 1992).

In contrast to the uncertainty associated with computer models, there is clear evidence from satellite and from surface monitoring that stratospheric ozone is being depleted, probably because of emissions of chlorine- and bromine-containing substances, so that globally the column of ozone from the surface through the atmosphere is thinning at about 3% per decade, a rate that seems to be increasing (Houghton et al., 1992). Because the chemicals believed to be causing this thinning are long-lived in the stratosphere, this thinning process is likely to continue for several decades even though the chemicals are being phased out of production. The consequence of this thinning is that ultraviolet (UV) radiation at the ground, normally severely attenuated by stratospheric O_3, will increase. There is, however, an inadequate network of surface monitoring stations to define the UV climate at the ground (i.e., taking into consideration effects of cloud, dust, etc.) or to determine whether UV has increased significantly. Even if annual mean UV exposure has not changed, there are indications that thinning of stratospheric ozone, similar but less extreme than that observed in the Antarctic, can occur episodically at temperate latitudes.

Air quality in the troposphere is also changing. In particular, in many highly populated regions, emissions of hydrocarbons and exhaust gases from motor vehicles provide the precursors for photochemical production of tropospheric ozone. There is evidence that background tropospheric ozone concentrations have increased by about 10% per decade in Europe over the past 20 years, and that the European background has approximately doubled since the beginning of the century (Houghton et al., 1992). Although there are no equivalent long-term records, it is likely that the tropospheric O_3 concentration in the United States and Southeast Asia has also increased. Perhaps more important than the mean background concentration of O_3 is the regular occurrence on a regional scale of high concentrations of tropospheric O_3 (typically 5 to 10 times background) associated with stagnant weather systems in summer. These concentrations have been demonstrated to reduce the growth and yield of many agricultural crops.

The atmospheric environment of many world crops of the future will undoubtedly contain more CO_2 than present, is likely to be somewhat warmer, and rainfall (and hence humidity) and cloudiness will probably also change, though in a direction unknown at present. Whether UV will increase much on a seasonal basis remains unclear, but episodes of increased UV at the ground, associated with patches of thinned stratospheric O_3, may occur, especially in temperate latitudes. Without doubt, in regions with large population densities, episodes of high concentrations of tropospheric O_3 and of other pollutants associated with fossil-fuel burning will become more common.

In this paper we review our knowledge of some of the interacting effects of these changing atmospheric conditions on agricultural crops, and we draw attention to particular gaps where further research is needed. In an attempt to structure what could otherwise be a very disparate summary, we will concentrate on two major gases, CO_2 and O_3, and will describe work that has investigated the effects of other stresses combined with them. In each case, we first briefly review responses to these gases alone.

CARBON DIOXIDE

Several recent reviews of responses of crops to CO_2 have been published (Goudriaan and Unsworth, 1990; Lawlor and Mitchell, 1991; van de Geijn et al., 1993; chapters 4 and 5 of this publication), and there is no advantage in repeating the detail of these here. At the leaf level, the two most well-known responses to elevated CO_2 are an increase in the rate of net photosynthesis, P_N, and a decrease in stomatal conductance, g_s. The increase in P_N is much larger in C_3 species (50 to 100% when CO_2 concentration doubles) than in C_4 species (10%), but a substantial decrease in g_s (30 to 40%) is observed in both types of species. There is some evidence of photosynthetic acclimation to elevated CO_2, so that increased rates of P_N do not persist at the leaf level in long-term studies with some species, but there is no evidence of acclimation in g_s. It is not clear whether increased CO_2 also indirectly affects plant growth, e.g., by changing rates of leaf initiation, expansion or longevity, but such effects could have a large influence on crop productivity.

Much of our understanding of effects of CO_2 on plants has been gained from studies with individual leaves. There have been far fewer studies of long-term effects of elevated CO_2 on canopy-scale photosynthesis and transpiration. Most of those reported have been in closed- or open-top chambers, though a few reports of field-scale exposures (FACE) are now becoming available. Drake and Leadley (1991) summarized the data available at that time and concluded that: (1) canopy photosynthesis increases in elevated CO_2 when there is a sink available for the carbon; (2) the relative effect of CO_2 is greatest at highest temperatures; and (3) elevated CO_2 alters many interacting factors, such as canopy architecture and partitioning of assimilates, that mediate gas exchange of canopies and ecosystems.

As a consequence of increased net photosynthesis, dry matter production and yield are substantially increased by elevated CO_2. Several authors (e.g., Lawlor and Mitchell, 1991) have reviewed the literature and concluded that, provided there is adequate water, nutrients and pest control, yields of C_3 and C_4 crops growing in about 700 μmol/mol CO_2 would be about 30 to 40% and 9%, respectively, greater than present yields (350 μmol/mol CO_2) if all other climatic factors were unchanged.

Table 8.1 summarizes harvest indices and yield from recent work at Nottingham, England (Clifford et al., 1993; Azam Ali, pers. comm.) in which stands of groundnut (C_3) and sorghum (C_4) were grown at two CO_2 concentrations in a computer-controlled glasshouse system designed to control temperature, humidity and soil moisture to simulate tropical growing conditions. Although harvest index (seed or pod yield per plant) was not altered by CO_2, yield increases were consistent with other reports for unstressed C_3 and C_4 crops.

However, in much of the world, ideal crop growing conditions, with adequate water and nutrition, are wishful thinking, and the influence of elevated CO_2 on aspects such as water-use may be much more relevant. Reviews by Morison (1985) and Eamus (1991) make it clear that, on a leaf area basis, elevated CO_2 improves the water-use efficiency (WUE – the ratio of weight of dry matter produced to weight of water transpired) of plants. On a whole-plant basis, the increase in leaf area resulting from enhanced photosynthesis may cause the amount of water-used per plant to remain close to that of plants growing at current levels of CO_2, although the WUE as defined here increases. We return to this topic later.

CO_2 AND TEMPERATURE

As discussed earlier, the atmospheric environment of the future is likely to include correlated increases of atmospheric CO_2 concentration and temperature. Long (1991) reviewed the mechanisms by which temperature and CO_2 affect photosynthesis in C_3 species, and developed models of the response of leaf canopy carbon exchange to changes in these variables. He showed (Figure 8.1) that the interaction of carbon

Table 8.1. Harvest indices (HI, seed or pod yield per plant/total dry mass per plant) and seed or pod yields for stands of groundnut (*Arachis hypogaea* L.) and sorghum (*Sorghum bicolor* L.) grown in controlled-environment glasshouses at 350 μmol/mol or 700 μmol/mol CO_2 (from Clifford et al., 1993 and Azam Ali, pers. comm.)

Crop	HI		Pod/seed yield (g/plant)		Percentage yield increase at elevated CO_2
	350 μmol/mol CO_2	700 μmol/mol CO_2	350 μmol/mol CO_2	700 μmol/mol CO_2	
Groundnut	0.195	0.212	9.06	13.29	30.9
	±0.019	±0.038	±0.06	±1.70	±13.3
Sorghum	0.333	0.334	25.78	27.37	6.2
	±0.016	±0.015	±0.11	±0.85	±3.3

Figure 8.1. Dependence of light-saturated rates of leaf CO_2 uptake (A_{sat}) on leaf temperature for three atmospheric CO_2 concentrations (C_a, µmol/mol of CO_2 in air). Arrows indicate T_{opt}, the temperature at which A_{sat} is maximal for each value of C_a (from Long, 1991)

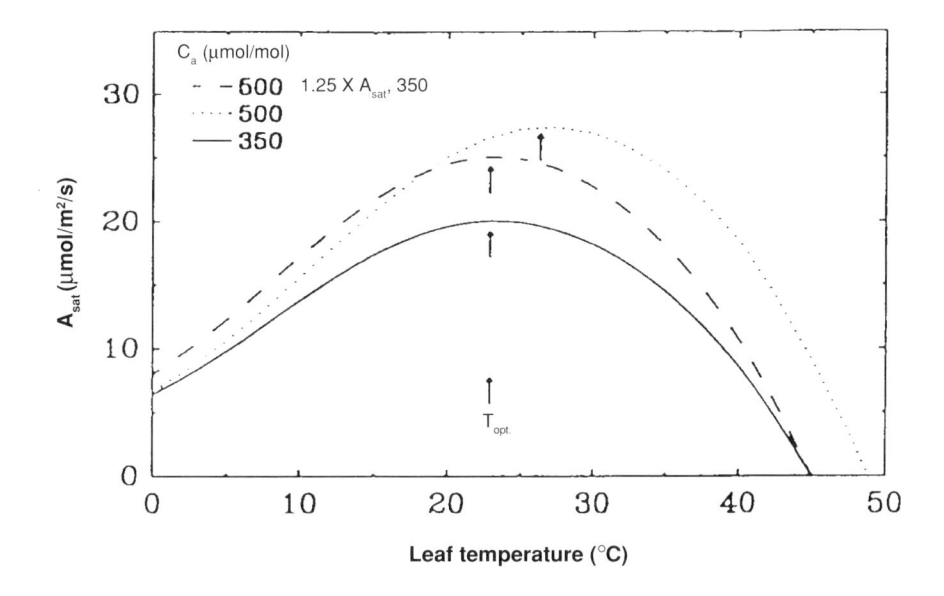

dioxide concentration and temperature causes the temperature optimum of the light-saturated rate of CO_2 uptake to increase as CO_2 increases. Without this interaction, an increase in global temperature of 2.5°C would increase the frequency of temperatures in temperate regions that are supra-optimal for light-saturated photosynthesis. However, if warming on this scale was coincident with a CO_2 concentration increase to above 500 µmol/mol, the temperature optimum would increase, and this inhibition would not occur. Long (1991) went on to develop a simple model of canopy CO_2 uptake (A_c) in response to temperature, taking into account the substantial fraction of the canopy leaf area that would be shaded and therefore not photosynthesizing at the saturated rate. Figure 8.2 shows the model sensitivity of daily canopy carbon uptake to temperature. Using a simple model of the diurnal variation of radiation and temperature at different latitudes, Long showed that the implications of Figure 8.2 are that daily net carbon assimilation responds much more to increased CO_2 at low latitudes than at high latitudes. These calculations are in good agreement with observations from long-term studies reported by Oechel and Strain (1985), Drake and Leadley (1991), and Kimball (1983).

Idso (1990) grew crops in field chambers in Phoenix, Arizona, at 300 and 600 µmol/mol CO_2. Figure 8.3 shows the ratio of net photosynthesis at the two CO_2 concentrations plotted against leaf temperature, and includes values calculated from Long's theoretical relationships, showing the good agreement at the leaf level. For a number of crops grown in open-top chambers in Phoenix, Arizona, Figure 8.4 (from Kimball

176   GLOBAL CLIMATE CHANGE AND AGRICULTURAL PRODUCTION

Figure 8.2. Simulated dependence of net canopy CO_2 uptake ($A_{c,\,tot}$), integrated over 24 h, on temperature at three atmospheric CO_2 concentrations (C_a). The simulation is for a canopy with a leaf area index (F) of 3.0 and a rate of horizontally to vertically projected leaf area (x) of 1.0, on 9 July, latitude 52°N and assuming cloudless conditions (from Long, 1991)

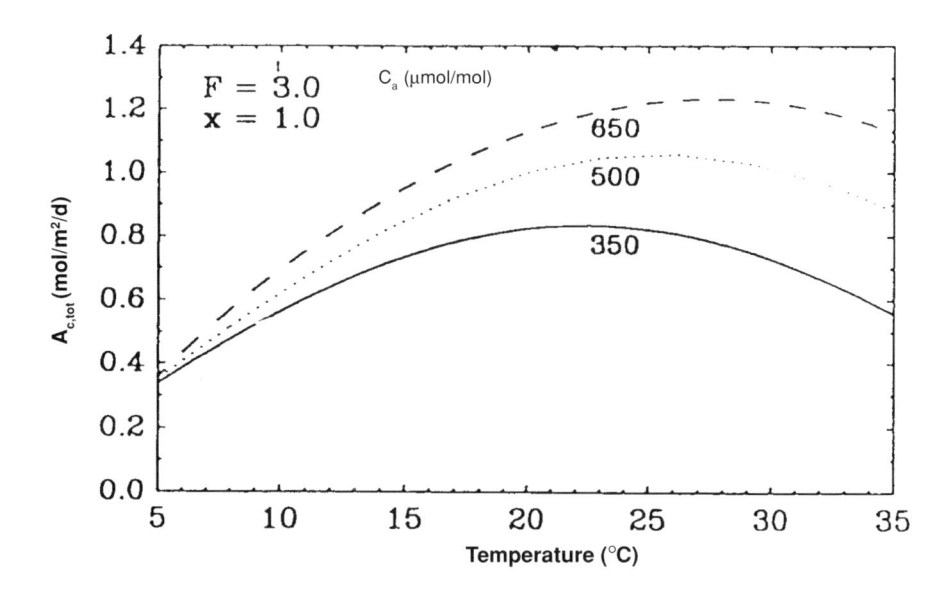

et al., 1993) summarizes the relative increase in growth (weekly dry matter production, comparing 650 μmol/mol CO_2 with 350 μmol/mol) with mean air temperature, indicating that, at low temperatures, elevated CO_2 actually *decreased* growth (a phenomenon predicted by Long's models).

Interactions between temperature and CO_2 also need to account for effect of temperature on development. Squire and Unsworth (summarized in Goudriaan and Unsworth, 1990) pointed out that determinate crops with discrete elements to their life cycle develop faster in higher temperatures, and so the stage of seed filling is shortened, limiting the benefits of elevated CO_2. Figure 8.5 illustrates results from a model of winter wheat growth that was modified for CO_2 response (though in a less sophisticated manner than Long, 1991). When CO_2 concentration doubled, but daily weather data from a typical year were used, the potential grain yield was 27% larger than in a control run at 340 μmol/mol CO_2 (11.5 and 9.0 t/ha, respectively). The date of maturity was unchanged. When daily temperatures were increased by 3°C and CO_2 was doubled, more rapid development of the crop shortened the growing season, and the potential grain yield (10.4 t/ha) was only 15% larger than in the control, but maturity occurred 30 days earlier. The wheat model used in these simulations contains a number of elements that are sensitive to temperature. Recently, Kocabas (1993) completed a detailed investigation of the model sensitivity to changes in mean temperature at various stages of crop development (Kocabas *et al.*, 1993), and to changes

Figure 8.3. Variation of relative CO_2 assimilation rate (net photosynthesis at an intercellular CO_2 concentration of 600 μmol/mol divided by net photosynthesis at 300 μmol/mol) with leaf temperature for bell pepper, tomato and cottonwood tree. The dotted line is the relationship predicted by the analysis of Long (1991) (adapted from Idso, 1990)

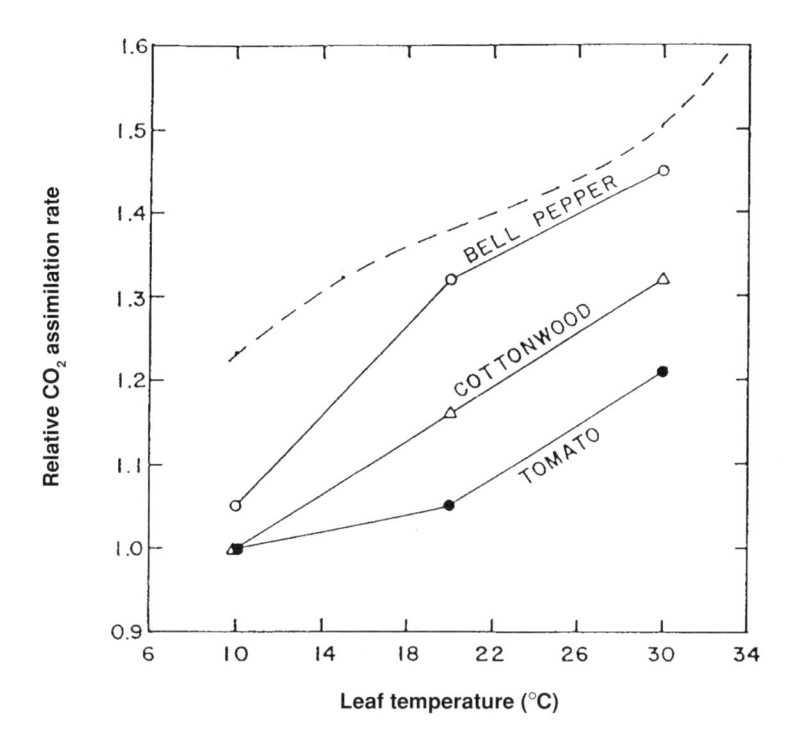

in variability of temperature in various developmental phases and over the whole season. The results indicated the particularly strong dependence of yield on temperature in the phase from emergence to double ridges in wheat, but suggested that changes in temperature variability of ±20% would not have any consistent effects on yield variability. It would be useful to incorporate Long's analysis into the wheat model to allow more precise simulation of the interactive effects of temperature and CO_2 on carbon assimilation and crop development.

Indeterminate crops, such as grass and sugar beet, continue to grow and produce yield as long as the temperature is above a minimum threshold, provided that other factors such as drought and nutrition do not limit growth. Consequently, simultaneous increases in temperature and CO_2 concentration increase the potential yield of such crops in temperate environments, where low temperatures at the beginning and end of the season usually limit productivity.

Perennial tree crops, which yield fruit, nuts and wood, pose complex problems in assessing the effects of CO_2 and temperature interactions, because their yield

178   GLOBAL CLIMATE CHANGE AND AGRICULTURAL PRODUCTION

Figure 8.4. Influence of mean air temperature on the growth modification factor (weekly dry matter production (W) at 650 µmol/mol CO_2 divided by W at 350 µmol/mol CO_2) for water hyacinth, azolla (water fern), carrot, radish and cotton (from Kimball *et al.*, 1993)

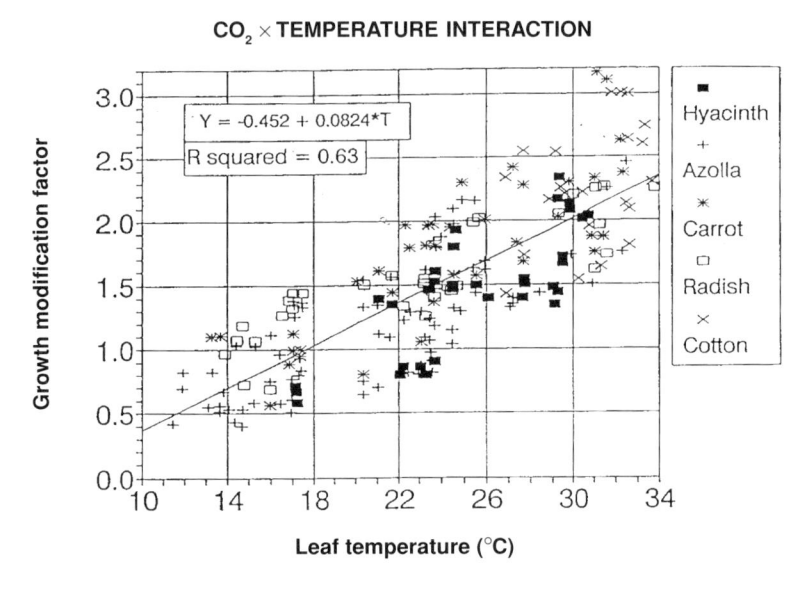

Figure 8.5. Calculations using the model ARCWHEAT (Porter, 1984), modified for CO_2 responses to illustrate potential production of winter wheat based on weather data from Rothamsted, England. The straight lines correspond to total dry matter production and grain yield predicted at 340 µmol/mol CO_2 and observed weather. The curves are predicted production and yield at 680 µmol/mol CO_2 and increases of daily mean temperature up to 4°C (from Squire and Unsworth, 1988)

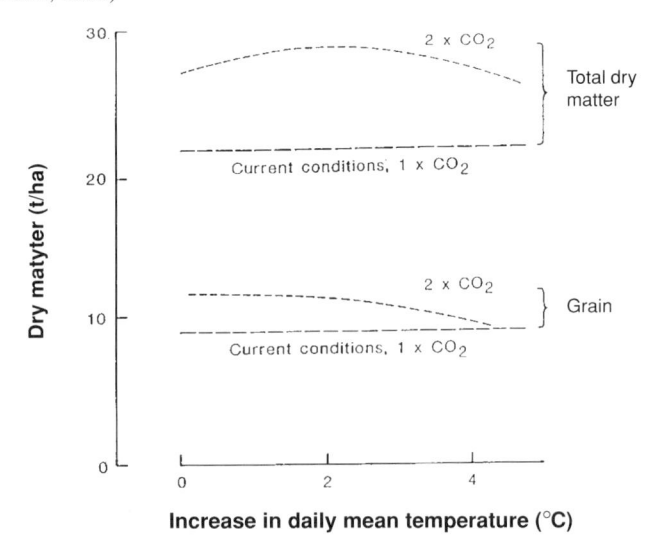

depends on a phased sequence of development over at least two years. Climatic warming could disrupt this sequence, for example, by providing insufficient winter chilling to synchronize spring bud break, or by advancing ovule development and leading to poor fruit set in spring (Cannell *et al.*, 1989). Cannell and Smith (1986) pointed out that climatic warming could either advance or delay bud burst in apples depending on the extent to which chilling requirements were met. The long-term influence of CO_2 on tree growth, and its interaction with temperature responses, is very uncertain at present.

CO_2 AND WATER AVAILABILITY

The influence of CO_2 on the water-use of crops may well prove to be the most important benefit of increased CO_2 concentrations for agriculture. Morison (1993) reviewed the recent literature and discussed the effect at a range of scales, pointing out that there are now many studies, including annual cereal crops, legumes and woody perennials, which show that, contrary to earlier assumptions, although the absolute amount of dry matter produced decreases at all CO_2 concentrations as water availability is reduced, the *relative* effect of increasing CO_2 on plant growth increases with decreasing water supply. There are at least two mechanisms responsible for this relative enhancement: (1) the reduction in stomatal conductance, with a consequent beneficial effect on leaf water potential and hence on leaf expansion; (2) greater allocation of carbon to roots in elevated CO_2, and hence improved potential for exploring the soil volume for water and nutrients. In contrast, the increased production of leaf area in elevated CO_2 acts to increase the water-use per unit ground area of crops.

It is less clear whether increasing CO_2 will decrease water-use of crop canopies in the field over periods of days or months. This is not only because of the various mechanisms described above, but also because soil evaporation must be accounted for, and feedbacks between leaf temperature, atmospheric humidity and canopy evaporation apply. Consequently, the very rough estimate that a 33% increase in dry matter production combined with a 33% decrease in transpiration in response to doubling CO_2 concentration would increase water-use efficiency by a factor of 2 is unlikely to apply. Goudriaan and Unsworth (1990) used a crop model with feedbacks between vegetation and the atmosphere to estimate that daily plant canopy water-use efficiency would increase by about 35% when carbon dioxide concentration doubled. In a simpler model, but including soil evaporation, Morison (1993) calculated that seasonal water-use by crops with leaf area indices between 2 and 6 would decrease by only 10 to 15% in doubled CO_2 concentration. Nijs *et al.* (1989) grew ryegrass in 350 µmol/mol CO_2 and 600 µmol/mol CO_2. While the total water-use of the canopy was unaffected by growth in elevated CO_2, WUE was increased by 25% on a canopy basis or 87% on a leaf area basis. Results from this experiment indicated that, because the influence of elevated CO_2 on P_N and g_s was greatest early in the season, before the canopy closed, WUE probably varied seasonally, and this may be a general feature for annual crops.

Azam Ali (pers. comm.) used two methods to assess transpiration of crops of sorghum and groundnut growing at 350 and 700 μmol/mol CO_2. Seasonal monitoring of soil water content with a neutron probe, combined with metered irrigation and subsidiary measurements of soil evaporation, enabled cumulative crop transpiration to be calculated. Porometry on individual leaves, and measurements of leaf area distribution, enabled transpiration per unit leaf area to be calculated. In groundnut but not in sorghum, elevated CO_2 reduced transpiration per unit leaf area throughout the season, but more so in the earlier stages. Figure 8.6 shows the seasonal relationships between dry mass and transpired water for the two crops; the slope is water-use efficiency (WUE). For the C_3 groundnut, seasonal WUE was significantly larger at 700 μmol/mol CO_2 (4.4 ± 0.2 g/kg) than for 350 μmol/mol CO_2 (3.0 ± 0.2 g/kg); equivalent values for sorghum (4.9 ± 0.3 and 4.2 ± 0.3 g/kg, respectively) did not differ significantly. The figure also shows that, in spite of the increased WUE in groundnut, the total amount of water-used in the season was about the same (450 mm) in both CO_2 treatments.

On an even larger scale, de Bruin and Jacob (1993) estimated the influence of doubling CO_2 concentration on regional-scale transpiration, allowing for interactions between transpiration, leaf temperature and the planetary boundary layer. They concluded that, for vegetation where stomatal conductance decreased by 34% in response to elevated CO_2, but leaf area was unchanged, regional transpiration on a typical summer day would decrease by about 11% for short crops such as cultivated grass and 17% for tall crops such as plantation forests. If negative feedback from the planetary boundary layer had been ignored, these values would have been overestimated (15 and 24%, respectively).

CO_2 AND OZONE

When Krupa and Kickert (1989) attempted to review the interactive responses of crops to CO_2 and O_3, they found no publications describing experimental work. We summarize here the few papers that have been published since then, and comment on other work known to be in progress. We briefly review the effects of ozone alone later in this paper.

Allen (1990) made an estimate of the possible scale by which CO_2 might influence responses of soybeans to O_3. He used published results relating O_3 exposure to yield (Heagle et al., 1983) and a model of the sensitivity of stomatal conductance for soybean to CO_2 (Rogers et al., 1983). According to these models, doubling CO_2 concentration would reduce g_s by about 30% and thus decrease the stomatal uptake of O_3 by the crop. Allen estimated that this would result in a yield decrease of about 15% compared with yield in ambient air. This estimate takes no account of the large direct effects of CO_2 on soybean growth and assumes no interactions between the two gases.

Scientists at the USDA Beltsville Research Center have studied soybeans growing in open-top chambers in the field and exposed to CO_2 and O_3. Kramer et al. (1991), in a preliminary study, showed that the yield loss attributed to ozone at the site was

COMBINED EFFECTS OF CO_2, TEMPERATURE, UV-B AND O_3

Figure 8.6. (a) Seasonal relationship between dry mass and transpired water for crops of groundnut growing in controlled-environment glasshouses at 350 µmol/mol CO_2 and 700 µmol/mol CO_2. Regression lines are shown dotted for 350 µmol/mol and solid for 700 µmol/mol. The slopes of the regression line are the water-use efficiencies (WUEs). (b) As 6(a) but for a crop of sorghum (Azam Ali, pers. comm.)

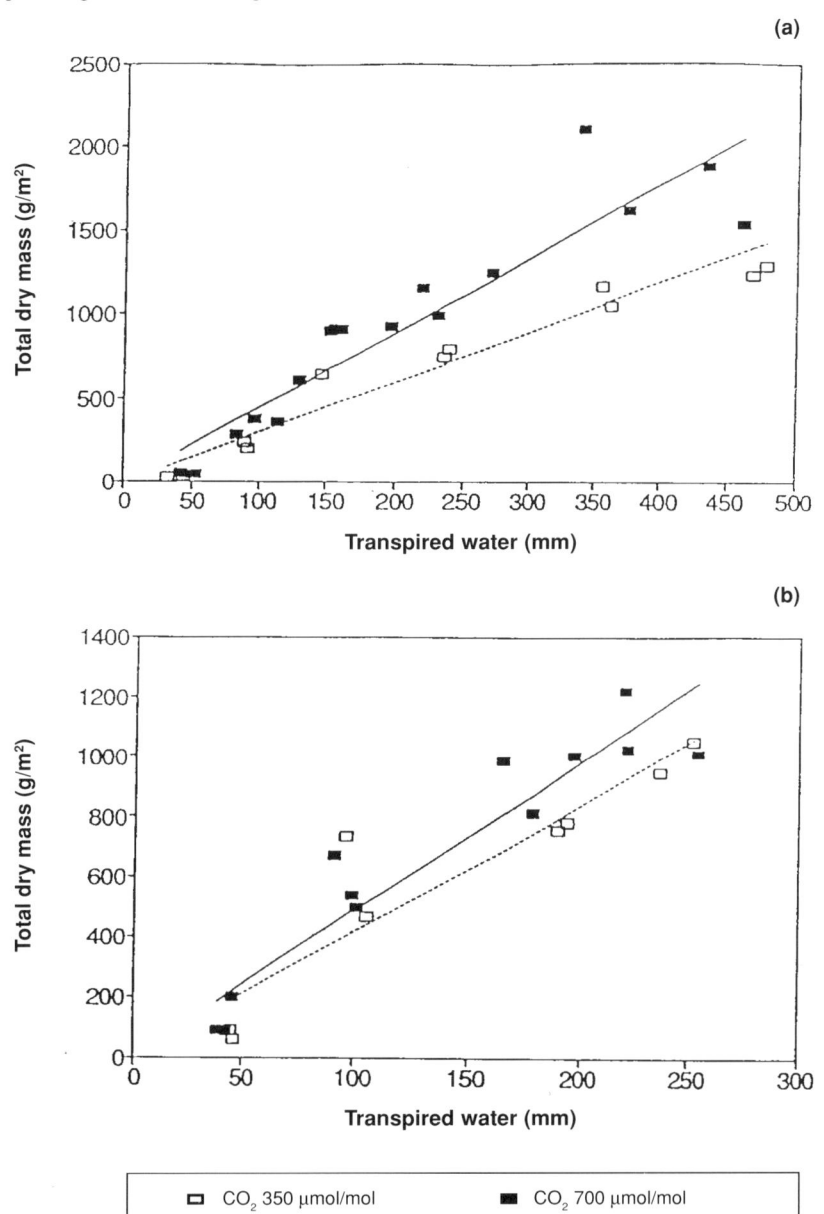

about 12%, but when CO_2 concentrations were increased to 500 μmol/mol the yield loss was only 6.7%. Mulchi *et al.* (1992) reported a more complex design with three O_3 treatments (charcoal-filtered, CF, non-filtered, NF, NF + 40 μmol/mol O_3) and three CO_2 treatments (ambient, + 50, + 150 μmol/mol). Increasing CO_2 concentration increased grain yield and grain oil content, and decreased protein content. Increasing O_3 reduced grain yield but did not significantly alter oil or protein content. Table 8.2 summarizes the interactive effects of CO_2 and O_3 on growth, yield and quality. Ozone caused a significant decrease in yield compared to the charcoal-filtered controls. When CO_2 was added to the ozone treatments, it partially counteracted the negative effect of O_3, so that yields in chambers with plus 150 μmol/mol CO_2 and NF air were similar to those in ambient CO_2 and CF air. The results suggested that most effects of CO_2 and O_3 were additive.

Barnes and Pfirrman (1992) grew radishes in a phytotron in Munich, Germany, at two CO_2 concentrations (385 and 765 μmol/mol) and two O_3 concentrations (20 nmol/mol and 73 nmol/mol), and studied gas exchange, growth and productivity, and mineral composition of the crop. In clean air the typical effects of elevated CO_2 on photosynthesis (increased) and stomatal conductance (decreased) were observed; the extra carbon assimilation stimulated root growth rate by 43% but there was no significant effect on shoot growth or leaf area. At ambient CO_2, the high O_3 treatment depressed photosynthesis by 26% (compared to the low treatment) and induced a slight reduction in g_s, with the net result that water-use efficiency declined. The reduction in carbon uptake was reflected in reduced growth, with roots more affected than shoots. Interactive effects of ozone and CO_2 were generally consistent with a reduction in ozone-induced responses in elevated CO_2. Early in the experiment, elevated CO_2 reduced the effect of O_3 on photosynthesis, but later this benefit disappeared, perhaps because long-term exposure to O_3 has an overall limiting effect on plant response to CO_2. The effects of CO_2 and O_3 on g_s appeared to be additive. Work is in progress in Munich on responses of grass swards to O_3, CO_2 and water availability (Payer *et al.*, 1993). Preliminary results indicate that the growth depression which is observed as water is withheld in ambient CO_2 is alleviated at higher CO_2 levels. Responses of yield and water-use efficiency to the mixtures of CO_2 and ozone seem to be additive rather than synergistic.

In Newcastle, England, work is in progress exposing winter and spring cultivars of wheat to two CO_2 concentrations and two O_3 concentrations (Barnes *et al.*, 1993). Early results indicate that after 50 days' exposure there was evidence of downregulation of photosynthesis in response to prolonged exposure to elevated CO_2; this was most pronounced in the winter cultivars. Long-term exposure to O_3 resulted in a decrease in the light-saturated rate of CO_2 assimilation, partial stomatal closure, and accumulation of water soluble carbohydrate and starch in leaves. These resulted in decreased growth, with root growth more severely affected than shoot growth. In plants exposed to elevated CO_2 and O_3, effects on stomatal conductance were less than additive, and CO_2 enhancement effects on photosynthesis and growth were reduced by O_3.

In summary, from the very limited experimental results available at the time of

Table 8.2. Direct and interactive effects of CO_2 and O_3 treatments on growth, yield components and grain quality characteristics of soybean growing in open-top field chambers (from Mulchi et al., 1992)

Chamber treatments		Shoot biomass[1] (g)	Leaf area[1] (dm^2)	SLW[1] (mg/cm^2)	Pods/ plant	Seeds/ plant	Weight of 100 seeds (g)	Grain yield (g)	Grain oil (%)	Grain protein (%)
CO_2	O_3									
CO_2 treatment means										
Ambient	—	55.3	55.7	2.88	51.7	128	16.9	436	20.4	40.4
+50 µmol/mol	—	61.1	58.1	3.07	57.8	140	17.7	497	22.3	39.3
+150 µmol/mol	—	59.2	58.0	2.99	60.9	144	17.6	509	22.1	39.2
LSD(0.05)		3.2	NS	NS	7.4	11	0.5	41	0.5	1.1
O_3 treatment means										
	CF	58.9	60.1	2.73	63.7	153	18.2	558	21.9	39.4
	NF	62.5	59.7	3.04	52.1	133	17.8	472	22.1	39.5
	NF+O_3	54.3	52.0	3.17	54.6	127	16.2	412	21.7	40.1
LSD(0.05)	3.2	7.5	0.36	7.4	11	0.5	0.5	NS	NS	
CO_2 and O_3 treatment										
Ambient	CF	57.6	61.4	2.31	56.2	147	17.5	513	21.4	40.1
	NF	59.2	57.5	3.07	50.9	122	17.6	432	21.7	40.4
	NF+O_3	49.2	48.2	3.26	48.2	117	15.5	363	21.2	40.8
+50 µmol/mol	CF	60.4	61.8	2.99	66.6	155	19.1	594	22.1	39.6
	NF	66.7	58.0	3.12	50.6	135	17.7	474	22.9	38.6
	NF+O_3	56.4	54.4	3.11	55.8	130	16.3	424	22.1	39.7
+150 µmol/mol	CF	58.6	57.1	2.88	68.4	157	18.1	568	22.4	38.6
	NF	61.7	63.6	2.94	54.7	142	18.0	510	21.9	39.4
	NF+O_3	57.4	53.3	3.14	59.6	134	16.8	448	22.0	39.8
	Av.	58.6	57.3	2.98	56.8	137	17.4	480	21.9	39.6
LSD(0.05)		5.7	NS	0.36	NS	NS	0.9	53	NS	NS
CV (%)		5.3	10.7	9.3	9.9	6.1	2.4	7.0	1.6	2.2

[1] Vegetative samples were collected on 15 August 1989.
NS = not significant; LSD = Least significant difference; CV = Coefficient of variance

writing, interactions between exposures to elevated O_3 and CO_2 seem to result in approximately additive effects on growth and yield. Both gases decrease stomatal conductance, and this results in less uptake of O_3 in elevated CO_2 treatments than in ambient air. Several authors have noted that this reduces (or at least delays) the damaging effects of O_3. In principle, the reverse is also true, i.e., increased O_3, by reducing stomatal conductance, negating some of the benefits to growth that elevated CO_2 confers. However, the O_3 exposure necessary to induce stomatal closure is generally sufficient to cause visible leaf injury, and this disbenefit is likely to be far more damaging than reduced CO_2 uptake through stomata. More field-based studies are necessary to assess fully the effects of the two gases in combination.

CO_2 AND UV-B RADIATION

Many studies of effects of UV-B radiation on crops are seriously flawed because UV exposures were either inadequately specified and/or were unrealistically large. Tevini (1993), in a thorough review of effects of UV-B radiation on plants, pointed out that, since the spectrum of UV from artificial light sources differs from that in the solar spectrum, and since many photobiological processes are strongly wavelength dependent, it is important to use weighting functions based on the action spectrum for specific responses before experiments in different exposure systems can be compared. Unfortunately, action spectra for many responses remain unknown. In addition, many plant responses to UV-B are larger when plants are growing at low light levels typical of growth chambers and some glasshouses than they are in full sunshine, probably because natural protective pigments are inadequately synthesized and repair processes are restricted in low light. For this reason, field-based studies of crops are most relevant for estimating responses of productivity and yield to UV-B. Even so, most field studies have used continuous UV-B exposures corresponding to 15 to 25% reductions in stratospheric O_3, i.e. roughly 30 to 50% increases in UV-B above present values. Such increases are much larger than are likely to occur as long-term means in agricultural regions under any likely depletion of stratospheric ozone in the next 30 years.

Effects of UV-B on yield of crops growing in the field have produced very variable results. Work in the late 1970s and early 1980s by Biggs and colleagues in Florida, using very high UV exposures, found yield reductions in only about half the crops studied. Work in Germany, also at high UV exposures, found no UV-B effects on cabbage, lettuce and rape. More recently, a large field-based study in Beltsville, Maryland, by Teramura, has revealed a high degree of intra-specific variability among soybean cultivars. The two most sensitive cultivars were grown for five seasons in the field in two UV treatments, corresponding to 16 and 25% ozone depletion simulations. In one cultivar (Essex) yield was reduced by 20 to 25% at the high UV treatment; in the other cultivar, yield generally increased by 10 to 22% in this treatment. Neither cultivar showed a consistent significant yield change at the lower UV exposure. The conclusion from these studies must be that, under any realistic scenario of increasing UV-B radiation, crop yields are unlikely to be altered to a significant extent. We therefore turn to the question of whether UV-B exposure alters the sensitivity of crops to other atmospheric changes.

Effects of the combination of elevated CO_2 and UV-B radiation on crop growth and yield have been addressed in very few studies. Teramura and his colleagues (Teramura et al., 1990) grew wheat, rice and soybean in a glasshouse experiment. Treatments in the factorial design were: ambient CO_2 (350 μmol/mol), ambient UV, elevated CO_2 (650 μmol/mol) and elevated UV-B (corresponding to a 10% decrease in stratospheric O_3 at the equator). Compared to the control, seed yield and total biomass increased significantly in elevated CO_2 for all three species. However, when UV-B and CO_2 were increased simultaneously, no increase in either seed yield (wheat and rice) or total biomass (rice) was observed compared to the control. In contrast, in soybean, the increases in seed yield and biomass induced by CO_2 alone were maintained in the high CO_2/high UV-B environment. Studies of leaf gas exchange indicated that UV-B reduced the apparent carboxylation efficiency in wheat and rice, but not in soybean.

Results of a CO_2/UV study on seedlings of sunflower and maize reported by Tevini (1993) are confounded by temperature changes in some treatments. A 25% increase in UV-B at 340 μmol/mol CO_2 changed sunflower and maize dry weights (W) by -14 and -24%, respectively (compared to ambient controls). The same UV-B increment combined with a +2°C temperature change altered W by +5% in sunflower and +31% in maize. Adding a doubled CO_2 concentration to the elevated UV and temperature regime altered W by +19 and +32% for sunflower and maize, respectively. The incomplete experimental design limits interpretation of this study, but the results suggest that temperature and CO_2 are much more important influences on growth than any likely realistic increases in UV-B.

OZONE

There are a number of recent reviews of the mechanism of action of O_3 on plants, and of crop responses to O_3 (e.g., Tingey and Andersen, 1991; Heck et al., 1988); consequently, we will only briefly summarize the main features. The response of plants to O_3 may be viewed as the culmination of a sequence of physical, biochemical and physiological events. The O_3 diffuses from the air into the leaf through stomata, and these exert control on O_3 uptake. Plants are able to detoxify O_3 or its metabolites and can repair or compensate for O_3 impacts, so visible signs of ozone injury may not occur if the rate of O_3 uptake is sufficiently small. When resources have to be devoted to repair or compensation mechanisms, or when O_3 uptake is too large for full repair to be achieved, growth and yield may be reduced in the absence of visible injury by O_3. The principal modes of action of O_3 on plants are through injury to proteins and membranes, reduction in photosynthesis (i.e., carbon gained), changes in allocation of carbohydrate, and acceleration of senescence. The main result of O_3 exposure is therefore a reduction in the capacity of the plant to accumulate photosynthate through loss in photosynthetic capacity and increased foliage senescence.

Methods for exposing crops to ozone in experiments designed to develop yield/exposure responses are well established. The most commonly used technique is the open-top field chamber, which enables crop stands to be grown in either charcoal-filtered air or in air to which known amounts of O_3 have been added. Coordinated programmes in the United States and Europe have investigated crop responses to O_3 in open-top chambers, and relationships between yield and seasonal mean O_3 concentration are available for many of the major world crops. The seasonal mean index, however, has been shown to be inadequate in relating ozone exposure to effects, primarily because of its inability to consider duration of exposure and its consideration of all concentrations as equal in their effect. Both duration and concentration are important in the effect of ozone (Hogsett et al., 1988). In particular, studies have demonstrated that ozone's effect is cumulative and that higher concentrations are more important than lower concentrations in causing an effect (Hogsett et al., 1988; Musselman et al., 1994). Consequently, the cumulative, peak-weighted indices are more appropriate because of their biological relevancy. Examples of such indices include SUM06 (in units of nmol/mol-h) which cumulates all concentrations equal to or greater than 0.06 nmol/mol, or SIGMOID which cumulates all concentrations during the season and weights all concentrations according to a sigmoid function. The US Environmental Protection Agency (EPA) Draft Ozone Criteria Document (1995) reviewed and tabulated the latest information concerning effects of ozone on crop yields using both the seasonal mean, and two of the cumulative, peak-weighted indices. The data for crop yield response are primarily from the National Crop Loss Assessment Network (NCLAN). Yield reductions in the NCLAN studies are calculated relative to a typical background O_3 concentration of SUM06 of 0 or a 7-h seasonal mean of 25 nmol/mol. Yield losses of 10% or less in 50% of the crops (12 crop species, 54 studies including well-watered and drought studies) would occur with SUM06 exposure concentrations of 26.4 nmol/mol-h in three months, or 49 nmol/mol 7-h seasonal mean. Some species/cultivars were particularly sensitive; 11% would be expected to have a yield reduction of 10% at a 7-hr seasonal mean of less than 35 nmol/mol. Similarly, 18% of the crops/cultivars are expected to have yield reductions of 10% or less at 3-month SUM06 concentrations of less than 10 nmol/mol-h. Both of these exposure values are quite low compared to the decade-average (1982-1991) exposure (3 months) from all monitoring sites across the United States (SUM06 = 29.5 nmol/mol-h; 7-h mean = 54 nmol/mol). In general, grain crops are less sensitive than others, but within-species variability and sensitivity may be greater than between species. Krupa and Kickert (1989) reviewed various published lists of the sensitivity of crop yields to O_3; onion, spinach, potato, alfalfa and cotton appear to be particularly sensitive crops, although it is possible to find resistant cultivars of all these species. Legumes range between sensitive and intermediate, and cereal and grass crops tend to be relatively resistant.

There is also a very large literature describing biotic and abiotic factors that modify plant response to O_3 (US EPA, 1995). We will concentrate in this paper on only a few interacting factors that are particularly relevant in terms of future changes in the atmospheric environment.

OZONE AND TEMPERATURE

When interactions between ozone exposure and temperature were reviewed by the US EPA (1986) it was concluded that there were variable and conflicting results. More recent analysis (US EPA, 1995) suggests that many of these studies in controlled environments confounded effects of temperature with changes in vapour pressure deficit (VPD) because relative humidity was kept constant while temperature was increased. Since VPD can have profound effects on stomatal responses, evaporation rates and leaf expansion, it is not surprising that these early results have led to confusion. Recently, Todd et al. (1991) designed an experiment with tomato seedlings in which differences in VPD at different temperatures were minimized during O_3 exposure. This study showed that, out of 11 growth variables measured, the only significant modifications by temperature of the effects of O_3 were on stem fresh weight and specific leaf area (leaf area/leaf dry weight). The authors suggested that VPD probably plays a more important role in determining sensitivity to O_3 than temperature. It is important to have more studies to clarify this issue, as many field-based studies of ozone responses use open-top chambers in which temperatures are increased by a few degrees above ambient. If either the changed temperature or the changed VPD in such chambers alters the sensitivity of the crops to O_3, there would be serious consequences for the application of much of our knowledge of the O_3 sensitivity of crops that has been gained from field-chamber experiments.

A second important O_3/temperature interaction, particularly affecting perennial species, including trees, is winter hardiness. Several studies have shown that exposures to O_3 reduce the frost-hardiness; Davison et al. (1988) reviewed the subject. Whilst most of this research has been with woody species, Barnes et al. (1988) showed that daily exposures to about 80 nmol/mol O_3 for 7 days significantly reduced the survival of pea seedlings after exposure to nighttime temperatures of about –4°C. It seems likely that, as with sulphur dioxide, exposure to ozone could alter the frost hardiness of a number of agricultural crops. This response would be most likely to be significant in the event of an early autumn frost on early-sown crops such as winter cereals. There are suggestions that exposure of trees in summer to ozone alters the hardiness the following winter. Eamus and Murray (1991) pointed out that, even in severe winters, there are brief periods of mild temperatures that induce partial dehardening of woody species. They speculated that O_3 increases tree predisposition to dehardening in such conditions and consequently puts trees at greater risk from low temperatures. There are likely to be significant differences between species in this phenomenon. For example, in Florida, Eissenstat et al. (1991) found that, although O_3 reduced frost hardiness of citrus and avocado, the effects were small and the likelihood of significantly changing frost resistance is slight.

OZONE AND WATER AVAILABILITY

The availability of soil water is one of the strongest influences on crop growth and productivity. When available, irrigation is commonly used to alleviate water stress, but in conditions of large evaporative demand, even irrigated crops may experience water stress. The most immediate effects of water stress are on plant-water potential, and may lead to reductions in leaf expansion rates and altered partitioning of dry matter, with increased assimilate partitioning to roots. When stomatal closure occurs, there is a resulting reduction in transpiration and CO_2 uptake and an increase in leaf temperature.

WATER STRESS AND O_3 RESPONSES

There has long been a general belief that water stress reduces the magnitude of any adverse effects of O_3, i.e., leaf injury, growth and yield reductions. This belief was based on observations that stomatal closure, induced by drought, reduced the rate and quantity of O_3 absorbed by leaves (US EPA, 1986). However, more recent results show that the O_3/water availability interaction is complex, and probably depends on the magnitude and timing of the water stress (which has seldom been adequately defined in air pollution experiments). One difficulty in quantifying the interaction is the very large effect that water stress alone may have on yield, so that any influence of O_3 is sometimes masked.

Temple et al. (1985) studied the effects of O_3 and water stress on cotton growing in California. In this typically hot, dry season, the water-stressed plants wilted frequently, and yielded much less than the well-watered controls; nevertheless, when compared to yields of plants in a background O_3 concentration of 25 nmol/mol, and expressed as a percentage, the estimated yield loss at a seasonal mean O_3 concentration of 50 nmol/mol was 7% in the well-watered plants and 2% in the water-stressed crop. Clearly the absolute yield loss from drought was much larger than the yield loss attributed to O_3 in either treatment.

Heagle et al. (1983) reviewed six studies of soybean responses to O_3 and water stress, and concluded that in only three were there significant interactions, i.e., the clear negative relationships between yield and O_3 exposure observed with well-watered plants were much reduced with water stressed plants. More recently, Heggestad and Lesser (1990) analysed three years of data for four soybean cultivars. They concluded that, in most cases, the relationships between yield and O_3 concentration had similar slopes for the water-stressed and well-watered treatments. If this conclusion applies more generally, the extra yield losses which would result from O_3 exposure could be estimated if a yield response to drought alone is known. Clearly, there is still uncertainty over the influence of water stress on the form of the yield response to O_3. This uncertainty is unlikely to be resolved until better experimental designs with more precise specifications of the degree of water stress are developed.

O_3 AND WATER-USE EFFICIENCY

For irrigated crops where water supplies may be limited or expensive, it is important to know whether O_3 would influence the water-use efficiency (WUE) and consequently the amount of water required in a season. Reich *et al.* (1985) exposed well-watered soybeans to 130 nmol/mol O_3 for 7 hours daily, and found a 25% decrease in WUE compared with controls in 10 nmol/mol O_3. Similar results were found for alfalfa (Temple *et al.*, 1988), probably indicating that carbon dioxide uptake was reduced more by O_3 than water loss. Tingey *et al.* (1994) also reported a significant decrease in WUE in soybeans exposed to episodic exposure regimes of 20-30 nmol/mol-h (3-month SUM06). The study found the O_3 did not close stomata, but rather affected CO_2 assimilation. A significant increase in leaf construction cost was reported, indicating that both carbon and water are used inefficiently by plants exposed to ozone. In contrast, Greitner and Winner (1988) found that O_3 exposure *increased* the WUE of radish and soybeans. This disagreement probably reflects variation with exposure dynamics, stage of development, and genetic variation between and within species.

Recently Barnes and Pfirrman (1992) investigated effects of O_3 and elevated CO_2 singly and in combination on the WUE of radish. Both gases reduced stomatal conductance, and the combination of gases reduced conductance still further. Figure 8.7 shows that O_3 reduced the instantaneous WUE, whereas CO_2 increased it (as discussed earlier). In the combination of O_3 and CO_2, there was initially no significant effect of O_3 (i.e., the response of WUE was similar to that observed in elevated CO_2), but as growth progressed, the WUE was significantly reduced in the combination treatment compared to in elevated CO_2 alone. This may be a response to accelerated senescence induced by O_3 exposure.

In summary, relatively severe water stress may reduce the yield losses attributable to O_3 in some crop species but, in general, the yield loss resulting from the water stress outweighs the benefits of the O_3 protection. For well-watered crops, O_3 exposure may change the water-use efficiency compared with crops in clean air, but the direction and magnitude of change probably depend on exposure and genetic factors. The influence of O_3 and CO_2 in combination on WUE is potentially important and merits further study.

OZONE AND UV-B RADIATION

Krupa and Kickert (1989) found no reports of O_3/UV-B interactions in their review. They assessed potential risks on a geographical basis, using distributions of major crops, tropospheric ozone concentrations and UV-B irradiance. They suggested that interactions might involve episodic exposure to the two stresses, so that peaks of O_3 would coincide with lower UV-B irradiance and vice versa. Runeckles and Krupa (1994) have developed this concept further, arguing that, as tropospheric O_3 increases, UV-B irradiance at the surface is reduced, because the O_3 absorbs some UV. In

Figure 8.7. Instantaneous water-use efficiency WUE (μmol CO_2 assimilated (A) per mol H_2O transpired (E) for radish (*Raphanus sativus* L. cv. Cherry Belle) plants exposed in climatic chambers to CO_2 and O_3. Treatments were 350 μmol/mol CO_2 + 20 nmol/mol O_3 (first columns) or + 80 nmol/mol O_3 (second columns); 750 μmol/mol CO_2 + 20 nmol/mol O_3 (third columns) or + 80 nmol/mol O_3 (fourth columns). Error bars show lsd at p=0.01 for each of the sampling times (from Barnes and Pfirrman, 1992)

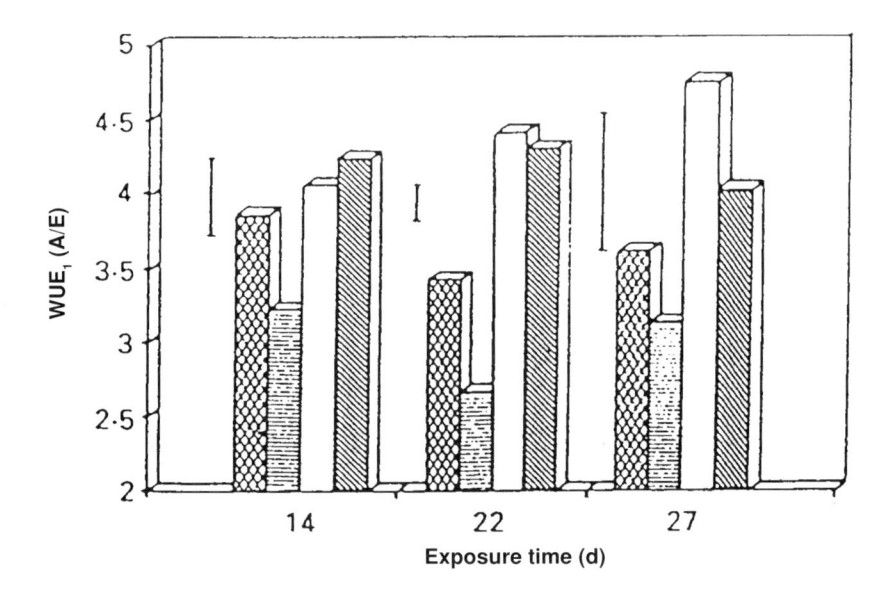

theory, this effect could be important, in spite of the relatively small contribution of tropospheric O_3 to the total atmospheric O_3 column, because scattering by aerosols and molecules in the troposphere increases the radiation path length (Bruhl and Crutzen, 1989). In an attempt to quantify this effect, Albar (1992) compared the measured solar spectral irradiance I (l) at the ground near Nottingham, England, on two days (12 and 18 July 1990) when the tropospheric O_3 concentration was 51 and 84 nmol/mol, respectively, and stratospheric O_3 was constant. Figure 8.8 shows that I (l) was about 20 to 40% greater in the UV-B waveband (280-320 nm) on the low O_3 day than on the high O_3 day, apparently supporting the hypothesis of Runeckles and Krupa (1994). However I (l) at longer wavelengths was also greater (by about 20 to 25%), and this increase is most probably a consequence of less aerosol (dust) being present on the low O_3 day. Since this aerosol effect would also apply in the UV-B waveband, it seems likely that the UV-B increase *directly* attributable to reduced tropospheric O_3 was no more than 10% to 15%. Variations in aerosol from day to day seem likely to be more important than variations in tropospheric O_3 in modulating the intensity of UV-B at the ground, but further observations would be valuable.

Figure 8.8. Variation of percentage increase in solar spectral irradiance $((I(\lambda, 51)-I(\lambda, 84))/I(\lambda, 84))$ with wavelength λ between days when the tropospheric ozone concentration was 84 nmol/mol O_3 and 51 nmol/mol O_3 (from Albar, 1992)

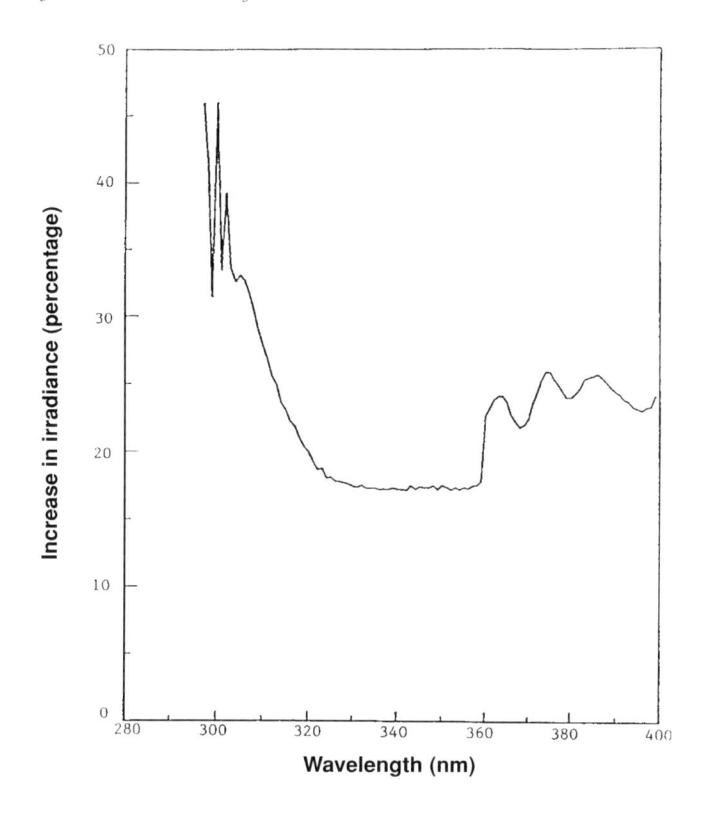

As mentioned earlier, many investigations of plant responses to UV-B are seriously flawed by inadequate specification and/or excessive UV exposure. An improved field-based system at Raleigh, North Carolina, using open-top chambers for studying crop responses to UV and O_3 (Booker et al., 1992a) has been used for three seasons of research on soybeans (Miller et al., 1994). The system exposed soybean crops from emergence to maturity to UV treatments (ranging from ambient to about twice ambient biologically effective UV-B) and to O_3 treatments giving seasonal mean 12 h/d O_3 concentrations from 14 to 83 nmol/mol.

The ozone treatments resulted in reductions of photosynthesis, accelerated senescence, and reduced yield, in agreement with many other published studies. In contrast, the UV treatments, even at this relatively large UV irradiance, did not induce any significant changes in photosynthesis or yield, and there were no UV/O_3 interactions (Booker et al., 1992b).

Fiscus et al. (1994) attempted to reconcile the lack of response to UV reported in the Raleigh experiments with other reports which indicate that increased UV-B causes physiological dysfunction and reductions in crop yields. They concluded that three factors make the conclusions of other field and glasshouse studies, at best, hard to interpret and possibly misleading: a frequent failure to monitor UV-B adequately; a tendency to underestimate UV exposures when relying on model calculations which do not allow sufficiently for effects of dust in the atmosphere; and no adjustment of UV exposure for seasonal and daily weather changes (and hence a tendency towards unrealistically large exposures).

Although it would be useful to see further well-designed field studies of UV/O_3 interactions using a wide range of crop and natural species, on the evidence of the careful work of Booker, Miller, Fiscus and their colleagues, it seems unlikely that UV/O_3 interactions are of any importance for crop productivity, and it seems clear that O_3 poses a much greater threat to yields of many crops than any likely increases in UV-B radiation.

CONCLUDING REMARKS

In our introduction we emphasized the need for knowledge of how interacting environmental changes will affect crop productivity. Many examples in this paper show that misleading conclusions can be drawn by considering crop responses to single factors. However, the experimental demands of studying multiple factors are great, and it is necessary to define priorities. In our view, the following interactions should have high priority:

- Interactions of CO_2 and temperature. If the benefits of increased CO_2 in terms of increased potential carbon assimilation by C_3 species are not achievable at low temperatures, there are important implications for agriculture in cool climates. Equally, the merits of C_3 versus C_4 crops may need reassessing in warmer climates in view of the direct benefits of CO_2 to C_3 species.
- Canopy scale studies of effects of CO_2 on water-use efficiency, and on the interacting effects of O_3. Although it is clear that increased CO_2 enhances WUE at the leaf or single plant level, there are very few studies relevant to field-grown crops. Equally, the very limited studies of effects of O_3 and CO_2 on WUE suggest that responses are non-additive, and this is an important issue to resolve for agriculture in Europe, the eastern United States and other regions where tropospheric O_3 concentrations influence crop productivity.

It is disappointing that, in spite of considerable interest in the effects of UV-B radiation on plants, much of the research in the past decade has been seriously flawed by employing very unrealistic UV-B exposures and/or by using highly artificial growth conditions. On the evidence of the few well-designed studies of crop responses to UV-B that we have reviewed here, we do not believe that UV-B poses serious threats to crops, and have not identified any important interactions between UV-B and other stressors that would make further research of this type a particularly high priority.

ACKNOWLEDGEMENTS

We are grateful to Ossama Albar, Sayed Azam Ali, Fitz Booker, Zahide Kocabas and their colleagues for allowing us access to unpublished results, and we thank Pat Sommer for her assistance in preparing the typescript.

REFERENCES

Albar, O.F. 1992. *The Spectral Distribution of Solar Ultraviolet Radiation at the Ground*. PhD Thesis. University of Nottingham, UK.

Allen, L.H. 1990. Plant responses to rising carbon dioxide and potential interactions with air pollutants. *J. Environmental Quality* **19**: 15-34.

Barnes, J.D. and Pfirrman, T. 1992. The influence of CO_2 and O_3, singly and in combination, on gas exchange, growth and nutrient status of radish (*Raphanus sativus* L.). *New Phytologist* **121**: 403-412.

Barnes, J.D., Reiling, K., Davison, A.W. and Renner, C.J. 1988. Interaction between ozone and winter stress. *Environmental Pollution* **53**: 235-254.

Barnes, J.D., Bannister, G. and Ollerenshaw, J. 1993. Interactive effects of CO_2 and/or O_3 on wheat (*Triticum aestivum* L.). *UK Committee for Air Pollution Effects Research Annual Meeting, Bangor, Wales* (Abstract).

Booker, F.L., Fiscus, E.L., Philbeck, R.B., Heagle, A.S., Miller, J.E. and Heck, W.W. 1992a. A supplemental ultraviolet-B radiation system for open-top field chambers. *J. Environmental Quality* **21**: 56-61.

Booker, F.L., Miller, J.E. and Fiscus, E.L. 1992b. Effects of ozone and UV-B radiation on pigments, biomass and peroxidase activity in soybean. In: *Tropospheric Ozone and the Environment: II. Effects, Modelling and Control*. R.L. Berglund (ed.). Air and Waste Management Association, Pittsburgh, PA. pp. 489-503.

Bruhl, C. and Crutzen, P.J. 1989. On the disproportionate role of tropospheric ozone as a filter against solar UV-B radiation. *Geophysics Research Letters* **16**(7): 703-706.

Cannell, M.G.R. and Smith, R.I. 1986. Climatic warming, spring budburst and frost damage on trees. *J. Applied Ecology* **23**: 177-191.

Cannell, M.G.R., Grace, J. and Booth, A. 1989. Possible impacts of climatic warming on trees and forests in the United Kingdom: a review. *Forestry* **62**: 337-364.

Clifford, S.C., Stronach, I.M., Mohamed, A.D., Azam-Ali, S.N. and Crout, N.M.J. 1993. The effects of elevated atmospheric carbon dioxide and water stress on light interception, dry matter production and yield in stands of groundnut (*Arachis hypogaea* L.). *J. Experimental Botany* **44**: 1763-1770.

Davison, A.W., Barnes, J.D. and Renner, C.J. 1988. Interactions between air pollutants and cold stress. In: *Gaseous Air Pollution and Plant Metabolism*. S. Schulte-Hostede, N.M. Darrall, L.W. Blank and A.R. Wellburn (eds.). Elsevier Science, London. pp. 307-328.

de Bruin, H.A.R. and Jacob, C.M.J. 1993. Impact of CO_2 enrichment on the regional evapotranspiration of agro-ecosystems, a theoretical and numerical modelling study.

In: *CO₂ and Biosphere*. J. Rozema, H. Lambers, S.C. van de Geijn and M.L. Cambridge (eds.). Kluwer Academic Publishers, Dordrecht. pp. 307-318.

Drake, B.G. and Leadley, P.W. 1991. Canopy photosynthesis of crops and native plant communities exposed to long-term elevated CO_2 treatment. *Plant, Cell and Environment* **14**: 853-860.

Eamus, D. 1991. The interaction of rising CO_2 and temperatures with water-use efficiency. *Plant, Cell and Environment* **14**: 843-852.

Eamus, D. and Murray, M. 1991. Photosynthetic and stomatal conductance responses of Norway Spruce and Beech to ozone acid mist and frost in a conceptual model. *Environmental Pollution* **72**: 23-44.

Eissenstat, D.M., Syverson, J.P., Dean, T.J., Yelenosky, G. and Johnson, J.D. 1991. Sensitivity of frost resistance and growth in citrus and avocado to chronic ozone exposure. *New Phytologist* **118**: 139-146.

Fiscus, E.L., Miller, J.E. and Booker, F.L. 1994. Is UV-B a hazard to soybean photosynthesis and yield?: Results of an ozone/UV-B interaction study and model predictions. In: *Stratospheric Ozone Depletion/UV-B Radiation in the Biosphere*. R.H. Biggs and M.E.B. Joyner (eds.). Springer-Verlag, Berlin. pp. 135-147.

Goudriaan, J. and Unsworth, M.H. 1990. Implications of increasing carbon dioxide and climate change for agricultural productivity and water resources. In: *Impact of Carbon Dioxide, Trace Gases, and Climate Change on Global Agriculture*. B.A. Kimball, N.J. Rosenberg and L.H. Allen, Jr. (eds.). ASA Spec. Pub. No. 53. American Society of Agronomy, Madison, Wisconsin. pp. 111-133.

Greitner, C.S. and Winner, W.E. 1988. Increases in delta 13C values of radish and soybean plants caused by ozone. *New Phytologist* **108**: 489-494.

Heagle, A.S., Heck, W.W., Rawlings, J.O. and Philbeck, R.B. 1983. Effects of chronic doses of ozone and sulfur dioxide on injury and yield of soybeans in open-top field chambers. *Crop Science* **23**: 1184-1191.

Heck, W.W., Taylor, O.C. and Tingey, D.T. 1988. *Assessment of Crop Loss from Air Pollutants*. Elsevier Applied Science, London.

Heggestad, H.E. and Lesser, V.M. 1990. Effects of ozone, sulfur dioxide, soil water deficit and cultivar on yields of soybean. *J. Environmental Quality* **19**: 488-495.

Hogsett, W.E., Tingey, D.T. and Lee, E.H. 1988. Ozone exposure indices: concepts for development and evaluation of their use. In: *Assessment of Crop Loss from Air Pollutants*. W.W. Heck, O.C. Taylor and D.T. Tingey (eds.). Elsevier and Applied Science, London. pp. 107-138.

Houghton, J.T., Callander, B.A. and Varney, S.K. 1992. *Climate Change 1992*. Cambridge University Press, Cambridge.

Idso, S.B. 1990. Interactive effects of carbon dioxide and climate variables on plant growth. In: *Impact of Carbon Dioxide, Trace Gases, and Climate Change on Global Agriculture*. B.A. Kimball, N.J. Rosenberg and L.H. Allen Jr. (eds.). ASA Spec. Pub. No. 53. American Society of Agronomy, Madison, Wisconsin. pp. 61-69.

Kimball, B.A. 1983. Carbon dioxide and agricultural yield: An assemblage and analysis of 430 prior observations. *Agronomy J.* **75**: 779-788.

Kimball, B.A., Mauney, J.R., Nakayama, F.S. and Idso, S.B. 1993. Effects of increasing atmospheric CO_2 on vegetation. In: *CO_2 and Biosphere*. J. Rozema, H. Lambers, S.C. van de Geijn and M.L. Cambridge (eds.). Kluwer Academic Publishers, Dordrecht. pp. 65-75.

Kocabas, Z. 1993. *Sensitivity of Crop Models to Climate Variables and Modelling Techniques*. University of Nottingham. PhD.

Kocabas, Z., Mitchell, R.A.C., Craigon, J. and Perry, J.N. 1993. Sensitivity analyses of AFRCWHEAT1 crop model: the effect of changes in radiation and temperature. *J. Agric. Sci. (Cambridge)* **120**: 149-158.

Kramer, G.F., Lee, E.H. and Rowland, R.A. 1991. Effects of elevated CO_2 concentration on the polyamine levels of field-grown soybean at three O_3 regimes. *Environmental Pollution* **73**(2): 137-152.

Krupa, S.V. and Kickert, R.N. 1989. The greenhouse effect: impacts of ultraviolet-B (UV-B) radiation, carbon dioxide, and ozone on vegetation. *Environmental Pollution* **61**: 263-393.

Lawlor, D.W. and Mitchell, R.A.C. 1991. The effects of increasing CO_2 on crop photosynthesis and productivity: a review of field studies. *Plant, Cell and Environment* **14**: 807-818.

Long, S.P. 1991. Modification of the response of photosynthetic productivity to rising temperature by atmospheric CO_2 concentrations: has its importance been underestimated? *Plant, Cell and Environment* **14**: 729-739.

Miller, J.E., Booker, F.L., Fiscus, E.L., Heagle, A.S., Pursley, W.A., Vozzo, S.F. and Heck, W.W. 1994. Ultraviolet-B radiation and ozone effects on growth, yield, and photosynthesis of soybean. *J. Environmental Quality* **23**(1): 83-91.

Morison, J.I.L. 1985. Sensitivity of stomata and water-use efficiency to high CO_2. *Plant, Cell and Environment* **8**: 467-474.

Morison, J.I.L. 1993. Response of plants to CO_2 under water limited conditions. In: *CO_2 and Biosphere*. J. Rozema, H. Lambers, S.C. van de Geijn and M.L. Cambridge (eds.). Kluwer Academic Publishers, Dordrecht. pp. 193-209.

Mulchi, C.L., Slaughter, L., Saleem, M., Lee, E.H., Pausch, R. and Rowland, R. 1992. Growth and physiological characteristics of soybean in open-top chambers in response to ozone and increased atmospheric CO_2. *Agriculture, Ecosystems and Environment* **38**: 107-118.

Musselman, R.C., McCool, P.M. and Lefohn, A.S. 1994. Ozone descriptors for an air quality standard to protect vegetation. *Journal of the Air and Waste Management Association*. **44**: 1383-1390.

Nijs, I., Impens, I. and Behaeghe, T. 1989. Leaf and canopy responses of *Lolium perenne* to long-term elevated atmospheric carbon dioxide concentration. *Planta* **177**: 312-320.

Oechel, W.C. and Strain, B.R. 1985. Native species responses to increased atmospheric carbon dioxide concentration. In: *Direct Effects of Increasing Carbon Dioxide on Vegetation*. B.R. Strain and J.D. Cure (eds.). US Department of Energy, Washington DC. pp. 119-154.

Payer, H., Firsching, K., Steiner, K. and Djelassi, K. 1993. Factorial effects of CO_2, O_3 and irrigation on the yield and water-use efficiency of experimental grass swards grown in closed chambers. *UK Committee for Air Pollution Effects Research Annual Meeting, Bangor, Wales* (Abstract).

Porter, J. 1984. A model of canopy development in winter wheat. *J. Agric. Sci. (Cambridge)* **102**: 383-392.

Reich, P.B., Schoettle, A.W. and Amundson, R.G. 1985. Effects of low concentrations of O_3, leaf age and water stress on leaf diffusive conductance and water-use efficiency in soybean. *Physiologia Plantarum* **63**: 58-64.

Rogers, H.H., Bingham, G.E., Cure, J.D., Smith, J.M. and Surano, K.A. 1983. Responses of selected plant species to elevated carbon dioxide in the field. *J. Environmental Quality* **12**: 569-574.

Runeckles, V.C. and Krupa, S.V. 1994. The impact of UV-B radiation and ozone on terrestrial vegetation. *Environmental Pollution* **83**: 191-213.

Squire, G.J. and Unsworth, M.H. 1988. *Effects of CO_2 and Climate Change on Agriculture. Report to U.K. Department of the Environment.* University of Nottingham, England.

Temple, P.J., Taylor, O.C. and Benoit, L.F. 1985. Cotton yield responses to ozone as mediated by soil moisture and evapotranspiration. *J. Environmental Quality* **14**: 55-60.

Temple, P.J., Benoit, L.F., Lennox, R.W., Reagan, C.A. and Taylor, O.C. 1988. Combined effects of ozone and water stress on alfalfa growth and yield. *J. Environmental Quality* **17**: 108-113.

Teramura, A.H., Sullivan, J.H. and Ziska, L.H. 1990. Interaction of elevated ultraviolet-B radiation and CO_2 on productivity and photosynthetic characteristics in wheat, rice, and soybean. *Plant Physiology* **94**: 470-475.

Tevini, M. 1993. *UV-B Radiation and Ozone Depletion: Effects on Humans, Animals, Plants, Microorganisms, and Materials.* Lewis, Boca Raton, Florida.

Tingey, D.T. and Andersen, C.P. 1991. The physiological basis of differential plant sensitivity to changes in atmospheric quality. In: *Ecological Genetics and Air Pollution.* G.E. Taylor, J.L.F. Pitelka and M.T. Clegg (eds.). Springer-Verlag, Berlin. pp. 209-235.

Tingey, D.T. and Taylor, G.E.J. 1982. Variation in plant response to ozone: a conceptual model of physiological events. In: *Effects of Gaseous Air Pollution in Agriculture and Horticulture.* M.H. Unsworth and D.P. Ormrod (eds.). Butterworth Scientific, London. pp. 113-138.

Tingey, D.T., Hogsett, W.E., Rodecap, K.D., Lee, E.H. and Moser, T.J. 1994. The impact of O_3 on leaf construction cost and carbon isotope discrimination. In: *Immissionsokologische Forschun im Wnadel der Zeit: Festschrift fur Robert Guderian.* W. Kuttler and M. Jochimsen (eds.). Westarp Wissenschaften: Essener Okologische Schriften. pp. 195-206.

Todd, A.G., Ormrod, D.P., Hale, B.A. and Goodyear, S.N. 1991. Temperature effects on tomato response to ozone at constant vapor pressure deficit. *Biotronics* **20**: 43-52.

US EPA. 1986. Effects of ozone and other photochemical oxidants on vegetation. In: *Air Quality Criteria for Ozone and Other Photochemical Oxidants.* Research

Triangle Park, North Carolina. Environmental Criteria and Assessment Office. pp. 6/1-6/298.
US EPA. 1995. *Air Quality Criteria for Ozone and other Photochemical Oxidants.* Draft edition. Research Triangle Park, North Carolina. Environmental Criteria and Assessment Office.
van de Geijn, S.C., Goudriaan, J. and Berendse, F. (eds.). 1993. *Climate Change; Crops and Terrestrial Ecosystems.* CABO-DLO, Wageningen. 144 p.

9. The Potential Effects of Climate Change on World Food Production and Security

GÜNTHER FISCHER
International Institute for Applied Systems Analysis, Laxenburg, Austria

KLAUS FROHBERG
Institut für Agrarentwicklung in Mittel- und Osteuropa, Halle, Germany

MARTIN L. PARRY
Jackson Environment Institute, University College, London, UK

CYNTHIA ROSENZWEIG
Goddard Institute for Space Studies and Columbia University, USA

Since the late 1950s, global agricultural output has increased at rates and to levels that are unprecedented in human history. Much of the productivity increase is attributed to the breeding of high-yielding crop varieties, intensive use of inorganic fertilizers and pesticides, expansion of irrigation, and capital-intensive farm management.

In the 1970s, the euphoria surrounding the 'Green Revolution' was questioned in the wake of the energy crisis and growing awareness of long-term environmental consequences. Concern over soil erosion, groundwater contamination, soil compaction and decline of natural soil fertility, and destruction of traditional social systems, led to a reappraisal of what were then considered to be the most advanced agricultural production techniques. Since then, agricultural research has expanded its scope to include sustainable and resource-efficient cropping systems and farm management practices.

Since the beginning of the 1980s yet another threat to agriculture has attracted much attention. Many climatologists predict significant global warming in the coming decades due to increasing atmospheric carbon dioxide and other trace gases. As a consequence, major changes in hydrological regimes have also been forecast to occur. The magnitude and geographical distribution of such climate-induced changes may affect our ability to expand food production as required to feed a population of more than 10 000 million people projected for the middle of the next century. Climate change could have far-reaching effects on patterns of trade among nations, development, and food security.

Beyond what is known about greenhouse gases and the climate system, however, lie great uncertainties: How much warming will occur, at what rate, and according to what geographical and seasonal pattern? What secondary processes will the warming trend induce, and what might be the physical and biological impacts of such processes? Will some areas benefit while other areas suffer, and who might the winners and losers be? And, if such damages are unavoidable, what can be done to adapt or modify our systems so as to minimize or overcome them? These are important and complex questions, and we have only begun to understand them and to develop methods for their analysis.

Recent research has focused on regional and national assessments of the potential effects of climate change on agriculture. For the most part this work has treated each region or nation in isolation, without relation to changes in production elsewhere and without paying attention to climate change effects on the world market and their feedbacks. Assessments of potential impacts have been achieved in national studies completed in the United States (Adams *et al.*, 1990, 1994; Smith and Tirpak, 1989), Australia (Pearman, 1988), and the United Kingdom (UK Department of the Environment, 1991). Regional studies have been conducted in high-latitude and semi-arid agricultural areas (Parry *et al.*, 1988). These regional and national studies have been summarized in the IPCC Working Group II Reports (IPCC, 1990b, 1996).

In 1989 the US Environmental Protection Agency (EPA), with additional support provided by the US Agency for International Development (USAID), commissioned a three-year study on the effects of climate change on world food supply. The present study is an initial attempt to arrive at an integrated global assessment of the potential effects of climate change on agriculture and the world food system. The collaborative project was jointly managed by the Goddard Institute of Space Studies (GISS) and the Environmental Change Unit (ECU) in collaboration with the International Institute for Applied Systems Analysis (IIASA) and involved about 50 scientists worldwide.

The aim of this paper is to provide a brief description of the study and an analysis of results obtained from a set of simulation runs carried out with the Basic Linked System of National Agricultural Models (Fischer *et al.*, 1988) constructed by the Food and Agriculture Program (FAP) at IIASA. IIASA's research provided a framework for analysing the world food system, viewing national agricultural systems as embedded in national economies, which in turn interact with each other at the international level.

STUDY METHODS

The implementation of the study involved a four-step procedure:
1. Selection of climate change scenarios.
2. Estimation of site-specific potential changes in crop yields.
3. Aggregation of crop modelling results to estimates of potential national/regional productivity changes.
4. Dynamic simulation of climate change yield impacts on the world food system.

Details of the methods are described in Rosenzweig *et al.* (1995), Rosenzweig and Parry (1994) and Rosenzweig and Iglesias (1994).

CLIMATE CHANGE SCENARIOS

Scenarios of climate change were developed in order to estimate their effects on crop yields and food trade. A climate change scenario is defined as a physically consistent set of changes in meteorological variables, based on generally accepted projections of CO_2 (and other trace gases) levels. The range of scenarios analysed is intended to capture the range of possible effects and to set limits on the associated uncertainty. One set of scenarios for this study was created by changing observed data on current climate (1951-1980) according to the results of doubled CO_2 simulations of three General Circulation Models (GCMs) (Table 9.1).

The temperature changes of these GCM scenarios (4.0-5.2°C) are at or near the upper end of the range (1.5-4.5°C) projected for doubled CO_2 warming by the IPCC (IPCC, 1990a, 1992). The GISS and GFDL scenarios, however, are near the mean temperature change (3.8°C) of recent doubled CO_2 experiments documented for atmospheric GCMs with a seasonal cycle and a mixed-layer ocean (IPCC, 1992).

GCMs currently provide the most advanced means of predicting the potential future climatic consequences of increasing radiatively active trace gases. They have been shown to simulate current temperatures reasonably well, but do not reproduce current precipitation accurately; and their ability to reproduce current climate varies considerably from region to region (IPCC, 1990a). Of special importance for agricultural climate change impacts, there is a notable lack of consensus among GCMs in prediction of regional soil moisture changes (Kellogg and Zhao, 1988). Furthermore, GCMs have so far not been able to produce reliable projections of changes in climate variability, such as alterations in the frequencies of drought and storms, even though these could significantly affect crop yields.

For the crop modelling part of this study, climate changes from doubled CO_2 GCM simulations are utilized with an associated level of 555 ppm CO_2; the assumed timing of the simulations with the world food model is that these conditions will occur in year 2060. Rates of future emissions of trace gases and the point in time when their effects will be fully realized are not certain. Because other greenhouse gases besides

Table 9.1 GCM doubled CO_2 climate change scenarios

GCM	Year[1]	Resolution (lat × long)	CO_2 (ppm)	Change in average global temp. (°C)	precip. (%)
GISS	1982	7.83°×10°	630	4.2	11
GFDL	1988	4.4°×7.5°	600	4.0	8
UKMO	1986	5.0°×7.5°	640	5.2	15

[1] When calculated.
Note: GFDL: Geophysical Fluid Dynamics Laboratory; UKMO: United Kingdom Meteorological Office.

CO_2, such as methane (CH_4), nitrous oxide (N_2O), and the chlorofluorocarbons (CFCs), are also increasing, an 'effective CO_2 doubling' has been defined as the combined radiative forcing of all greenhouse gases having the same forcing as doubled CO_2 (usually defined as ~600 ppm). Level of CO_2 is important when estimating potential impacts on crops, because crop growth and water use have been shown to benefit from increased levels of CO_2 (Cure and Acock, 1986). A CO_2 level of 555 ppm was associated with the effective doubled CO_2 climate projections for use in the crop modelling simulations. This was based on the GISS GCM transient trace gas scenario A described in Hansen *et al.* (1988), in which the simulated climate had warmed to the effective doubled CO_2 level of about 4°C by 2060. This level assumes that non-CO_2 trace gases contribute ~15% of the change in radiative forcing from 300 to ~600 ppm.

Two other climate change scenarios were tested. The GISS transient scenario consists of a separate GCM run with gradually increasing atmospheric CO_2 levels. The CO_2 concentrations in the GISS transient scenario A were assumed to be 405 ppm, 460 ppm and 530 ppm, respectively, in the decades of the 2010s, 2030s and 2050s. Crop modelling experiments were conducted separately for these three time periods.

Another scenario, termed GISS-A, utilized the climate changes projected for the 2030s from the GISS transient run with 555 ppm CO_2 for the crop model simulations. This scenario was used to test the consequences of a lower sensitivity of the climate system to increasing atmospheric greenhouse gas concentrations. The GISS-A scenario projects a global temperature rise of 2.4°C and a 5% increase in precipitation.

ESTIMATION OF SITE-SPECIFIC POTENTIAL CHANGES IN CROP YIELDS

Crop models and a decision support system developed by the International Benchmark Sites Network for Agrotechnology Transfer (IBSNAT, 1989) were used to estimate how climate change and increasing levels of carbon dioxide may alter yields of major crops. The simulations represented both major production areas and vulnerable regions at low, mid and high latitudes. The IBSNAT models simulate crop growth and yield formation as influenced by genetics, climate, soils and management practices. Models used were for wheat, maize, rice and soybean.

The crop models account for the beneficial physiological effects of increased CO_2 concentrations on crop growth and water use (Peart *et al.*, 1989). Most plants growing in experimental environments with increased levels of atmospheric CO_2 exhibit increased rates of net photosynthesis and reduced stomatal openings, thereby reducing transpiration per unit leaf area while enhancing photosynthesis. Crop modelling simulation experiments were conducted for baseline climate (1951-1980) and GCM doubled CO_2 climate change scenarios with and without the physiological effects of CO_2.

The study also tested the efficacy of farm-level adaptations to climate change, including change in planting date, change of cultivar, irrigation, fertilizer, and change of crop. These measures were then grouped into two levels of adaptation: Level 1 implies little change to existing agricultural systems reflecting relatively easy and low-cost farmer response to a changing climate. Level 2 implies more substantial changes to agricultural systems possibly requiring resources beyond the farmer's means. It must be noted that costs of adaptation and future water availability for irrigation under the climate change scenarios were not considered in the study.

AGGREGATION OF CROP MODELLING RESULTS

Data on crop yield changes expected for different scenarios of climate change had to be compiled for all crop commodities and geographical groupings represented in the IIASA/FAP world food model, the Basic Linked System (BLS).

Crop model results for wheat, rice, maize and soybean from the selected sites were aggregated by weighting regional yield changes, based on current production, to estimate changes in national yields. The regional yield estimates represent the current mix of rainfed and irrigated production, the current crop varieties, nitrogen management and soils. Production data were gathered by scientists participating in the study and from the FAO, the USDA Crop Production Statistical Division, and the USDA International Service.

Changes in national yields of other crops and commodity groups and of regions not simulated with crop models were estimated based on similarities to modelled crops and growing conditions, and previous published and unpublished climate change impact studies. Estimates were made of yield changes for the three GCM scenarios with and without direct effects of CO_2. The yield changes with the direct effects of CO_2 were based on the mean responses to CO_2 for the different crops in the crop model simulations.

In total, 12 climate change yield impact scenarios were developed: for each of three climate models (GISS, GFDL, UKMO) four scenarios are specified (GCM without direct effects of CO_2 on yields; with direct effects of CO_2 on yields; with direct effects of CO_2 and Adaptation Level 1; with direct effects of CO_2 and Adaptation Level 2).

SIMULATION OF CLIMATE CHANGE YIELD IMPACTS ON THE WORLD FOOD SYSTEM

The agricultural production components of the national models in the BLS were modified to incorporate the projected changes in yields and crop productivity. Impacts were assessed for the period 1990 to 2060, with population growth and technology trends projected to that year. The simulation results obtained under each of the climate impact scenarios are compared to a 'neutral' point of departure, a BLS reference scenario which assumes that policies, especially those affecting economic and technological development remain more or less unchanged. The reference scenario projects a possible future based on an internally consistent set of assumptions. The difference between the reference scenario and the climate change simulations is the climate-induced *dynamic* effect. Note that the results of this study are considered to be long-term projections, not forecasts. The considerable length of the projection period makes it impossible to avoid judgement concerning some of the assumptions in the system.

CROP MODELS AND YIELD SIMULATIONS

The IBSNAT crop models were used to estimate how climate change and increasing levels of carbon dioxide may alter yields of world crops at 112 sites in 18 countries. (Figure 9.1). The crop models used were CERES-Wheat (Ritchie and Otter, 1985; Godwin et al., 1989), CERES-Maize (Jones and Kiniry, 1986; Ritchie et al., 1989), CERES-Rice (Godwin et al., 1993) and SOYGRO (Jones et al., 1989).

The IBSNAT models are comprised of parameterizations of important physiological processes responsible for plant growth and development, evapotranspiration, and partitioning of photosynthate to produce economic yield. The simplified functions enable prediction of growth of crops as influenced by the major factors that affect yields, i.e., genetics, climate (daily solar radiation, maximum and minimum temperatures, and precipitation), soils, and management practices. The models include a soil moisture balance submodel so that they can be used to predict both rainfed and irrigated crop yields. The cereal models simulate the effects of nitrogen fertilizer on crop growth, and these were studied in several countries in the context of climatic change. For the most part, however, the results of this study assume optimum nutrient levels.

The IBSNAT models were selected for use in this study because they have been validated over a wide range of environments (e.g., Otter-Nacke et al., 1986) and are

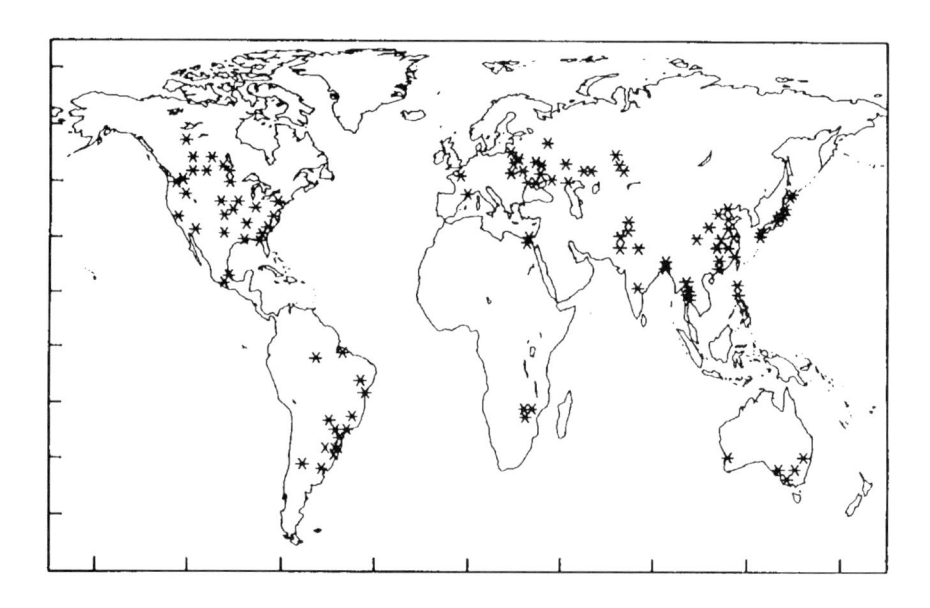

Figure 9.1. Crop modelling sites

not specific to any particular location or soil type. The validation of the crop models over different environments also improves their ability to estimate effects of changes in climate. Furthermore, because management practices, such as the choice of varieties, planting date, fertilizer application and irrigation, may be varied in the models, they permit experiments that simulate adaptation by farmers to climate change.

PHYSIOLOGICAL EFFECTS OF CO_2

Ratios were calculated between measured daily photosynthesis and evapotranspiration rates for a canopy exposed to high CO_2 values, based on published results (Allen et al., 1987; Cure and Acock, 1986; Kimball, 1983), and the ratios were applied to the appropriate variable in the crop models on a daily basis (see Peart et al., 1989 for a detailed description of methods). The photosynthesis ratios (555 ppm CO_2/330 ppm CO_2) for soybean, wheat, rice and maize were 1.21, 1.17, 1.17 and 1.06, respectively. Changes in stomatal resistance were set at 49.7/34.4 s/m for C_3 crops and at 87.4/55.8 s/m for C_4 crops, based on experimental results by Rogers et al. (1983). As simulated in this study, the direct effects of CO_2 may bias yield changes in a positive direction, since there is uncertainty regarding whether experimental results will be observed in the open field under conditions likely to be operative when farmers are managing crops. Plants growing in experimental settings are often subject to fewer environmental stresses and less competition from weeds and pests than are likely to be encountered in farmers' fields. Recent field free-air release studies have found overall positive CO_2 effects on cotton under current climate conditions (Hendry, 1993).

CROP YIELD SIMULATIONS

Crop modelling simulation experiments were performed for baseline climate (1951-1980), and GCM doubled CO_2 climate change scenarios with and without the physiological effects of CO_2. This involved the following tasks:
- For the countries studied, geographical boundaries were defined for the major crop production regions; agricultural systems (e.g., rainfed and/or irrigated production, number of crops grown per year) were described, and data on regional and national rainfed and irrigated production of major crops were gathered.
- Observed climate data for representative sites within these regions were obtained for the baseline period (1951-1980), or for as many years of daily data as were available; the soil, crop variety, and management inputs necessary to run the crop models at the selected sites were specified.
- The crop models were validated with experimental data from field trials, to the extent possible.
- The crop models were run with baseline data, and GCM climate change scenarios, with and without direct effects of CO_2 on crop growth. Rainfed and/or irrigated simulations were carried out as appropriate to current growing practices.
- Alterations in farm-level agricultural practices that would lessen any adverse

consequences of climate change were identified and evaluated, by simulating irrigated production and other adaptation responses, e.g., shifts in planting date and substitution of crop varieties.

FARM-LEVEL ADAPTATIONS

In each country, the agricultural scientists used the crop models to test the possible responses to the worst climate change scenario (this was usually, but not always, the UKMO scenario). These adaptations included change in planting date, change of cultivar, irrigation, fertilizer and change of crop. Irrigation simulations in the crop models assumed automatic irrigation to field capacity when plant available water dropped to 50%, and an irrigation efficiency of 100%. Not all adaptation possibilities were simulated at every site and country: choice of adaptations to be tested was made by the participating scientists, based on their knowledge of current agricultural systems (Table 9.2).

The adaptation simulations were not comprehensive because not all possible combinations of farmer responses were tested at every site. Spatial analyses of crop, climatic and soil resources are needed to test fully the possibilities for crop substitution. Neither the availability of water supplies for irrigation nor the costs of adaptation were considered in this study; these are both critical needs for further research.

Table 9.2 Adaptations tested in crop modelling study

Country	Crop tested	Change of planting date	Change of cultivar/crop	Additional irrigation	Additional N fertilizer
Argentina	m	x	xc,xz	x	
Australia	r,w	xx	x	xx	
Bangladesh	r		x		
Brazil	w,m,s	xx	x,xc	x,xx,xxx	x
Canada	w	x		x,xx	
China	r	x	xp,xz		
Egypt	m,w	x	x	x	
France	m,w	x,xz	x	x	
India	w			x	
Japan	r,w,m	xx		xx	
Mexico	m	x	xc	xxx	x
Pakistan	w	x		x	
Philippines	r	xp	xp		
Thailand	r		x		
Uruguay	b	x	x.	x	x
USA	w,m,s	x	x	x	
Former USSR	w	x,xz	x		
Zimbabwe	m	xp		x,xp	

w=wheat, m=maize, r=rice, s=soybean, b=barley.
xc=hypothetical new cultivar; xp=new cultivar and change in planting date; xx=irrigation and change in planting date; x=simple change
xxx=irrigation and increase nitrogen fertilizer; xz=suggested shift in crop production zone.

EFFECTS ON CROP YIELDS

Depending on present conditions, global warming and CO_2 enrichment can have positive or negative impacts. Simulated yield increases in the mid and high latitudes are caused primarily by:
- *Positive physiological effects of CO_2.* At sites with cooler initial temperature regimes, increased photosynthesis more than compensated for the shortening of the growing period caused by warming.
- *Lengthened growing season and amelioration of cold temperature effects on growth.* At some sites near the high-latitude boundaries of current agricultural production, increased temperatures extended the frost-free growing season and provided regimes more conducive to greater crop productivity.

The primary causes of decreases in simulated yields are:
- *Shortening of the growing period.* Higher temperatures during the growing season speed annual crops through their development (especially the grain-filling stage), causing less grain to be produced. This occurred at all sites except those with the coolest growing-season temperatures in Canada and the former USSR.
- *Decrease in water availability.* This is due to a combination of increases in evapotranspiration rates in the warmer climate, enhanced losses of soil moisture and, in some cases, a projected decrease in precipitation in the climate change scenarios.
- *Poor vernalization.* Vernalization is the requirement of some temperate cereal crops, e.g., winter wheat, for a period of low winter temperatures to initiate or accelerate the flowering process. Low vernalization results in low flower bud initiation and ultimately reduced yields. Decreases in winter wheat yields at some sites in Canada and the former USSR were caused by lack of vernalization.

CROP YIELDS WITHOUT ADAPTATION

Table 9.3 shows modelled wheat yield changes for the GCM doubled CO_2 climate change scenarios (the yield changes include results from both rainfed and irrigated simulations, weighted by current percentage of the respective practice). Climate changes without the direct physiological effects of CO_2 cause decreases in simulated wheat yields in all cases, while the direct effects of CO_2 mitigate the negative effects primarily in mid and high latitudes.

The magnitudes of the estimated yield changes vary by crop. Global wheat yield changes weighted by national production are positive with the direct CO_2 effects, while maize yield is most negatively affected, reflecting its greater production in low-latitude areas where simulated yield decreases are greater. Maize production declines most with direct CO_2 effects, probably due to its lower response to the physiological effects of CO_2 on crop growth. Simulated soybean yields are most reduced without the direct effects of CO_2, but are least affected in the less severe GISS and GFDL climate change scenarios when direct CO_2 effects are simulated. Soybean

Table 9.3. Current production and change in simulated wheat yield under doubled CO_2 climate change scenarios, with and without the direct effects of CO_2[1]

Country	Current production				Change in simulated yields (%)					
	Yield	Area	Prod.	%	climate change alone			with physiological effects		
	(tonnes/ha)	('000 ha)	('000 tonnes)		GISS	GFDL	UKMO	GISS	GFDL	UKMO
Australia	1.38	11 546	15574	3.2	−18	−16	−14	8	11	9
Brazil	1.13	2788	3 625	0.8	−51	−38	−53	−33	−17	−31
Canada	1.88	11 365	21412	4.4	−12	−10	−38	27	27	−7
China	2.53	29 092	73 527	15.3	−5	−12	−17	16	8	−0
Egypt	3.79	572	2 166	0.4	−36	−28	−54	−31	−26	−51
France	5.93	4 636	27 485	5.7	−12	−28	−23	4	−15	−9
India	1.74	22 876	39 703	8.2	−32	−38	−56	3	−9	−33
Japan	3.25	237	722	0.2	−18	−21	−40	−1	−5	−27
Pakistan	1.73	7 478	12918	2.7	−57	−29	−73	−19	31	−55
Uruguay (barley)	2.15	91	195	0.0	−41	−48	−50	−23	−31	−35
USA	2.72	26 595	64 390	13.4	−21	−23	−33	−2	−2	−14
Former USSR										
winter	2.46	18 988	46 959	9.7	−3	−17	−22	29	9	0
spring	1.14	36 647	41 959	8.7	−12	−25	−48	21	3	−25
World[2]	2.09	231	482	72.7	−16	−22	−33	11	4	−13

[1] Results for each country represent the site results weighted according to regional production. The world estimates represent the country results weighted by national production.
[2] World area and production × 1 000 000.

responds positively to increased CO_2, but is the crop most affected by the high temperatures of the UKMO scenario.

The differences among countries in simulated crop yield responses to climate change without the direct effects of CO_2 are primarily related to differences in current growing conditions. Higher temperatures tend to shorten the growing period at all locations tested. At low latitudes, however, crops are currently grown at higher temperatures, produce lower yields, and are nearer the limits of temperature tolerances for heat and water stress. Warming at low latitudes thus results in accelerated growing periods for crops, more severe heat and water stress, and greater yield decreases than at higher latitudes. In many mid- and high-latitude areas where current temperature regimes are cooler, increased temperatures, while still shortening grain-filling periods, thus exerting a negative influence on yields, do not significantly increase stress levels. At some sites near the high-latitude boundaries of current agricultural production, increased temperatures can benefit crops otherwise limited by cold temperatures and short growing seasons, although the extent of soil suitable for expanded agricultural production in these regions was not studied explicitly. Potential for expansion of cultivated land is embedded in the BLS world food trade model and is reflected in shifts in production calculated by that model.

The GISS and GFDL climate change scenarios produced yield changes ranging from +30 to −30%. Effects under the GISS scenario are, in general, more adverse than under the GFDL scenario to crop yields in parts of Asia and South America, while effects under the GFDL scenario result in more negative yields in the United States and Africa and less positive results in former USSR. The UKMO climate change scenario, which has the greatest warming (5.2°C global surface air temperature increase), causes average national crop yields to decline almost everywhere.

CROP YIELDS WITH ADAPTATION

The adaptation studies conducted by the scientists participating in the project suggest that ease of adaptation to climate change is likely to vary with crop, site and adaptation technique. For example, at present, many Mexican producers can only afford to use small doses of nitrogen fertilizer at planting; if more fertilizer becomes available to more farmers some of the yield reductions under the climate change scenarios might be offset. However, given the current economic and environmental constraints in countries such as Mexico, a future with unlimited water and nutrients is unlikely (Liverman *et al.*, 1994). In contrast, switching from spring to winter wheat at the modelling sites in the former USSR produces a favourable response (Menzhulin *et al.*, 1994), suggesting that agricultural productivity may be enhanced there with the relatively easy shift to winter wheat varieties.

Adaptation Level 1, simulating minor changes to existing agricultural systems, compensated for the climate change scenarios incompletely, particularly in the developing countries. Adaptation Level 2, implying major changes to current agricultural systems, compensated almost fully for the negative climate change impacts in the GISS and the GFDL scenarios. With the high level of global warming as projected by the UKMO climate change scenario, neither Level 1 nor Level 2 adaptation fully overcame the negative climate change effects on crop yields in most countries, even when the direct CO_2 effects were taken into account.

MODELLING THE WORLD FOOD SYSTEM

The world food system consists of many actors, some powerful and others dependent. Producers and consumers interact through national and international markets. While there is a trend towards internationalization in the world food system, only ~15% of world cereal production currently crosses national borders (Fischer *et al.*, 1990). National governments shape the system by imposing regulations and by investments in agricultural research, improvements in infrastructure, and education. Although the system does not guarantee stability nor equity, food and fibre have been produced over time in increasingly efficient ways. These efficiencies have generated long-term real declines in prices of major food staples.

THE BASIC LINKED SYSTEM OF NATIONAL AGRICULTURAL MODELS

The Basic Linked System of National Agricultural Policy Models (BLS) consists of some 35 national and/or regional models: 18 national models, 2 models for regions with close economic cooperation (EU and Eastern Europe + former USSR[1]), 14 aggregate models of country groupings, and a small component that accounts for statistical discrepancies and imbalances during the historical period. The individual models are linked together by means of a world market module. A detailed description of the entire system is provided in Fischer *et al.* (1988). Earlier results obtained with the system are discussed in Parikh *et al.* (1988) and in Fischer *et al.*, (1990,1994).

The BLS is a general equilibrium model system. This necessitates that all economic activities are represented in the model. Financial flows as well as commodity flows within a country and at the international level are consistent in the sense that they balance.

The country models are linked through trade, world market prices and financial flows. The system is solved in annual increments, simultaneously for all countries. It is assumed that supply does not adjust instantaneously to new economic conditions. Only supply that will be marketed in the following year is affected by possible changes in the economic environment. A first round of exports from all the countries is calculated for an initial set of world prices, and international market clearance is checked for each commodity. World prices are then revised, using an optimizing algorithm, and again transmitted to the national models. Next, these generate new domestic equilibria and adjust net exports. This process is repeated until the world markets are

---

[1] The political changes as well as changes in national boundaries of the very recent past are not captured in the BLS, although the model formulation has been adjusted, away from centrally planned economies to more market-oriented behaviour.

cleared in all commodities. At each stage of the iteration the domestic markets are in equilibrium. Since these steps are taken on a year-by-year basis, a recursive dynamic simulation results.

An upper bound on land available for cropping and for use as pasture is determined by the availability of land resources as well as economic conditions; e.g., by the economic returns to land. The physical resource limits in developing countries were derived from a FAO assessment of potential arable land (FAO/UNFPA/IIASA, 1983; FAO, 1988). The responsiveness of how much land can be cultivated due to changing economic conditions is rather low since time and investment are needed to bring new land into cultivation.

Technological development is assumed to be largely determined by exogenous factors. Technical progress is included in the models as biological technical progress in the yield functions of both crops and livestock. Rates of technical progress were estimated from historical data and, in general, show a decline over time. Mechanical technical progress is part of the function determining the level of harvested crop area and livestock husbandry. Induced (endogenous) technical progress is not considered for any of these cases or for non-agricultural production.

Information generated by simulating with the BLS consists of a number of variables. At the world-market level these include prices, net exports, global production and consumption. At the country level, the information generated includes: producer and retail prices, level of production, use of primary production factors (land, labour and capital), intermediate input use (feed, fertilizer and other chemicals), level of human consumption, stocks and net trade, gross domestic product and investment by sector, population number and labour force, welfare measures such as equivalent income, and the level of policy measures as determined by the government (e.g., taxes, tariffs).

BLS REFERENCE PROJECTIONS WITHOUT CLIMATE CHANGE

A set of reference scenarios was designed for studying the relative effects of climate change in relation to technology, population and economic growth. The standard reference scenario we describe, termed scenario REF-M, assumes 'business as usual' in the sense that no radical shifts in technological and political trends are included. Protectionist policy measures in agriculture, however, are assumed to be lowered to half the observed historical levels. The transition to such reduced protection of the agriculture sectors is implemented between the beginning of the simulation period and year 2020. Thereafter the respective policy settings are kept constant.

Another reference projection, scenario REF-H, assumes faster economic growth than the standard reference scenario. In the BLS, the dynamics of economic growth can be influenced by adjusting the rate of investment to the amount of technical progress. In scenario REF-H, growth of Gross Domestic Product (GDP) is especially higher for Asian countries compared to scenario REF-M. Similarly, a lower growth simulation run, scenario REF-L, has been created by curtailing investments in favour of consumption.

Additional reference projections with the BLS were based on the standard reference scenario REF-M. Two projections deal with the sensitivity of the world food system with respect to land development and availability of agricultural inputs. In scenario REF-MA, expansion of arable land after year 2000 is constrained to half the expansion in scenario REF-M. Another simulation run, scenario REF-MF, limited the use of fertilizers by implementing a tax on fertilizers of 50% in developed countries and of 33% in developing countries. In addition, a ceiling on fertilizer use per hectare was enforced. Both these scenarios were simulated to test the impact of possible agricultural policies geared towards reducing greenhouse gas emissions from agriculture (specifically, CO_2 release from deforestation and N_2O emissions from nitrogen fertilizers).

Finally, reference projection REF-MP uses the economic assumptions of the standard reference run REF-M but alters the demographic projections from medium population growth to low population growth.

All the simulations are carried out on a yearly basis from 1980 to 2060.

ASSUMPTIONS ABOUT POPULATION

Population growth rates were obtained from the UN World Population Prospects, Medium Variant (UN, 1989) in all simulations except for scenario REF-MP where the Low Variant was used. Since the UN projected national population levels only up to the year 2025, the remainder of the projection period was covered by growth rates compiled from long-term population projections of the World Bank (Table 9.4) (World Bank, 1990). Labour participation rates are taken from projections of the International Labour Organization. The allocation of total labour force between agriculture and non-agricultural sectors responds to relative prices and incomes.

During the last two decades, population at the global level has been growing at an average rate of about 1.8% per annum, from 3 600 million in 1970 to about 5 300 million people in 1990. In the BLS standard reference projection, the average annual population growth rate is projected to decline gradually from 1.7% per annum during 1980 to 2000 to around 0.5% per annum for the period 2040 to 2060. This would bring global population numbers to 6 100 million in year 2000 and about 10 300 million in year 2060. Most of the increase in population numbers occurs in developing countries, increasing their share in world population from around 73% in 1980 to 78% in year 2000 and some 86% in year 2060. In 2060, almost 9 000 million people are projected to live in developing countries, of which approximately 5 000 million are in Asia, and 2 200 million in Africa. With lower population growth in scenario REF-MP, population reaches 7 200 million in year 2020, and some 8 600 million in year 2060, about 17% less than in scenario REF-M.

ECONOMIC GROWTH

Growth rates in most of the national models of the BLS are determined based on three elements: (a) capital accumulation through investment and depreciation, relat-

Table 9.4. Estimated population and average annual growth, year 1980-2060

Region	1980 mill.	2000 mill.	2020 mill.	2060 mill.	1980-2000 % p.a.	2000-2020 % p.a.	2020-2040 % p.a.	2040-2060 % p.a.
MEDIUM VARIANT								
WORLD	4 378	6 125	7 883	10 315	1.7	1.3	0.8	0.5
Developed	1 186	1 340	1 445	1 470	0.6	0.4	0.1	−0.0
North America	810	915	980	970	0.6	0.4	0.1	−0.1
Western Europe and other developed market economies	428	472	501	505	0.5	0.3	0.1	−0.1
Eastern Europe + former USSR	374	424	465	500	0.6	0.5	0.3	0.1
Pacific OECD countries	134	151	156	145	0.6	0.2	−0.1	−0.2
Developing	3 193	4 786	6 437	8 843	2.1	1.5	1.0	0.6
Africa	412	757	1 274	2 240	3.1	2.6	1.7	1.1
Latin America	351	522	695	878	2.0	1.4	0.9	0.3
West Asia	188	325	485	777	2.8	2.0	1.4	1.0
South Asia	898	1 396	1 889	2541	2.2	1.5	0.9	0.6
Centrally Planned Asia	1 086	1 419	1 637	1 843	1.4	0.7	0.4	0.2
Pacific Asian countries	258	366	458	564	1.8	1.1	0.7	0.4
LOW VARIANT								
WORLD	4 380	5 968	7 215	8 565	1.6	1.0	0.5	0.3
Developed	1 185	1 319	1 364	1 348	0.5	0.2	−0.0	−0.0
Developing	3 195	4 649	5 851	7 217	1.9	1.2	0.6	0.4

ed to a savings function that depends on lagged GDP levels as well as balance of trade and financial aid flows, (b) dynamics of the labour force as a result of demographic changes, and (c) technical progress. Table 9.5 presents some indicators of economic development derived from the simulation results of the reference projections.

All regions, developed and developing alike, show a declining rate of GDP growth over time, which is consistent with historical developments. The declining rates of population growth (and the related decline in the growth of the labour force) as well as a general slowdown in productivity increases contribute to this development.

In the standard reference scenario, GDP at the global level increases at an average 2.4% annually during 1980 to 2020. Economic growth declines to about 1.5% per annum during 2020 to 2040. Overall, global GDP increases 4.4 times during the simulation period compared to a 2.4-fold increase in population numbers. This results in an average annual increase in GDP per caput of 1.4 and 1.1% in developed

Table 9.5. Economic growth indicators under different reference projections (average annual percentage change)

	REF-L*			REF-M*			REF-MP*			REF-H*		
	World	Deve-loped	Deve-loping	World	Deve-loped	Deve-loping	World	Deve-loped	Deve-loping	World	Deve-loped	Deve-loping
GDP[1]												
1980-2020	2.2	2.0	3.1	2.4	2.2	3.3	2.3	2.1	3.1	2.7	2.2	4.0
1980-2060	1.7	1.5	2.3	1.8	1.6	2.4	1.7	1.6	2.3	2.2	1.7	3.2
GDP/CAP[2]												
1980-2020	0.8	1.5	1.3	1.0	1.7	1.5	1.1	1.8	1.6	1.3	1.8	2.2
1980-2060	0.7	1.3	1.0	0.8	1.4	1.1	0.9	1.4	1.2	1.2	1.5	1.9
AGRICULTURE[3]												
1980-2020	1.5	0.7	2.0	1.5	0.7	2.0	1.4	0.7	1.9	1.6	0.7	2.1
1980-2060	1.2	0.5	1.5	1.2	0.5	1.6	1.1	0.4	1.4	1.2	0.5	1.6
FOOD PRODUCTION												
1980-2020	1.5	1.0	1.8	1.5	1.0	1.9	1.4	0.9	1.7	1.6	1.0	2.0
1980-2060	1.1	0.7	1.4	1.1	0.7	1.4	1.0	0.6	1.2	1.2	0.7	1.5

[1] Gross Domestic Product.
[2] Gross Domestic Product per caput.
[3] Gross Domestic Product of agriculture.

* The terms REF-L, REF-M, REF-MP, REF-H are described in the section on *BLS reference projections without climate change*.

and developing countries, respectively, over the 80-year period from 1980 to 2060. It should be noted, however, that increases in per caput indicators are higher at regional and national levels compared to global figures owing to an aggregation effect induced by the demographic development, giving increasingly higher weights to poorer developing countries.

With faster economic growth, especially in developing countries such as in scenario REF-H, world GDP grows almost 5.6 times between year 1980 and 2060, compared to only 3.9 times as in the lower growth projection REF-L. Note that even with relatively minor variations on the basic economic assumptions, the resulting output in year 2060 varies by some 40%. Similarly, food production (measured in terms of net food energy, i.e., production less feed, seed and waste) as well as overall agriculture growth is projected to exceed population growth throughout the simulation period.

PRODUCTION, DEMAND AND TRADE

The standard reference projection, scenario REF-M, presents the perspective of a world in which the effective demand for food grows substantially owing to higher incomes and larger populations. Technological progress and economic development assumed in the reference scenario allow this increase in demand to be met at somewhat decreasing world market prices for agricultural products, consistent with historical trends. Table 9.6 shows global production of agricultural commodities in the standard reference scenario REF-M and in the higher income scenario REF-H.

Global trade in the reference scenario increases somewhat faster than global agricultural production. For cereals, the share of net exports in global production is estimated to increase from 13% in 1980 to 15% in 2060, with wheat and coarse grains showing an almost three-fold and rice a four-fold increase in trade levels. In general, the share of global trade in global production of commodity aggregates increases gradually over time indicating a growing specialization in production. Increasing demand in developing countries, due to rising incomes and growing populations, leads to a deterioration in the level of agricultural self-sufficiency for this group of countries, which changes from a net surplus of about 3% in 1979/81 into a 1% deficit by the year 2060 caused by increasing deficits in cereals, meat and milk.

RISK OF HUNGER

Finally, we ask where all this leaves the hungry. To evaluate the impact of alternative scenarios on the poor in different countries, it was necessary to generate a consistent hunger indicator in the BLS. Country-wise estimates of the number of undernourished persons have been made by FAO (1984, 1987). To recover the FAO method in a reduced form, suitable for use in the simulation models, a cross-country regression has been estimated explaining the share of people at risk of hunger by a measure of food energy availability relative to nutritional requirements. Food availability, in turn, depends on income and price levels.

Table 9.6. Global production in two BLS reference projections

Commodity	1980	Production in year Scenario REF-M[1] 2000	2020	2040	2060	Production in year Scenario REF-H[1] 2000	2020	2040	2060	Unit of measurement
Wheat	441	603	742	861	958	658	811	911	1056	million tonnes
Rice[2]	249	367	480	586	659	415	545	661	749	million tonnes milled equivalent
Coarse grains	741	1022	1289	1506	1669	1065	1349	1587	1772	million tonnes
Bovine+ovine meat	65	83	105	123	136	84	107	125	139	million tonnest carcass weight
Dairy	470	613	750	877	997	616	758	893	1021	million tonnes whole milk equivalent
Other meat	17	25	33	41	48	25	34	42	49	million tonnes protein equivalent
Protein feed	36	52	64	76	85	52	65	77	87	million tonnes protein equivalent
Other food	225	326	433	538	629	326	436	545	640	million US dollars 1970
Non-food	26	34	41	47	52	34	41	48	53	million US dollars 1970
Agriculture	370	522	676	821	942	533	696	848	977	million US dollars 1970

[1] The terms REF-M and REF-H are described in the section on *BLS reference projections without climate change*.
[2] Production is in milled rice equivalent; conversion factor from paddy is 0.667.

Table 9.7. People at risk of hunger[1] (millions)

	REF-M			REF-MP	REF-MA	REF-MF	REF-L	REF-H
	Year 1980	Year 2020	Year 2060	Year 2060	Year 2060	Year 2060	Year 2060	Year 2060
DEVELOPING	501	715	641	395	727	722	757	498
Africa	116	291	415	305	447	446	441	375
Latin America	36	39	24	13	34	33	32	20
West Asia	28	55	72	46	84	85	78	68
South Asia	265	319	128	30	160	157	202	35
Centrally Planned Asia[2]	26	33	0	0	0	0	0	0
Pacific Asian countries	30	6	2	0	3	2	4	0

[1] The term 'at risk of hunger' is used following the FAO methodology of the lack of food or income to achieve a dietary intake above 1.4 times the basal metabolic rate.

[2] Estimate does not include China.

In the standard reference scenario the incidence of hunger decreases markedly from an estimated 23% of population in developing countries (excluding China) in year 1980 to some 9% in year 2060. Yet, despite this remarkable improvement the estimated number of people at risk of hunger increases somewhat, from about 500 million[2] in 1980 to almost 720 million in year 2020, and some 640 million in the year 2060. The projected number of undernourished people in developing countries is shown in Table 9.7. No estimate was attempted for developed regions. Of course, the projected number of people at risk of hunger is sensitive to the scenario assumption, ranging from less than 400 million under the low population run, about 500 million assuming faster economic development, to about 760 million under the lower growth scenario.

While the estimates show an improvement of the food security situation both in relative and absolute terms in Asian countries, the African continent experiences a mixed outcome largely due to the dramatic population increase. The estimated share of people at risk of hunger in total African population declines from 28% in year 1980 to 18% by 2060. However, the number of hungry is projected to increase more than three-fold, from about 120 million people in 1980 to 415 million in 2060, thereby making Africa the region with the largest number of undernourished. With lower population growth, scenario REF-MP, the estimated level of people at risk of hunger is reduced by more than 25%; higher economic development, scenario REF-H, reduces the number of hungry by some 40 million, i.e., approximately 10%.

---

[2] FAO has recently estimated (FAO, 1993) that the number of undernourished in developing countries amounted to 941 million people in 1979/81 and to 843 million people in 1988/90. These estimates are based on a threshold food energy level of 1.54 times Basal Metabolic Rate (BMR). The BLS estimates assume a lower threshold of 1.4 times BMR.

AN ASSESSMENT OF THE WORLD FOOD SYSTEM UNDER ALTERNATIVE SCENARIOS

The evaluation of the potential impact of climate change on production and trade of agricultural commodities, in particular on food staples, is carried out by comparing the results of the climate change scenarios to the reference projections. Various aspects of these reference projections have been presented in the previous section.

The climate change yield impact scenarios devised within the project involve a large number of experiments that relate to:
- different GCM doubled CO_2 simulations;
- different assumptions with regard to impacts of climate change on plant growth and yield levels, such as physiological effects of 555 ppm CO_2, or time pace of impact;
- different assumptions regarding farm-level adaptation to mitigate yield impacts;
- policy changes to affect both the reference run and climate change experiments, e.g., population growth, trade policies, economic growth, and policies to reduce emission of greenhouse gases from agriculture, such as limitation of arable land expansion, rice cultivation, or use of chemical fertilizers.

Well over 70 experiments have been simulated. Results from six sets of simulation experiments are reported here:
1. Simulations *without* physiological effects of 555 ppm CO_2 on crop growth and yield.
2. Simulations *with* physiological effects of 555 ppm CO_2 on crop growth and yield.
3. Simulations with physiological effects of 555 ppm CO_2 on crop growth and yield, and adaptations to mitigate negative yield impacts at the farm-level that would not involve any major changes in agricultural practices (Adaptation Level 1).
4. Simulations with the physiological effects of 555 ppm CO_2 on crop growth and yield, and adaptations at the farm-level that, in addition to the former, would also involve major changes in agricultural practices (Adaptation Level 2).
5. Simulations with the physiological effects of 555 ppm CO_2 on crop growth and yield, but with temperature and precipitation changes projected by the GISS GCM for the 2030s.
6. Simulations with the GISS transient run A and associated gradual increases in CO_2 level.

METHODS

Data on crop yield changes estimated for the different scenarios of climate change were compiled for 34 countries or major regions of the world. Most models included in the BLS distinguish between yield and acreage functions. The yield response functions of major crops use the level of fertilizer application and a term related to technology as explanatory variables. While technical progress is specified relative to a time trend, the level of fertilizer application is derived from optimality conditions, i.e., by equating the marginal value product of fertilizer to its price. Yield variations caused by climate change were introduced into the yield response functions by means of a multiplicative factor applied to the relevant parameters in the mathematical representation. This implies that both average and marginal fertilizer productivity are affected by the imposed yield changes. Alternative schemes for introducing yield changes are conceivable, e.g., with an additive term in the response functions rather than a multiplier. More empirical knowledge with regard to the effect of climate-induced yield changes on marginal productivity is needed to select the most appropriate implementation.

Since additional country and/or crop specific information to suggest explicit modifications of extents suitable for crop cultivation due to impacts of climate change was not available[3], the land allocation is only indirectly influenced through the implied changes in overall performance of the agricultural sector as well as changing comparative advantage of the competing crop production activities. It should be noted, however, that the BLS is equipped to handle explicit area constraints in the resource allocation module of the agricultural production component.

The adjustment processes taking place in the different scenarios are the outcome of the imposed yield changes triggering changes in national production levels and costs, leading to changes of agricultural prices in the international national markets. This in turn affects investment allocation and labour migration between sectors as well as reallocation of resources within agriculture. Time is an important aspect in this assessment: the yield modifications due to climate change are assumed to start occurring in 1990, reaching their full impact in year 2060. This allows the economic actors in the national and international food system to adjust their behaviour over a 70-year period. Yet, the dynamic impacts in some of the scenarios are sizeable.

For the GISS transient scenario A, climate change yield impacts were phased in linearly between the climate 'snapshots', i.e., the yield change multiplier terms incorporated in the yield response functions of the BLS are being built up gradually as a function of time between 1990-2010, 2010-2030 and 2030-2050, so as to fully reach the specified impact levels respectively in years 2010, 2030 and 2050. Beyond year 2050 the yield change multiplier is extrapolated extending the trend of the period 2030-2050.

CLIMATE CHANGE YIELD IMPACTS WITHOUT ECONOMIC ADJUSTMENTS

Before assessing the impacts of introducing a set of climate-change-induced yield modifications through simulation with the BLS, we may ask what distortion such a change in agricultural productivity would imply for the world food system. This measure of distortion is termed the *static* climate change yield impact, as it measures

---

[3] For instance, an assessment of agro-ecological zones (AEZ) under altered climatic conditions as currently being conducted for several countries could provide such information.

the hypothetical effect of yield changes without adjustments of the economic system taking place over time. It refers to a state of the system that is *not* in equilibrium. As such it is only of theoretical interest, but helps in the understanding and quantification of the nature and magnitude of adjustments taking place due to changing economic conditions.

Table 9.8 shows *static* climate change yield impacts estimated for the world and for developed and developing countries. The estimates of *static* climate change yield impacts, without assuming direct physiological effects of 555 ppm CO_2 on crop growth and yields, represent a fairly pessimistic outlook, with decreases in crop productivity on the order of 20 to 30%. Such an assumption is not regarded as very probable and will not be further discussed in the analysis.

When direct physiological effects of CO_2 on yields are included, the magnitude and even the direction of the aggregate *static* impact at the world level varies with GCM climate scenario and with the assumptions regarding farm-level adaptation. In all cases the most negative effects are obtained in scenarios using the UKMO climate change scenario, which has the highest mean global warming, 5.2°C. Results derived from the GISS scenario show only small negative effects or even gains at the global level.

The impacts are, however, quite unevenly distributed. At the aggregate level, developed countries experience an increase in productivity in all but the UKMO scenario. In contrast, developing regions suffer a loss in productivity in all estimates presented here. Table 9.9 shows the continental-level results of scenarios assuming direct physiological effects of increased atmospheric CO_2 concentrations (555 ppm) and, where applicable, some farm-level adaptations (Adaptation Level 1). Under the GISS and GFDL climate projections, crop productivity in developed regions benefits substantially, especially in the former USSR and in the Pacific OECD countries (Australia, Japan and New Zealand). Impacts on developing regions are all negative, except for the group of Centrally Planned Asia that includes China. Under the GISS and UKMO GCM scenarios, countries in Central and South America are most affected. The GFDL GCM estimates are worst for West Asia, South Asia, and Africa. Static impacts derived for the GISS-A climate change scenario are mostly positive (except for the GISS-A scenario estimates which take into account the physiological effects of 555 ppm CO_2 but assume only modest climate sensitivity to increased atmospheric greenhouse gas concentrations) amounting to ~10% globally. Note, however, that the global increase is estimated to be more than twice the gain in developing countries.

CLIMATE CHANGE YIELD IMPACTS WITH ECONOMIC ADJUSTMENTS

The calculations above discuss an effect that would result if climate-induced yield changes were to occur without agronomic and economic adjustment. In the scenario assumptions, however, yield productivity changes are introduced gradually to reach their full impact only after a 70-year period, from 1990 to 2060. In scenarios with shortfalls in food production caused by climate change, market imbalances push international prices upwards and provide incentives for reallocation of capital and

Table 9.8. Simulated static impact using three GCMs (GISS, GFDL, UKMO)

	Cereals production % change			Crop production % change			GDP agriculture % change		
	GISS	GFDL	UKMO	GISS	GFDL	UKMO	GISS	GFDL	UKMO
WORLD TOTAL									
without phys. effect of CO_2	−22.1	−21.8	−22.4	−25.4	−24.4	−25.0	−33.6	−33.0	−33.5
with phys. effect of CO_2	−5.1	+2.8	−0.1	−9.0	+0.3	−2.8	−18.2	−8.9	−12.2
Adaptation Level 1	−1.7	+2.8	+0.9	−5.5	+0.3	−1.7	−12.9	−8.3	−10.1
Adaptation Level 2	+1.4	+4.6	+3.2	−1.1	+2.3	+1.0	−6.1	−3.3	−4.4
DEVELOPED									
without phys. effect of CO_2	−13.9	−6.1	−10.3	−21.3	−15.3	−18.6	−30.4	−27.1	−28.9
with phys. effect of CO_2	+2.6	+18.6	+10.6	−5.1	+9.6	+2.1	−15.8	−3.2	−9.8
Adaptation Level 1	+7.8	+18.6	+13.1	+0.1	+9.9	+5.0	−6.7	−0.1	−3.6
Adaptation Level 2	+7.8	+18.7	+13.1	+3.3	+9.9	+6.4	−2.8	+1.4	−0.8
DEVELOPING									
Without phys. effect of CO_2	−28.5	−25.3	−26.5	−28.6	−26.4	−27.1	−36.2	−34.3	−35.1
With phys. effect of CO_2	−11.2	−0.7	−3.7	−12.0	−1.8	−4.5	−20.1	−10.2	−13.0
Adaptation Level 1	−9.2	−0.7	−3.2	−10.0	−1.8	−3.9	−17.8	−10.1	−12.3
Adaptation Level 2	−3.6	+1.4	−0.1	−4.5	+0.6	−0.8	−8.7	−4.3	−5.6

Table 9.9. Static climate change impact (%), Adaptation Level 1, year 2060

	GISS			GFDL			UKMO			GISS-A		
	Cereals	Other	All crops	Cereals	Other	All crops	Cereals	Other	All crops	Cereals	Other	All crops
DEVELOPED												
North America	+2.7	+12.6	+5.9	−3.8	+3.6	−0.7	−10.8	−9.8	−10.1	+6.6	+22.0	+12.1
Western Europe and other developed market economies	+6.2	+13.5	+10.3	+4.3	+9.6	+7.1	+2.7	+7.3	+5.1	+18.7	+22.2	+20.5
Easten Europe + former USSR	+17.7	+25.9	+22.8	+2.6	+13.5	+8.6	−8.3	−0.3	−4.0	+14.0	+16.9	+15.9
Pacific OECD countries	+8.3	+13.9	+11.0	+8.5	+10.5	+9.1	+7.4	+8.0	+7.4	+17.6	+17.2	+17.0
DEVELOPING												
Africa	−20.6	+0.8	−3.0	−24.0	−4.7	−8.1	−25.6	−7.0	−10.3	−4.7	+7.4	+5.4
Latin America	−16.7	−6.1	−8.7	−14.5	+0.2	−3.2	−22.7	−15.8	−17.7	+3.1	+16.6	+13.8
West Asia	−12.2	−4.6	−6.5	−17.4	−9.2	−11.1	−22.5	−15.0	−16.9	+8.8	+13.5	+12.3
South Asia	−9.8	−4.0	−6.7	−10.7	−6.6	−8.4	−28.8	−26.1	−26.6	+1.8	+7.8	+5.0
Centrally Planned Asia	+3.3	+9.4	+6.9	+1.4	+7.2	+5.0	−0.8	+4.7	+2.6	+1.9	+13.2	+9.7
Pacific Asian countries	−14.9	−11.1	−11.4	−5.6	−3.7	−3.2	−14.8	−5.1	−8.1	+6.1	−3.2	−3.6

Table 9.10. Percentage change in world market prices, year 2060

	Cereals				All crops			
	GISS	GFDL	UKMO	GISS-A	GISS	GFDL	UKMO	GISS-A
without phys. effect of CO_2	306	356	818	81	234	270	592	70
with phys. effect of CO_2	24	33	145	−21	8	17	90	−25
Adaptation Level 1	13	22	98		2	10	67	
Adaptation Level 2	−4	2	36		−8	−3	25	

human resources. At the same time, consumers react to price changes and adjust their patterns of consumption.

Table 9.10 contains changes in world market prices for cereals and an overall crop price index, as observed in the climate change scenarios relative to the standard reference projection. When direct physiological effects of CO_2 on plant growth and yields are not included, major increases in world market prices result in four- to nine-fold increases of cereal prices depending on GCM scenario. Apart from the scientific evidence of the beneficial physiological effects of elevated CO_2 levels on crop yields, such increases would elicit strong public reaction and policy measures to mitigate the negative yield impacts. Hence, the outcome for scenarios without the physiological effects of CO_2 on yields are probably unrealistically extreme.

When the physiological effects of 555 ppm CO_2 on yields are included in the assessment, cereal prices increase on the order of 24 to 145% relative to the standard reference projection. The index of crop prices increases by 8 to 90%, depending on GCM climate change scenario. Changes caused by GISS and GFDL scenarios are rather modest, resulting in an increase of about 25 to 33% in cereal prices, and an increase of less than 20% in overall crop prices. Only the yield impacts derived under a climate change as projected by the UKMO GCM produced large agricultural price increases. On the other hand, under the GISS-A climate scenario, where impacts are dominated by positive physiological effects of CO_2, major price decreases occur.

Price changes are further reduced when farm-level adaptation is considered. The crop price index rises less than 10% in both GISS and GFDL simulation runs. The UKMO projection, however, still produces a two-thirds crop price increase. Figure 9.2 compares the level of crop prices to that generated in the BLS standard reference projection. Results are shown for three simulation runs with physiological effects of 555 ppm CO_2 and farm-level adaptation: GISS and GFDL doubled CO_2 runs and a simulation run based on GISS transient scenario A. Note that in the GISS doubled CO_2 and GISS transient run, positive crop impacts dominate for about half the simulation period. In the GISS transient run the crop price index falls initially (around year 2010) by as much as 10% below the level of the reference run. Then, as negative impacts in developing countries increase and beneficial impacts in developed countries level off, prices return to and finally exceed the price index of the reference scenario.

With adaptation measures involving major changes in agricultural practices, i.e., Adaptation Level 2, prices would even fall below reference run levels in the GISS and GFDL scenarios. Note that the assumptions underlying Adaptation Level 2, sometimes requiring major investments, may not be economically viable. Scenarios with low-cost adaptation measures, i.e., Adaptation Level 1, appear to be more realistic.

Table 9.11 highlights the estimated *dynamic* impacts of climate change on agriculture resulting after 70 years of simulation with economic adjustment. According to these calculations, which include an optimistic assessment of direct physiological effects of 555 ppm CO_2 on crop yields, the impact on global agriculture GDP would be between −2 and +1% in all but the UKMO scenarios where it ranges between −2 and −6%. Developed countries are likely to experience some increase in agricultural

Figure 9.2. Impact of climate change on crop prices (direct CO_2 effects and farm-level adaptation taken into account)

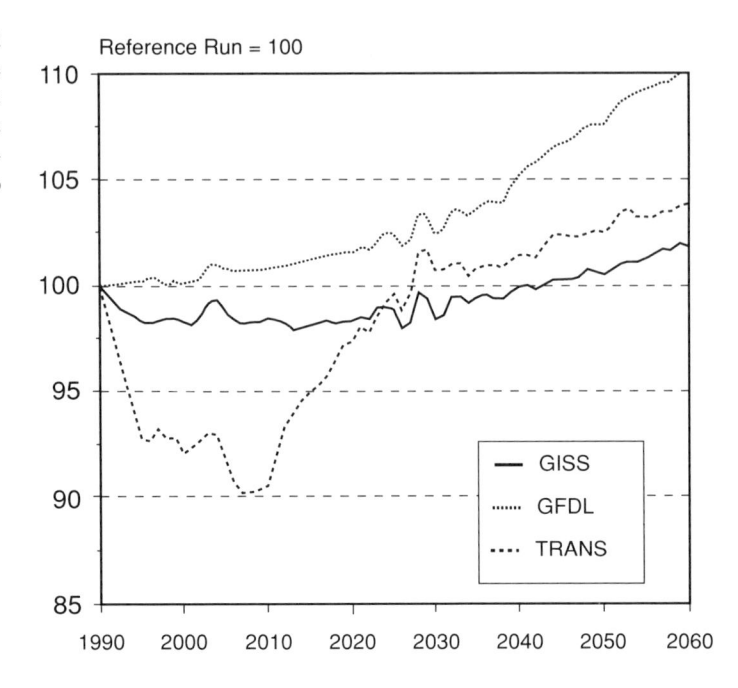

output. On the contrary, developing countries are projected to suffer a production loss in most scenarios. Table 9.12 lists the simulated regional impacts considering physiological effects of CO_2 on crop growth and some farm-level adaptation, Adaptation Level 1. It also includes results from the GISS-A scenario. Among developed regions, simulated positive impacts on agricultural output are largest for Europe, the former USSR and the Pacific OECD countries. Dynamic impacts in developing regions are mostly negative except for Centrally Planned Asia which benefits in all these scenarios.

It is important to note that these changes in comparative advantage between developed and developing regions are likely to accentuate the magnitude of the static impacts suggested by the analysis without economic adjustment. Winners are likely to gain more, and losers to lose even more. We can distinguish two prototypical situations in these scenario results. (1) When global supply is only marginally affected, there is little impact on prices. Then the shift in relative productivity from developing to developed regions dominates the adjustment process. For instance, in the GISS and GFDL scenarios with farm-level adaptation, agriculture production shifts somewhat from developing to developed countries taking account of the differences in projected yield changes. (2) When global crop yields are strongly affected, as in the UKMO scenario, the supply gap is so substantial that massive price increases result. These in turn provide production incentives to both regions to recover more than half the production forgone due to climate change (according to static crop model estimates).

Table 9.11. Impact of climate change with economic adjustment, year 2060

	Cereals production % change			Crop production % change			GDP agriculture % change		
	GISS	GFDL	UKMO	GISS	GFDL	UKMO	GISS	GFDL	UKMO
WORLD TOTAL									
without phys. effect of CO_2	−10.9	−12.1	−19.6	−11.5	−12.8	−18.0	−10.2	−11.7	−16.4
with phys. effect of CO_2	−1.2	−2.8	−7.6	−0.5	−1.7	−6.4	−0.4	−1.8	−5.4
Adaptation Level 1	0.0	−1.6	−5.2	+0.2	−1.0	−5.0	+0.2	−1.2	−4.4
Adaptation Level 2	+1.1	−0.1	−2.4	+1.1	+0.2	−2.3	+1.0	0.0	−2.0
DEVELOPED									
without phys. effect of CO_2	−3.9	−10.1	−23.9	+3.8	−5.5	−12.7	+1.1	−6.2	−12.5
with phys. effect of CO_2	+11.3	+5.2	−3.6	+15.6	+7.6	−0.9	+11.6	+5.1	−1.9
Adaptation Level 1	+14.2	+7.9	+3.8	+17.6	+9.1	+4.0	+13.3	+6.5	+1.8
Adaptation Level 2	+11.0	+3.0	+1.8	+15.1	+8.6	+2.2	+11.8	+6.5	+1.3
DEVELOPING									
without phys. effect of CO_2	−16.2	−13.7	−16.3	−16.6	−12.8	−19.8	−13.9	−13.5	−17.7
with phys. effect of CO_2	−11.0	−9.2	−10.9	−5.8	−4.9	−8.2	−4.4	−4.0	−6.6
Adaptation Level 1	−11.2	−9.2	−12.5	−5.6	−4.4	−8.1	−4.1	−3.7	−6.4
Adaptation Level 2	−6.6	−5.6	−5.8	−3.6	−2.7	−3.9	−2.6	−2.2	−3.1

Table 9.12. Dynamic impact of climate change (%), Adaptation Level 1, year 2060

	Cereal production				Crop production				GDPA[1]			
	GISS	GFDL	UKMO	GISS-A	GISS	GFDL	UKMO	GISS-A	GISS	GFDL	UKMO	GISS-A
DEVELOPED
North America	+10.6	+5.9	−5.2	+4.1	+9.3	+4.8	−3.2	+2.0	+7.5	+3.2	−3.0	+0.8
Western Europe and other developed market economies	+6.5	+7.7	+12.2	+14.7	+10.7	+6.7	+11.7	+14.5	+8.0	+5.2	+7.9	+10.9
Eastern Europe + former USSR	+24.6	+7.6	+6.0	+19.9	+30.7	+12.7	−1.3	+20.1	+26.8	+11.2	−2.7	+17.3
Pacific OECD countries	+19.6	+31.7	+53.2	+4.2	+16.3	+24.0	+52.0	−1.1	+4.0	+5.5	+12.1	+0.2
DEVELOPING
Africa	−23.7	−24.5	−16.2	−8.2	−4.1	−9.3	+1.4	−3.3	−2.2	−7.6	+1.4	−2.8
Latin America	−25.0	−17.8	−14.5	−10.3	−13.8	−2.6	−11.1	+4.0	−10.9	−3.2	−7.9	+1.3
West Asia	−13.6	−17.0	−18.6	+2.8	−7.9	−11.5	−12.4	+6.3	−5.6	−8.7	−9.5	+5.8
South Asia	−11.9	−8.5	−26.8	−3.5	−7.8	−7.2	−25.1	+2.0	−6.2	−5.2	−20.0	+2.2
Centrally Planned Asia	+4.1	+2.6	+2.1	+1.6	+3.8	+2.5	+1.7	+3.0	+3.3	+2.1	+1.4	+1.9
Pacific Asian countries	−12.3	+0.3	−1.6	−12.8	−14.4	−2.3	−2.3	−11.3	−12.8	−3.1	−3.6	−9.1

[1] Gross Domestic Product from Agriculture

The magnitude of the impacts and these different responses at aggregate regional level are shown in Figure 9.3 and Figure 9.4, for cereals and total crops, respectively. Figure 9.5 illustrates the impacts projected by cereal commodity (with physiological effects of 555 ppm CO_2 and farm-level adaptation). Accordingly, global wheat production is less likely to suffer negative climate impacts than other cereal commodities; it increases in GISS and GFDL scenarios and declines less than other cereals in the UKMO scenario. This is likely due to the location of major wheat-producing regions at mid and high latitudes, where yield declines are projected to be lower.

Net imports of cereals into developing countries increase under all scenarios. The change in cereal imports, relative to the standard reference projection, is largely determined by the magnitude of the estimated yield change, the change in relative productivity in developing and developed regions, the change in world market prices, and changes in real incomes of consumers in developing countries. For example, under the GISS climate change scenarios, productivity is altered in favour of developed countries with relatively small changes in incomes and prices, resulting in pronounced increases of net cereal imports into developing countries.

With less agricultural production in developing countries and higher prices on international markets, the estimated number of people at risk of hunger is likely to increase. This occurs in all but one scenario (Table 9.13). The largest increase is to be expected from the UKMO scenario without CO_2 physiological effects; the smallest change, a decline of 2%, occurred in the GISS scenario considering physiological effects of increased CO_2 and Adaptation Level 2.

CONCLUSIONS

The distortions of the world food system simulated in the climate change scenarios fall well within the range of estimates obtained from the different reference projections. For instance, the decline in cereal production even under the worst climate change scenario based on the UKMO GCM experiment (assuming physiological effects of increased atmospheric CO_2 concentrations and some farm-level adaptation), amounts to less than half the difference between the cereal production levels simulated in the higher and lower economic growth reference projections, scenarios REF-H and REF-L. However, the ability of the world food system to absorb negative yield

Table 9.13. Impact of climate change on people at risk of hunger, year 2060

	Additional million people				% change			
DEVELOPING (excl. China)	GISS	GFDL	UKMO	GISS-A	GISS	GFDL	UKMO	GISS-A
Without phys. effect of CO_2	721	801	1446	265	112	125	225	41
With phys. effect of CO_2	63	108	369	−84	10	17	58	−13
Adaptation Level 1	38	87	300		6	14	47	
Adaptation Level 2	−12	18	119		−2	3	19	

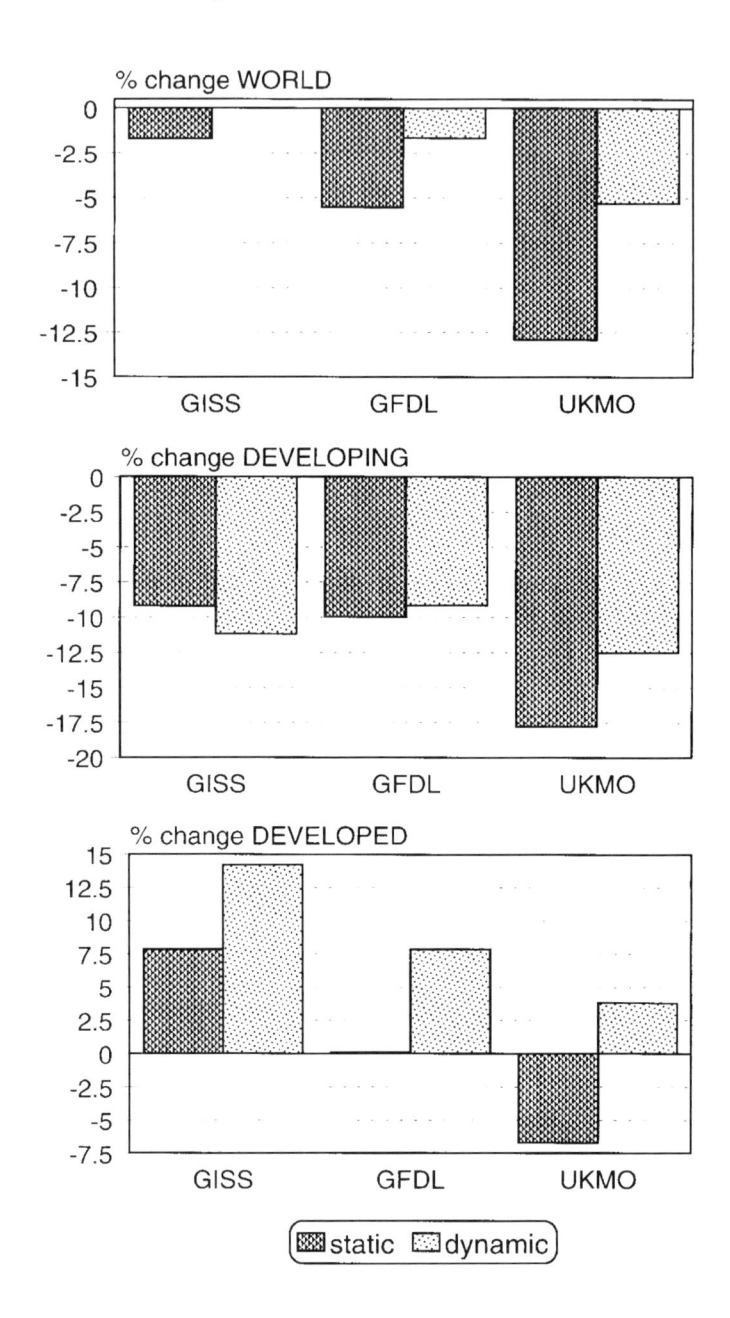

Figure 9.3. Impact of climate change on regional cereal production, year 2060, with physiological effects of 555 ppm CO_2 and farm-level Adaptation 1

EFFECTS ON WORLD FOOD PRODUCTION AND SECURITY 229

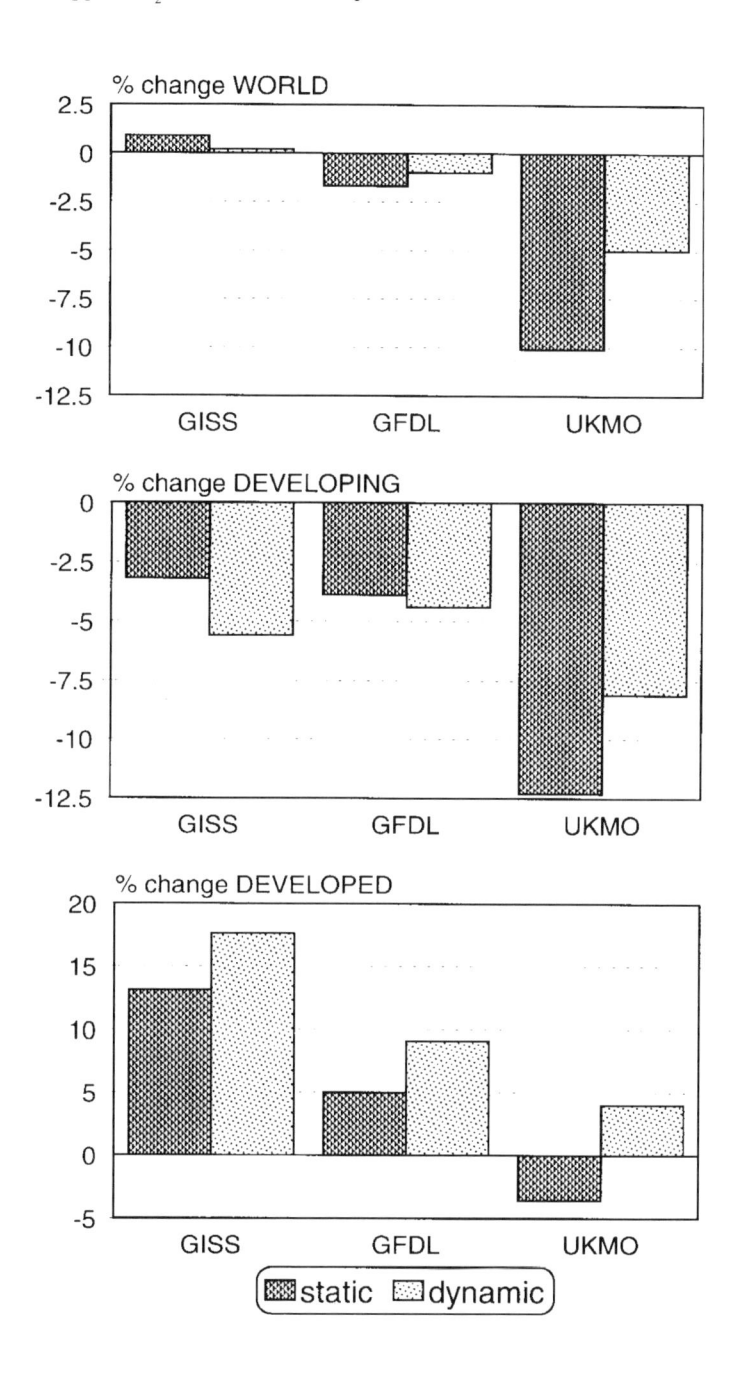

Figure 9.4. Impact of climate change on regional crop production, year 2060, with physiological effects of 555 ppm CO_2 and farm-level Adaptation 1

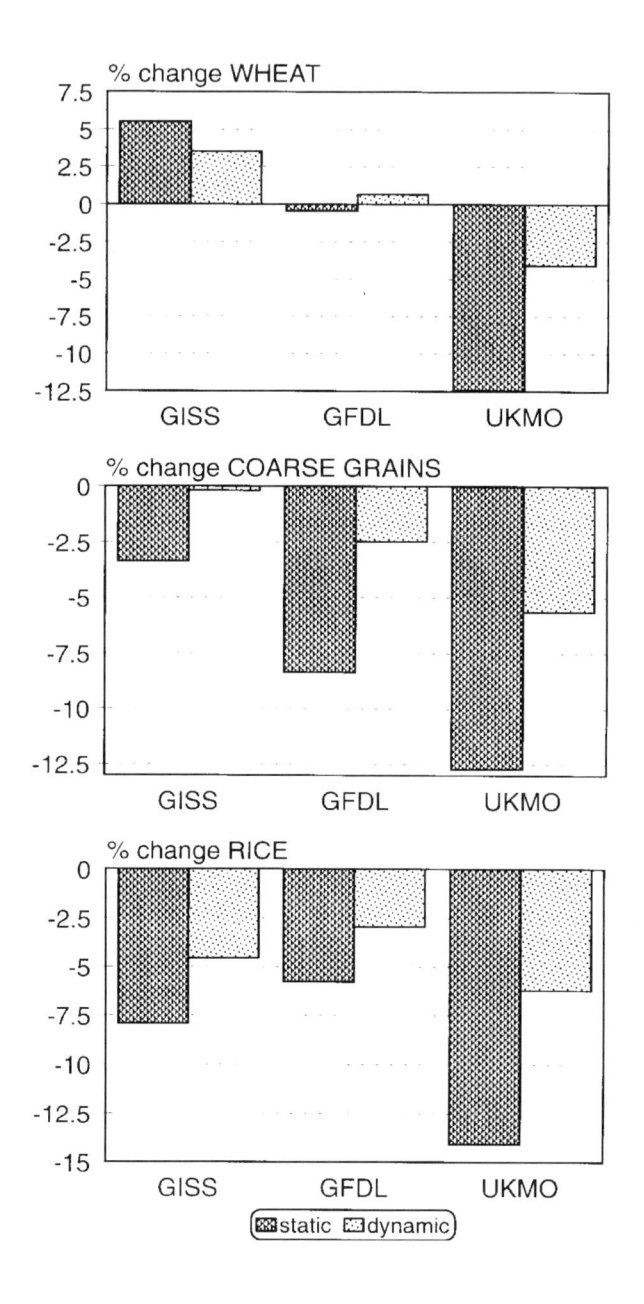

Figure 9.5. Impact of climate change on cereal commodities, year 2060, with physiological effects of 555 ppm CO_2 and farm-level Adjustment 1

impacts decreases with the magnitude of the impact. Economic adaptation can largely compensate for moderate yield changes such as the GISS and GFDL scenarios, but not greater ones such as the UKMO scenario.

The effects of changes in climate on crop yields are likely to vary greatly from region to region across the globe. The results of the scenarios tested in this study indicate that the effects on crop yields in mid- and high-latitude regions appear to be positive or less adverse than those in low-latitude regions, provided the potentially beneficial direct physiological effects of CO_2 on crop growth can be fully realized. From a development perspective, the most serious concern relates to the apparent difference in incremental yield impacts between developed and developing countries. The scenario results suggest that if climatic change were to retard economic development beyond the direct effects on agriculture in the poorer regions, especially in Africa, then overall impacts could be sizeable.

In all climate change scenarios, relative productivity of agriculture changes in favour of developed countries, with implications on resource allocation. Economic feedback mechanisms are likely to emphasize and accentuate the uneven distribution of climate change impacts across the world, resulting in net gains for developed countries in all but the UKMO scenarios and a noticeable loss to developing countries. As a result, net imports of cereals into developing countries increase in all scenarios, on the order of 20 to 50% compared to trade in the reference scenario.

Including direct physiological effects of CO_2 on crop yields, world cereal production is estimated to decrease between 1 to 3% under GISS and GFDL scenarios, and 7% in the UKMO climate scenario based projections. Assuming adaptation to climate change at farm-level, cereal production would still be reduced between 0 to 2% and 5%, respectively, for the GISS/GFDL and UKMO scenarios. The largest negative changes would occur in developing countries, averaging around -10%. This loss of production in developing countries, together with rising agricultural prices, is likely to increase the number of people at risk of hunger, in the order of 5 to 15% in the less severe climate scenarios, and ~50% in the UKMO based projections. Under a possibly more realistic climate change scenario with a lower climate sensitivity to increasing greenhouse gas concentrations, aggregate crop productivity at global level increases by ~10% until year 2060, i.e., the end of the simulation period considered in the analysis. Impacts are assessed to be positive in almost all of the 10 aggregate world regions reported in the study. However, the simulated percentage increases in developed countries are about twice the increases in developing regions.

The analysis also shows that the yield impacts do not only vary with geographic region, but are also unequally distributed over time. Results of scenarios based on the GISS transient scenario A demonstrate that benefits from physiological effects due to increasing atmospheric CO_2 levels may outweigh negative impacts from changing temperature and precipitation regimes at least in the near-term. The yield-increasing factors in that scenario dominate possible negative impacts until year 2020. Understanding the biophysical processes of CO_2 and climate change effects on crops remains an important research area.

It must be realized, however, that the ability to estimate climate change yield impacts on world food supply, demand and trade is surrounded by large uncertainties regarding important elements, such as the magnitude and spatial characteristics of climate change, the range and efficiency of adaptation possibilities, the long-term aspects of technological change and agricultural productivity, and even future demographic trends. Also, the adoption of efficient adaptation techniques is far from certain. In developing countries there may be social, economic or technical constraints, and adaptive measures may not necessarily result in sustainable production over long time-frames.

Determining how countries, particularly developing countries, can and will respond to reduced yields and increased costs of food is a critical research need arising from this study. Will such countries be able to import large amounts of food? Will the burden for adaptation be passed on to the poorest? From a political and social standpoint, the results of the study indicate the potential for a decrease in food security in developing countries. The study suggests that the worst situation arises from a scenario of severe climate change, low economic growth, continuing large population increases, and little farm-level adaptation. In order to minimize possible adverse consequences, like production losses, food price increases, environmental stresses, and an increase in the number of people at risk of hunger, the way forward is to encourage the agricultural sector to continue to develop crop breeding and management programmes for heat and drought conditions, in combination with measures taken to preserve the environment, to use resources more efficiently, and to slow the growth of the human population of the world. The latter step would also be consistent with efforts to slow emissions of greenhouse gases, and thus the rate and eventual magnitude of global climate change.

In the face of these uncertainties, both national and international organizations should encourage the development of new approaches likely to be effective in preparing for climate change. Agricultural research would benefit from increased attention to both macroclimate and microclimate in all experiments and variety trials. Another climate change impact potentially significant for future agricultural production is soil organic matter loss due to soil warming. Considering the vulnerability of agricultural production to the occurrence of climate extremes, research should be directed to determine what are the heat-tolerance limits of currently grown and of alternative crops and varieties. At what threshold values of air or soil temperature do severe problems begin? What agronomic methods are the best to moderate the thermal regime affecting crop growth?

To the extent that the progressive greenhouse effect cannot be prevented in practice, policies should be devised to facilitate the adjustment of agriculture to the likelihood of environmental change. Such adjustments may include modification of agronomic practices, adoption of crops known to be heat-resistant and drought-resistant, increased efficiency of irrigation and water conservation, and improved pest management. Such adjustments are worthy of being implemented in any case, be it with or without climatic change.

Although some countries in the temperate zone may reap some benefits from climate change, many countries in the tropical and subtropical zones appear to be more vulnerable. Particular hazards are the possibly increased flooding of low-lying areas, the increased frequency and severity of droughts in semi-arid areas, and potential decreases in attainable crop yields. It happens that the latter countries tend to be the poorest and the least able to make the necessary economic adjustments. Much of the expected change in global climate is due to the past and present activities of the industrial countries; so it is their responsibility to commit themselves to, and to play an active role in, a comprehensive international effort to prepare for the likely consequences.

REFERENCES

Adams, R.M., Rosenzweig, C., Peart, R.M., Ritchie, J.T., McCarl, B.A., Glyer, J.D., Curry, R.B., Jones, J.W., Boote, K.J. and Allen, L.H. Jr. 1990. Global climate change in US agriculture. *Nature* **345**(6272): 219-224.

Adams, R.M., Fleming, R.A., Change, C-C., McCarl, B.A. and Rosenzweig, C. 1994. A reassessment of the economic effects of global climate change on U.S. agriculture. *Climatic Change* **30**(2): 147-167.

Allen, L.H. Jr., Boote, K.J., Jones, J.W., Jones, P.H., Valle, R.R., Acock, B., Rogers, H.H. and Dahlman, R.C. 1987. Response of vegetation to rising carbon dioxide: photosynthesis, biomass and seed yield of soybean. *Global Biogrochemical Cycles* **1**: 1-14.

Cure, J.D. and Acock, B. 1986. Crop responses to carbon dioxide: A literature survey. *Agricultural and Forest Meteorology* **38**: 127-145.

FAO. 1984. *Fourth World Food Survey*. Rome.

FAO. 1987. *Fifth World Food Survey*. Rome.

FAO. 1988. *World Agriculture Toward 2000*. Rome.

FAO. 1993. *Agriculture: Towards 2010*. Rome.

FAO/UNFPA/IIASA. 1983. *Potential Population Supporting Capacity of Lands in the Developing World*. FAO, Rome.

Fischer, G., Frohberg, K., Keyzer, M.A. and Parikh, K.S. 1988. *Linked National Models: A Tool for International Policy Analysis*. Kluwer Academic Publishers, Dordrecht.

Fischer, G., Frohberg, K., Keyzer, M.A., Parikh, K.S. and Tims, W. 1990. *Hunger – Beyond the Reach of the Invisible Hand*. IIASA, Laxenburg.

Fischer, G., Frohberg, K., Parry, M.L. and Rosenzweig, C. 1994. Climate change and world food supply, demand and trade. Who benefits, who loses? *Global Environmental Change* **4**(1): 7-23.

Godwin, D., Ritchie, J.T., Singh, U. and Hunt, L. 1989. *A User's Guide to CERES-Wheat – V2.10*. International Fertilizer Development Center, Muscle Shoals, AL.

Godwin, D., Singh, U., Ritchie, J.T. and Alocilja, E.C. 1993. *A User's Guide to CERES-Rice*. International Fertilizer Development Center, Muscle Shoals, AL.

Hansen, J., Fung, I., Lacis, A., Rind, D., Russell, G., Lebedeff, S., Ruedy, R. and

Stone, P. 1988. Global climate changes as forecast by the GISS 3-D model. *J. Geophysical Research* **93**(D8): 9341-9364.

Hendrey, G.R. 1993. *Free-air CO_2 Enrichment for Plant Research in the Field*. C.K. Smoley, Boca Raton, Florida.

IBSNAT [International Benchmark Sites Network for Agrotechnology Transfer Project] 1989. *Decision Support System for Agrotechnology Transfer Version 2.1 (DSSAT V2.1)*. Department of Agronomy and Soil Science. College of Tropical Agriculture and Human Resources. University of Hawaii, Honolulu.

IPCC. 1990a. *Climate Change: The IPCC Scientific Assessment*. J.T. Houghton, G.J. Jenkins and J.J. Ephraums (eds.). Cambridge University Press, Cambridge.

IPCC. 1990b. *Climate Change: The IPCC Impacts Assessment*. W.J.McG. Tegart, G.W. Sheldon and D.C. Griffiths (eds.). Australian Government Publishing Service, Canberra.

IPCC. 1992. *Climate Change 1992*. The Supplementary Report to the IPCC Scientific Assessment. J.T. Houghton, B.A. Callander and S.K. Varney (eds.). Intergovernmental Panel on Climate Change. Cambridge University Press, Cambridge.

IPCC. 1996. *Climate Change 1995: Impacts, Adaptations and Mitigation of Climate Change: Scientific-Technical Analyses*, R.T. Watson, M.C. Zinyowera and R.H. Moss (eds), Contribution of Working Group II to the Second Assessment Report of the Intergovernmental Panel on Climate Change, Cambridge University Press, Cambridge.

Jones, C.A. and Kiniry, J.R. 1986. *CERES-Maize: A Simulation Model of Maize Growth and Development*. Texas A&M Press, College Station.

Jones, J.W., Boote, K.J., Hoogenboom, G., Jagtap, S.S. and Wilkerson, G.G. 1989. *SOYGRO V5.42: Soybean Crop Growth Simulation Model. User's Guide*. Department of Agricultural Engineering and Department of Agronomy, University of Florida, Gainesville, FL.

Kellogg, W.W. and Zhao, Z.-C. 1988. Sensitivity of soil moisture to doubling of carbon dioxide in climate model experiments. I. North America. *J. Climate* **1**: 348-366.

Kimball, B.A. 1983. Carbon dioxide and agricultural yield: an assemblage and analysis of 430 prior observations. *Agronomy J.* **75**: 779-788.

Liverman, D., Dilley, M., O'Brien, K. and Menchaca, L. 1994. Possible impacts of climate change on maize yields in Mexico. In: *Implications of Climate Change for International Agriculture: Crop Modeling Study*. C. Rosenzweig and A. Iglesias (eds.). US Environmental Protection Agency. EPA 230-B-94-003. Washington DC.

Menzhulin, G.V., Koval, L.A. and Badenko, A.L. 1994. Potential effects of global warming and carbon dioxide on wheat production in the former Soviet Union. In: *Implications of Climate Change for International Agriculture: Crop Modeling Study*. C. Rosenzweig and A. Iglesias (eds.). US Environmental Protection Agency. EPA 230-B-94-003. Washington DC.

Otter-Nacke, S., Godwin, D.C. and Ritchie, J.T. 1986. *Testing and Validating the CERES-Wheat Model in Diverse Environments*. AgRISTARS YM-15-004-7. Johnson Space Center No. 20244. Houston.

Parikh, K.S., Fischer, G., Frohberg, K. and Gulbrandsen, O. 1988. *Toward Free Trade in Agriculture*. Martinus Nijhoff, The Hague.

Parry, M.L., Carter, T.R. and Konijn, N.T. (eds.). 1988. *The Impact of Climate Variations on Agriculture. Vol. 1: Assessments in Cool Temperate and Cold Regions. Vol. 2: Assessments in Semi-arid Regions.* Kluwer, Dordrecht.

Pearman, G. 1988. *Greenhouse: Planning for Climate Change.* CSIRO, Canberra.

Peart, R.M., Jones, J.W., Curry, R.B., Boote, K. and Allen, L.H. Jr. 1989. Impact of climate change on crop yield in the Southeastern U.S.A. In: *The Potential Effects of Global Climate Change on the United States.* J.B. Smith and D.A. Tirpak (eds.). US Environmental Protection Agency, Washington DC.

Ritchie, J.T. and Otter, S. 1985. Description and performance of CERES-Wheat: A user-oriented wheat yield model. In: *ARS Wheat Yield Project.* W.O. Willis (ed.). Department of Agriculture, Agricultural Research Service. ARS-38. Washington DC.

Ritchie, J.T., Singh, U., Godwin, D. and Hunt, L. 1989. *A User's Guide to CERES-Maize – V 2.10.* International Fertilizer Development Center, Muscle Shoals, AL.

Rogers, H.H., Bingham, G.E., Cure, J.D., Smith, J.M. and Surano, K.A. 1983. Responses of selected plant species to elevated carbon dioxide in the field. *J. Environmental Quality* **12**: 569-574.

Rosenzweig, C. and Iglesias, A. 1994. *Implications of Climate Change for International Agriculture: Crop Modeling Study.* US Environmental Protection Agency. EPA 230-B-94-003. Washington DC.

Rosenzweig, C. and Parry, M.L. 1994. Potential impacts of climate change on world food supply. *Nature* **367**: 133-138.

Rosenzweig, C., Allen, L.H. Jr., Harper, L.A., Hollinger, S.E. and Jones, J.W. (eds.). 1995. *Climate Change and Agriculture: Analysis of Potential International Impacts.* ASA Special Publication No. 59. American Society of Agronomy, Madison, WI. 382 p.

Smith, J.B. and Tirpak, D.A. (eds.). 1989. *The Potential Effects of Global Climate Change on the United States.* Report to Congress. EPA-230-05-89-050. US Environmental Protection Agency, Washington DC.

UK Department of the Environment. 1991. *The Potential Effects of Climate Change in the United Kingdom.* Climate Change Impacts Review Group. HMSO, London.

UN. 1989. *World Population Prospects 1988.* New York.

World Bank. 1990. *World Population Projections.* 1989-1990 Edition. Johns Hopkins University Press, Baltimore.

10. Climate Change, Global Agriculture and Regional Vulnerability

JOHN REILLY
Natural Resources and Environment Division, Economic Research Service, USDA, Washington DC, USA

The potential impacts of climate change on agriculture are highly uncertain. The large number of studies conducted over the past few years for many different sites across the world show few, if any, robust conclusions of either the magnitude or direction of impact for individual countries or regions. Where apparent consensus exists it frequently appears to occur because only one or two studies have been conducted using a single climate scenario. Many such studies have focused on doubled (2×CO_2) General Circulation Model (GCM) equilibrium scenarios. These do not begin to describe the variety of climatic conditions any particular region is likely to experience as the actual climate changes over time.

Potential future climate changes are also made more uncertain because of the recently recognized role of sulphate aerosols which may partly offset the warming expected from increased concentrations of CO_2, methane, nitrous oxide and other radiatively active trace gases. The significant spatial variation in sulphate aerosol concentrations means that the regional pattern of climate change may be quite different from that simulated on the basis of CO_2 increase alone. The short lifetime of aerosols in the atmosphere (a few days) means that if the use of high sulphur coal in India or China increases or efforts to control sulphur emission in the United States or Europe are intensified, the spatial pattern of climate change could change significantly within a relatively short period of time due to changes in the aerosol cooling effect.

Different impact methodologies also yield widely varying results of the direct impacts of climate change on crop yields and agricultural production even when examining the same region and the same climate scenarios. The socio-economic environment, agricultural technology and natural resource base will also necessarily undergo profound changes over the next 100 years whether agriculture meets the many challenges of feeding the world's growing population or fails to do so.

The robust conclusion that does emerge from impact studies is that climate change has the potential to change significantly the productivity of agriculture at most locations. Some currently highly productive areas may become much less productive. Some currently marginal areas may benefit substantially while others may become unproductive. Crop yield studies show regional variations of +20, 30 or more per cent in some areas and equal size losses in other areas. Most areas can expect change and will need to adapt, but the direction of change, particularly of precipitation, and required adaptations cannot now be predicted. Nor may it ever be possible to predict them with confidence. Current evidence suggests that poleward regions where agriculture is limited by short growing seasons are more likely to gain while subtropical and tropical regions may be more likely to suffer drought and losses in productivity. However, these broad conclusions hardly provide the basis for mapping out a long-term strategy for agricultural adaptation. Thus, policy must retain flexibility to respond as conditions change.

A further issue is how do climate change impacts on agricultural production fit within the other pressing challenges facing agriculture in different regions of the world. Is climate change a minor threat, likely to be undetected among the many changes that will reshape the agricultural sectors of the world's economies? Or is it another critical challenge to an agricultural sector straining to cope with a growing population, resource degradation, tighter constraints on available resources, and exhaustion of technological capabilities to expand production using existing land and water resources?

It is useful to place some of the 2×CO_2 agricultural projections in the context of other future projections. If we accept long-term demographic trends, the largest absolute addition to the world's population will occur during the decade of the 1990s, the growth rate having already slowed from that of the 1950s and 1960s. By the time 2×CO_2 climate scenarios are expected to be realized (some time around 2100 or later), the world population will have stabilized and agricultural research will no longer face the challenge of increasing productivity to keep up with a growing population.

Therefore there is a need for more specific analysis about how climate will change over the next 10, 20 or 30 years rather than over the next 100. It also provides a caution not to consider our response to climate change apart from our response to the immediate needs of agriculture: feeding a growing population where presently an estimated 740 million people still suffer from hunger and malnutrition while maintaining the productivity of basic agricultural resources and meeting the demands placed on agriculture to minimize damage to the environment.

This paper will: (1) briefly discuss the major methodologies used to estimate impacts of climate change as different models lead to substantially different estimates of climate change impacts; (2) review the broad literature reporting results of crop yield studies of climate change conducted for many different areas (how much (or little) do we know?); (3) review the set of estimates that has been made for global agricultural production and what it means for regional agricultural impacts; (4) discuss the issue of vulnerability, adding a precise definition, while reviewing some of the vulnerability concepts that have been used in the literature; and (5) review specific issues of adaption – how can the world's agricultural system, or more to the point, those populations highly dependent on agriculture, make themselves less likely to suffer loss from climate change.

IMPACT ASSESSMENT METHODOLOGIES

Climate change presents a challenge for researchers attempting to quantify its impact due to the global scale of likely impacts, the diversity of agriculture systems, and the decades' long time scale. Current climatic, soil and socio-economic conditions vary widely across the world. Each crop and crop variety has specific climatic tolerances and optima. It is not possible to model world agriculture in a way that captures the details of plant response in every location. The availability of data with the necessary geographic detail is currently the major limitation rather than computational capability or basic understanding of crop responses to climate. A specific problem has been how to take the detailed knowledge of plant response into aggregate assessments of regional assessments. In general, compromises are necessary in developing quantitative analyses at regional scales.

There are two basic approaches for evaluating crop and farmer response to changing climate: (1) structural modelling of the agronomic response of plants and the economic/management decisions of farmers based on theoretical specifications and controlled experimental evidence; and (2) reliance on the observed response of crops and farmers to varying climate.

For the first approach, sufficient structure and detail are needed to represent specific crops and crop varieties for which responses to different conditions are known through detailed experiments. Similar detail on farm management allows direct modelling of the timing of field operations, crop choices, and how these decisions affect costs and revenues. These approaches typically model a representative crop or farm. Both in the case of economic models of farm decisions and in the case of crop response models, the original purpose was to improve understanding of how the crop grows or how a farmer manages. In the case of models of a representative farm, one might hope to offer prescriptive advice for the farmer – where farm operations differ from the profit maximizing (or cost minimizing) model results, it provides guidance for how farmers might improve farm performance. In both cases, the idealized representation of the crop and farm operation tends to give results that differ markedly from the actual experience on farms operating under real world conditions. This may reflect the fact that farmers do not operate as profit maximizers (they could improve their performance) or that the models fail to consider some of the factors that the farmer takes into account such as risk or lack of immediate employment alternatives. Because of the idealized nature of these models, many analysts consider that these provide evidence of the *potential production* or *potential profitability*. Imposing climate change on these models gives estimates of how potential production may change due to climate change. Using these results as indicative of how climate will actually affect agriculture thus rests on the assumption that the change in the potential represents the change likely to be actually experienced. Many approaches of this type have used detailed crop response models requiring daily weather records. For aggregate analyses inferences must be made from relatively few sites and crops to large areas and diverse production systems because of the complexity of the models and the need for detailed data on weather over a decade or more. This is the basic approach of Fischer *et al.* reported elsewhere in this volume.

The work of Leemans and Solomon (1993) is in a similar vein, choosing much simpler representations of crop/climate interactions but is still related to basic agronomic representation of crop growth in response to temperature and precipitation. The advantage of their approach is that, because of the minimal amounts of climatic data required (mean monthly data on temperature and precipitation), the crop models can be applied at a resolution of $0.5° \times 0.5°$ latitude-longitude grids.

The second approach, relying on observed response of crops and farmers, provided some of the earliest estimates of the potential effects. The simplest example of this approach is to observe the current climatic boundaries of crops and to redraw these boundaries for a predicted changed climate (e.g., Rosenzweig, 1985). In a similar vein, researchers have applied statistical analysis of data across geographic areas to separate climate from other factors (e.g., different soil quality, varying economic conditions) that explain production differences across regions and have used these to estimate the potential agricultural impacts of climate change (e.g., Mendelsohn *et al.*, 1994). An advantage of using direct evidence from observed production is that the data reflect how farmers operating under commercial conditions and crops growing under such conditions actually respond to geographically varying climatic conditions. Here, the most recent work uses extremely reduced form models (e.g., Mendelsohn *et al.*, 1994) although estimation of more detailed structural models is possible. Darwin *et al.* (1995) use revealed evidence from geographic variation in climate in a global model, allocating production and input use to climatically determined land classes based on current production patterns. Climate change impacts are then simulated by altering the distribution of land classes and assuming that when an area's land class changes, its underlying production level changes to that of the new land class.

The Darwin *et al.* (1995) approach links the basic agricultural productivity of land classes, described by a production function, with a computable general equilibrium model of the world economy. Thus, actual production in a region or land class depends on the final market clearing prices. The model also treats interactions with other sectors of the economy, most importantly sectors that compete for land and water. My interest in this section is in contrasting approaches used to estimate the initial impact of climate on agricultural production. As demonstrated by Fischer *et al.* (1994), Reilly *et al.* (1994) and Adams *et al.* (1988), given an initial climate shock on productivity, there are a number of ways to introduce this shock into a variety of different types of economic models to generate estimates of the market impact and realize production under new equilibrium prices.

The advantage of these approaches is that the response of crops and farmers is based on actual response under current operating conditions rather than an idealized view of how crops and farmers respond. The basic caveat associated with this approach is that one must have faith that land currently producing one set of outputs can change to the new set of outputs once climate changes. Whether these types of approaches accurately capture the productivity impact depends on how well they control for other factors (such as soil quality) and whether farmers can adjust their production as climate changes. This latter consideration leads to the interpretation that these approaches capture the long-run equilibrium response to climate change and may not capture adjustment costs associated with changing to new crops and production practices.

CROPS RESPONSE ESTIMATES FOR DIFFERENT REGIONS OF THE WORLD

Table 10.1 summarizes the results of the large number of studies of the impact of climate change on potential crop production. While the table does not provide the detail on the range of specific studies, methods and climate scenarios evaluated, it provides an indication of the wide range of estimates. The general conclusion of global studies, that tropical areas may more likely suffer negative consequences, is partly supported by the results in the table. For example, Latin America and Africa show primarily negative impacts. However, very few studies have been conducted in these regions. For Europe, the United States and Canada, and for Asia (including China) and the Pacific Rim, where many more studies have been conducted, the results generally range from severe negative effects (−60, −70%, or complete crop failure) to equally large potential yield increases.

The wide ranges of estimates are due to several, as yet unresolved, factors. Differences among climate scenarios are important and these can generate wide ranges of impacts even when using identical methods for the same regions. For example, a study of the potential impact on rice yields conducted for most of the countries of South and Southeast Asia and for China, Japan, and Korea using the same crop model found yield changes for India to range from -3 to +28%, for Malaysia from +2 to +27%, for the Philippines from −14 to +14%, and for mainland China from −18 to −4% (Matthews *et al.*, 1994a, b) depending on whether the GISS, GFDL or UKMO climate scenario was used.

The impacts across sites can vary widely within a region. Thus, how many and which sites are chosen to represent a region and how the site-specific estimates are aggregated can have important effects on the results. Studies for the United States and Canada demonstrate the wide range of impacts across sites with total or near-total crop failure every year projected for wheat and soybeans at one site in the United States (Rosenzweig *et al.*, 1994) but wheat yield increases of 180 to 230% for other sites in the United States and Canada (Rosenzweig *et al.*, 1994; Brklacich *et al.*, 1994; Brklacich and Smit, 1992).

Whether and how changes in a crop variety are specified in a study can have a large impact. Studies conducted of wheat response in Australia found impacts ranging from −34 to +65% for the same climate scenario and site depending on which known and currently grown wheat cultivar was specified in the crop model (Wang *et*

Table 10.1. Regional crop yield for 2×CO_2 GCM equilibrium climates

Region	Crop	Yield impact (%)	Countries studied/comments
Latin America	maize	−61 to increase	Argentina, Brazil, Chile, Mexico. Range is across GCM scenarios, with and without the CO_2 effect.
	wheat	−50 to −5	Argentina, Uruguay, Brazil. Range is across GCM scenarios, with and without the CO_2 effect.
	soybean	−10 to +40	Brazil. Range is across GCM scenarios, with CO_2 effect.
Former Soviet Union	wheat grain	−19 to +41 −14 to +13	Range is across GCM scenarios and region, with CO_2 effect.
Europe	maize	−30 to increase	France, Spain, N Europe. With adaptation, CO_2 effect. Longer growing season; irrigation efficiency loss; northward shift.
	wheat vegetables	increase or decrease increase	France, UK, N Europe. With adaptation, CO_2 effect. Longer season: northward shift, greater pest damage; lower risk of crop failure.
North America	maize wheat	−55 to +62 −100 to +234	USA and Canada. Range across GCM scenarios and sites with/without CO_2 effect.
	soybean	−96 to +58	USA. Less severe or increase in yield when CO_2 effect and adaptation considered.
Africa	maize	−65 to +6	Egypt, Kenya, South Africa, Zimbabwe. With CO_2 effect, range across sites and climate scenarios.
	millet	−79 to −63	Senegal. Carrying capacity fell 11-38%.
	biomass	decrease	South Africa; agrozone shifts.
South Asia	rice maize wheat	−22 to +28 −65 to −10 -61 to +67	Bangladesh, India, Philippines, Thailand, Indonesia, Malaysia, Myanmar. Range over GCM scenarios, and sites; with CO_2 effect; some studies also consider adaptation.
Mainland China and Taiwan	rice	−78 to +28	Includes rainfed and irrigated rice. Positive effects in NE and NW China, negative in most of the country. Genetic variation provides scope for adaptation.
Asia (other) and Pacific Rim	rice	−45 to +30	Japan and South Korea. Range is across GCM scenarios. Generally positive in northern Japan; negative in south.
	pasture	−1 to +35	Australia and New Zealand. Regional variation.
	wheat	−41 to +65	Australia and Japan. Wide variation, depending on cultivar.

Source: Reilly *et al.* (1996).

al., 1992). Similarly, Matthews et al. (1994a, b) concluded that the severe yield losses in South, Southeast and East Asia for rice in many scenarios was due to a threshold temperature effect that caused spikelet sterility but that genetic variation with regard to the threshold likely provided significant opportunity to switch varieties as temperatures rose. Thus, an impact analysis that narrowly specifies a crop variety is likely to generate a much different estimated impact than an analysis that specifies responses on the basis of the genetic variation across existing cultivars. Some studies have attempted to evaluate how future crop breeding may change the range of genetic variability available in future varieties (Easterling et al., 1993).

Finally, the estimated amount of adaptation likely to be undertaken by farmers varies. Fundamental views about how the farm sector responds to changing conditions (of any kind) shape the choice of methodological approach, and these methodological approaches can give apparently widely different estimates of impact. Specification of the crop variety in a crop response model illustrates this difference. For some analysts, the prospect that farmers will not change the variety of crop grown over the next 100 years as climate, technology, prices and other factors change is so remote that they choose to represent change among varieties as an essentially autonomous response of the farm sector. Other analysts choose more specific crop variety characteristics, viewing even crop variety change as neither automatic nor without cost. For example, different varieties of wheat produce flours with different characteristics and the cultural practices for growing spring and winter wheat differ. Similarly, studies of impacts on Japanese rice production estimate negative impacts for the southern parts of the country because of the climate tolerances of Japonica rice which is preferred over Indica varieties in Japan (Seino, 1993).

The differences from simply whether or not one assumes farmers will adopt the better adapted variety are large but these differences are potentially magnified many fold because the series of potential adaptations are broad with some requiring more specific recognition, action and investment by farmers. How do farmers choose a planting date – by planting at the same time each year regardless of weather conditions or by planting when soil temperatures are sufficient for crop growth, when the rainy season starts, or when the fields can be tilled? If the decision is partly keyed to weather conditions then the farm decision-making process will lead to some amount of autonomous adjustment to climate change. Similarly, will the changes in tillage and irrigation practices, crop rotation schemes, crops, and crop processing and harvesting that are likely to occur over the next 100 years due to many factors also reflect changes in climate that are occurring simultaneously, or will farmers be unable to detect climate change and therefore fail to adapt these systems, becoming and remaining ill-adapted to the climate conditions occurring locally? If they adapt to current conditions (but cannot confidently look ahead) how maladjusted will their long-lived investments be after 3, 5, 10 or 20 years of continuous changes in climate?

Table 10.2 provides the range of estimates for the United States that have been generated based on quite different methodologies and assumptions about the extent to which adaptation will occur. While the table covers only the United States, it is

Table 10.2. Estimates of the impact of climate change on United States agriculture, percentage change

Climate scenario	Mendelsohn et al. (1994); without CO$_2$ effect		Darwin et al. (1995); without CO$_2$ effect				Rosenzweig and Parry (1994); USA average yield effects across crops	
			Effects on farm income		Effects on cereal prod.			
	1 Area wgts.	2 Rev. wgts.	3 Farm-level adaptation	4 Full adjustment	5 No adjustment	6 Full adjustment	7 No adjustment	8 Adjustment and CO$_2$
GISS	−1.8	+2.0	+4.1	−7.8	−24.4	−3.0	−14 to −21	0 to +17
GFDL	−1.2	+4.2	−16.1	+4.3	−38.0	−2.0	−23 to −29	+9 to −10
UKMO	−4.5	+1.1	−4.4	−5.4	−38.4	−5.0	−25 to −58	+1 to −20
OSU	−3.6	−0.7	−10.0	+5.8	−33.3	−5.2		

Notes: Mendelsohn et al. are annualized impact on land values as a percentage of total value of crop and livestock production. The values of crop and livestock production are for 1990 from Darwin et al. (1995). For a description of the Mendelsohn et al. methodology see Mendelsohn et al. (1994). The simulations of the model for the GISS, GFDL, UKMO and OSU reported above were provided by personal communication with Mendelsohn, 29 March 1995. Darwin et al. results are computed from simulations reported in Darwin et al. (1995). Rosenzweig and Parry summarize the range of crop yield impacts used in their 1994 study for the United States. The US average crop yield shocks estimated by them were reported in Reilly et al. (1993). Specific crop yield studies which were, in part, the basis for these estimates were reported in Rosenzweig et al. (1994).

GISS: Goddard Institute for Space Studies.
GFDL: Geophysical Fluid Dynamics Laboratory.
UKMO: United Kingdom Meteorological Office.
OSU: Oregon State University.

likely that applying this range of approaches in other regions would also generate a similar range of estimates. The Mendelsohn *et al.* (1994) estimates (columns 1 and 2) are based on an econometric model estimated on cross-section data and reflect, according to the authors, long-run, full adjustment of USA agriculture to a climate change shock. The methodology does not allow consideration of how crop prices may change and thus may be most comparable to the initial crop yield shock used in other methodologies. Except for column 8, none of the reported estimates consider the direct effect of CO_2 on plant growth. Unfortunately, the wide ranging methodologies do not or have not generally reported results that are directly comparable, thus some interpretation is necessary.

The starkest difference in methodology is between Mendelsohn *et al.* (1994) and Rosenzweig and Parry (1994). Columns 1 and 2 reflect results from models estimated with different weights on the individual observations. Mendelsohn *et al.* (1994) suggest the column 2 estimates based on revenue weights are more appropriate because they reflect the economic value of crops. They suggest that the more negative estimates based on area weights (column 1) reflect the type of bias that may be introduced by focusing on cereal crops which generally have a lower value per hectare than many other crops such as fruits and vegetables. Contrasting the climate impact shock they estimate (column 2) with the types of yield shocks estimated by Rosenzweig and Parry (1994) (column 7) provides a dramatically different picture of the impact of climate change on US agriculture. Rosenzweig and Parry include some adjustments but, unfortunately, the yield shocks for the United States comparable to the Mendelsohn *et al.* study (climate change and adaptation with no CO_2 effect) have not been reported. However, in their study adaptation did not have a particularly powerful effect on mitigating losses as reported by Reilly and Hohmann (1993). The relatively more benign impacts for the USA in the Rosenzweig and Parry yield estimates (column 8, with the CO_2 and adaptation) are, in a large part, less severe because of the CO_2. Thus, different methodologies, including adaptation but not the CO_2 effect, apparently produce estimates of impact for four major climate scenarios in the order of −1 to +5% using a Mendelsohn *et al.* methodology but on the order of −10 to −25% using the Rosenzweig and Parry methodology. In deriving the −10 to −25% range, I assume that adaption in their study may have reduced losses by 5 to 10% whereas the CO_2 fertilization effect reduced losses by 75 to 100%, which is the relative importance of these two factors on a global basis as in their data as estimated by Reilly and Hohmann (1993).

The Darwin *et al.* (1995) study used an independently derived set of climate shocks, representing climate change as a change in land class where the productivity of each land class was estimated from current data. Their methodology for estimating the direct effect of climate was more akin to Mendelsohn *et al.* (1994), using the observed differences in production across geographically varying climate as the basis for the projections. Their results, columns 3-6, help explain and confirm some of the differences between the other two studies. The initial shock on United States cereal production in the Darwin *et al.* (1995) study (column 5) is similarly (and generally more) severe than the yield shocks estimated by Rosenzweig and Parry (1994) (column 7). However, Darwin *et al.* (1995) estimate that, by just considering the immediate farm-level adjustment (without price changes and without expansion of agricultural production into new areas), farmers could offset between 70 and 120% of the initial losses (i.e., comparing column 5 and column 3). Note that this comparison is between impacts on cereal production and impacts on farm income, which is comparable (given that the simulation in column 3 does not allow prices to change) except that farm income includes impacts to agriculture for livestock and non-grains production as well. Columns 4 and 6 provide the estimates after full-adjustment including changes in world prices and trade, for cereal production and farm income. Note that the farm income effects with full adjustment (column 4) are sometimes worse than the farm income effects with only farm-level adaptation (column 3) because the Darwin *et al.* (1995) study considers worldwide effects with international trade. Thus, the impacts that occur in the rest of the world under the GISS and UKMO climate scenarios lead the US to lose international comparative advantage once full adjustment of international markets is considered.

Together these three studies indicate the wide range of estimated impacts for the same region and same climate scenarios depending on the methodology used. Mendelsohn *et al.* (1994) and Darwin *et al.* (1995) use methodologies that they argue more completely consider adaptation and they find impacts after adaptation to be generally less than Rosenzweig and Parry (1994), but even between these two approaches there are significant differences in estimated impacts for some climate scenarios in comparable estimates (columns 2 and 3).

The above discussion identified four separate factors that contribute to widely varying estimates of regional impacts of climate change apart from how or whether the CO_2 effect on crops is included in the simulation. These factors – varying climate scenarios, wide variation across sites within a region, how genetic variability across known crop varieties is addressed within the crop-response-modelling approach, and differences across impact methodologies particularly in how different methods address the capability of farmers to adapt – appear to be of roughly equal magnitude in explaining the wide range of estimates.

GLOBAL STUDIES AND THEIR IMPLICATIONS FOR REGIONAL EFFECTS

Accurate consideration of national and local food supply and economic effects depends on an appraisal of changes in global food supply and prices. International markets can moderate or reinforce local and national changes. In 1988, for example, drought presented a more severe threat because it occurred coincidentally in several of the major grain-growing regions of the world. Reilly *et al.* (1994) demonstrate that considering country-level production impacts of climate change in the absence of consideration of the global impacts can generate highly misleading results. Agricultural exporting countries, whose productivity is reduced by climate change, may find themselves with a financial bonanza if world agricultural prices rise because of climate change. These same countries may suffer significant economic loss if climate change turns out to be generally beneficial to world agriculture even if agricultural productivity in their country benefits. This feature of the agricultural economy is well-known and reflects what is, in aggregate, an inelastic demand for food. This point, which is a fundamental observation of agricultural economists, means that absolutely no implications for food availability, price or farm financial success can be drawn from local and country-level estimates of production impacts of climate change unless one assumes that production changes around the world will generally balance to leave little impact on global production and prices. A country may attempt to carry out a set of policies that maintains a neutral effect on the country's agricultural sector *vis-à-vis* the rest of the world, but maintaining such policies will generally entail significant economic cost through subsidization of domestic agricultural production and/or consumption or through import or export controls. There are many different ways these costs may be borne (higher food prices, government expenditures, lost efficiency in the producing sector, lost export opportunities) depending on how the policies are structured.

There are now a number of different attempts to estimate the impacts of climate change on global agriculture, in part to consider the global impacts but more importantly to consider more accurately what the regional impacts could be recognizing that what happens to global agriculture due to climate change will likely be more important for the viability and economic success of local agriculture than what happens to local production potential itself. Kane *et al.* (1992) and Tobey *et al.* (1992) examined the sensitivity of agriculture to potential yield losses in major temperate grain-growing regions based on very stylized climate change impacts. They loosely linked the potential for yield losses in temperate regions to climate projections that showed increasing aridity in the continental mid-latitude areas. They made alternative assumptions about how agriculture might be affected in higher latitudes and in the tropics. They also developed scenarios that reflected the estimated yield impacts for different parts of the world that were summarized in the 1990 Intergovernmental Panel on Climate Change assessment (Parry *et al.*, 1990). The yield response estimates used by Rosenzweig and Parry (1994) also reported in Fischer *et al.* (this volume) were also the basis of Reilly *et al.* (1994) and in greater detail Reilly *et al.* (1993). Many of the general conclusions are similar between the studies indicating that, given a set of yield shocks, economic modelling of international markets in itself is not a major source of difference in results even though there are major differences in the modelling approaches. Rather these different economic modelling approaches focus on different aspects and degrees of detail of agricultural economic interactions among crops, livestock, land use and the rest of the economy.

Among the issues that give rise to uncertainty in these studies are the following factors:

1. The timing of expected climate change. For example, Rosenzweig and Parry

(1994) assume that the 4.0 to 5.2°C scenarios occur in 2060, but the most recent IPCC work suggests the mean estimate for 2060 is closer to 1.5°C and that the range of global temperature impacts by 2100 is likely to be between 1.0 and 5°C.

2. Aggregation from detailed sites. Detailed plant growth models, the basis for many studies, require daily temperature and precipitation records for a 10- to 30-year historical climate record and detailed soil data, limiting the number of sites for which data are readily available and that can be practicably assessed. An alternative approach (Leemans and Solomon, 1993; Carter et al., 1991) makes use of geographic information system databases that contain more extensive information on current climates across the world. These efforts have not been linked to an economic model. Results confirm the pattern of relative decreased crop potential in the tropical areas and increased potential in the northern areas but are not aggregated to determine the net global effect.

3. Coverage of agricultural activities. Simulation of crop response models has been limited to a few major crops for a region, usually important grain crops, with yield effects extended to other crops. Left out are the indirect impacts of climate change through impacts on insect, disease and weed pests; on soils; and on livestock production. Mendelsohn et al. (1994) argue that their statistical approach accounts for all agricultural activities, implicitly accounting for the full effects of climate.

4. Other resource changes and competition for resources from other sectors. Land and water resource allocation is a conspicuous limitation in global studies. Water demand for other uses will grow, water use may have reached or passed sustainable levels of use in some areas, irrigation is responsible for salinization and land degradation, and water pricing and water system management are far from efficient under current conditions (e.g., Umali, 1993; Moore, 1991). Climate change also will affect demand for resources from other sectors.

The Darwin et al. (1995) study addresses many of these considerations in a global model including eight world regions. An Applied Computable General Equilibrium Model (ACGE), land and climatic resource changes are based on a geographic information system; changing climate shifts the distribution of land across several agroclimatic land classes. Other resource-using sectors are included and are also affected by climate change. The model is a static one, imposing climate change on current economic and agricultural markets and thus does not directly address the issue of timing.

The global results (Table 10.3) are comparable to Rosenzweig and Parry in terms of direct supply impacts for the world in the 'no adaptation' case, but the study finds that adaptation is able to turn global losses to small global benefits (unrestricted case). Even when the model is constrained to continue to produce on existing amounts of land within each region and prices are not allowed to respond, adaptation mitigates a significant share of the losses. These results contrast with those of Rosenzweig and

Table 10.3. Percentage changes in the supply and production of cereals for the world by climate change scenario

	Supply		Production	
	No adaptation	Land use fixed	Land use fixed	No restrictions
World				
GISS	-22.6	-2.4	0.2	0.9
GFDL	-23.5	-4.4	-0.6	0.3
UKMO	-29.3	-6.4	-0.2	1.2
OSU	-18.6	-3.9	-0.5	0.2

Note: Changes in supply represent the additional quantities firms would be willing to sell *at 1990 prices* under the alternative climate. Changes in production represent changes in equilibrium quantities.

Source: Darwin *et al.* (1995).

Parry (1994), giving generally smaller impacts and possible benefits even without the CO_2 effect and in that they show adaptation to be quite important.

Again, the global results are important because they are the first step in considering whether a local economy's consumers will be able to purchase food if it is unavailable domestically, how local producers may be affected by changes in demand for their crops, or how the cost of a country's agricultural policies may change because of changing international market conditions.

REGIONAL VULNERABILITY

The previous sections documented the wide range of uncertainty in the potential direction and magnitude of climate change impact. While many new studies have been conducted, most have focused on specific climate scenarios associated with $2\times CO_2$ GCM scenarios or arbitrary changes in climatic conditions to provide evidence of the general sensitivity of agriculture and crop production to climate change. The wide range of estimates limits the ability to extend, interpolate or extrapolate from the specific climate scenarios used in these studies to 'more' or 'less' climate change or to draw implications for impacts beyond the sites where studies were conducted.

Given these uncertainties in both magnitude and direction of impact, a key issue is *vulnerability* to possible climate change. Vulnerability is used here to mean the *potential* for negative consequences that are difficult to ameliorate through adaptive measures given the range of possible climate changes that might reasonably occur. Defining an area or population as vulnerable is, thus, not a prediction of negative consequences of climate change; it is an indication that across the range of possible climate changes, there are some climatic outcomes that would lead to relatively more serious consequences for the region than for other regions.

Vulnerability has been used rather loosely in many discussions. Before discussing some of the research that has examined potential vulnerability, I introduce a more formal definition. For the sake of simplicity, consider that climate can be described as a single variable, C. We are uncertain about what value C will take at some future point but we can describe the probability, p, that C will take on a specific value by the probability density function f(C). Further consider that we are able to describe the sensitivity of agriculture, A, to changes in climate by the function g(C). We can then define the expected loss function, L(C) as f(C) x g(C). A population, region, or crop is relatively more vulnerable under this definition if the area under L(C) where damages occur is larger than for a comparison population, region, or crop. Thus, I use the term vulnerability to describe only that portion of L(C) where damages occur. For other purposes, it useful to consider expected (net) damage (or benefit), that is the mean of the values of loss function which is a probability weighted mean of the damages.

Two, purely illustrative, numerical examples are plotted in Figure 10.1. For these examples I have chosen to represent f(C) as a gamma distribution. In panel A, damages are represented by a quadratic function while in panel B, damages are represented by a logarithmic function. These choices illustrate just two of the ways that our expectations about the degree of future climate change and our understanding of the sensitivity of an agricultural system to climate change may interact. In these numerical illustrations, the system characterized by quadratic losses (Panel A) is more vulnerable to loss than the system described by logarithmic losses. Even though the quadratic sensitivity to climate leads to potentially larger losses at extreme temperature change, the system is less vulnerable because climate change is not likely to be that extreme in this example. In fact, the small region of beneficial warming (negative damages) in Panel A gives rise to a substantial possibility of beneficial effects of warming for the system described in this panel. In Panel B, in contrast, damages initially rise sharply but the rate of increase slows. This characterization of system sensitivity indicates damages across the entire range of expected temperature change. Even though damages do not have the potential to become as severe as in Panel A, the system is more vulnerable to damage because climate is more likely to be in the relatively higher damage range of the sensitivity function.

In practice, multiple dimensions of climate affect any agricultural system. The simple characterization in Figure 10.1 is meant to make the definition of vulnerability mathematically precise even though it is not possible at this time to estimate formally the multidimensional, joint distribution of important climate variables. Nor do we precisely know the damage function that relates changes in these climate variables to agricultural impacts. The advantage is to make explicit that we must consider both our expectations with regard to climate and damage sensitivity. To make the example concrete, a semi-arid area may be extremely sensitive to damage if it becomes more arid. But, if our expectation is that it is highly likely that the region will become wetter, the region is not vulnerable. Another region in a humid agroclimatic zone may be vulnerable if substantial warming and drying are likely for the area.

Up to this point, I have not been explicit with regard to what I propose to measure as a damage. The existing literature suggests several different possible measures and therefore several different dimensions of vulnerability. Many studies focus on crop *yields*. Evidence suggests that yields of crops grown where temperature could easily

Figure 10.1. Defining vulnerability

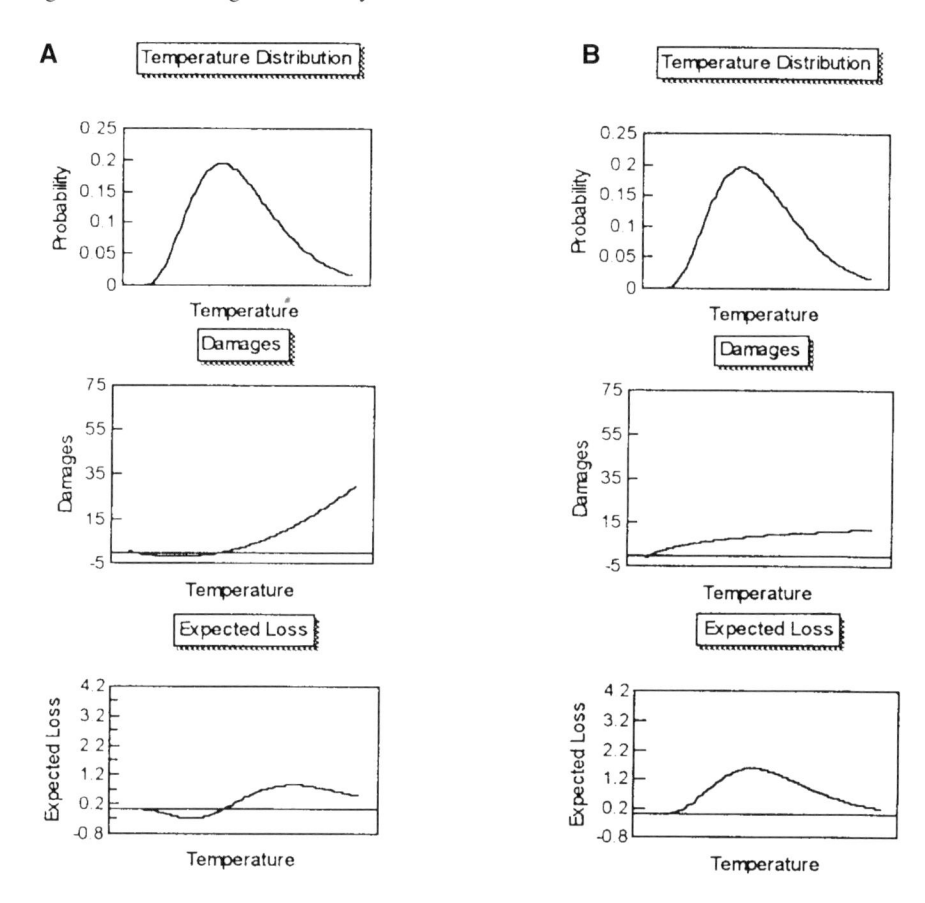

exceed threshold values during critical crop growth periods are more vulnerable to warming (e.g., rice sterility: Matthews *et al.*, 1994a, b).

Farmer or farm sector vulnerability may be measured in terms of impact on profitability or viability of the farming system. Farmers with limited financial resources and farming systems with few adaptive technological opportunities available to limit or reverse adverse climate change may suffer significant disruption and financial loss for relatively small changes in crop yields and productivity or these farms may be located in areas more likely to suffer yield losses. For example, Parry *et al.* (1988a, b) focused on semi-arid and cool temperate and cold agricultural areas as those that might be more clearly affected by climate change and climate variability.

Regional economic vulnerability reflects the sensitivity of the regional or national economy to farm sector and climate change impacts. A regional economy that offers only limited employment alternatives for workers dislocated by the changing profitability of farming and other climatically sensitive sectors may be relatively more vulnerable than those that are economically diverse. For example, Rosenberg (1993) examined the Great Plains area of the United States because of its heavy dependence on agriculture. Increasing aridity is expected in this region under climate change and thus it was considered to be potentially more economically vulnerable than other regions in the USA.

Hunger vulnerability has been used to mean the 'aggregate measure of the factors that influence exposure to hunger and predisposition to its consequences' involving 'interactions of climate change, resource constraints, population growth, and economic development' (Downing, 1992; Bohl *et al.*, 1994). Downing (1992) concluded that the semi-extensive farming zone, on the margin of more intensive land uses, appears to be particularly sensitive to small changes in climate. Socio-economic groups in such areas, already vulnerable in terms of self-sufficiency and food security, could be further marginalized. We likely ought not to look only at agriculturally dependent people. The means people have within society and the family to obtain food and how their allocation will change if production potential changes must be considered. A poor urban household may suffer due to production losses elsewhere in the region while the rural farmer may continue to eat. Or, women and children of rural peasant farms may go hungry, while 'excess' production from the region is sold. Assessing who has the means and rights to food during shortfalls is thus likely more critical in a climate vulnerability study than assessing how production may change. For hunger and famine in general, the relative importance of acquiring (versus producing) food has been demonstrated by Sen (1981, 1993).

Given the diverse currently existing conditions, the geographical variation likely to exist in any climate change scenario, and the wide uncertainty that must be associated with local prediction of future climates, some vulnerable agricultural areas and populations likely exist for nearly every region even if the expected value for the region is a net benefit. This makes vulnerability a relative concept – while there may be a few areas where even the most extreme climate change we can imagine would not generate losses, in general, the problem is to consider whether a particular region or population is relatively more vulnerable than others.

While perhaps most difficult to evaluate, vulnerability in terms of hunger and malnutrition ought to be the first concern. If so, then we can almost certainly eliminate the richer countries of the world as vulnerable. For poorer regions, it is the poorest members of these areas or those that could be made poor by climate change that are most at risk. The wide uncertainty with regard to local and regional climate change means it is difficult to rule out negative possibilities for any area. Thus, without even considering specific climate scenarios, we can assert that those who are currently poor, malnourished and dependent on local production for food are the most vulnerable in terms of hunger and malnutrition to climate change of the world's populations. Similarly, severe economic vulnerability is most likely where a large share of the population depend on agriculture, leaving little alternative employment opportunities. Again, we need not assess climate scenarios or projected yield changes to establish where these vulnerable populations live. Given these considerations, Table 10.4 presents some of the critical dimensions of areas of the world that might be used to

Table 10.4. Basic regional agricultural indicators and vulnerability

	Sub-Saharan Africa	Near East/ North Africa	South Asia	Southeast Asia	East Asia	Oceania	Former USSR	Europe	Latin America	USA, Canada
Ag. land (%)*	41	27	55	36	51	57	27	47	36	27
Cropland (%)*	7	7	44	13	11	6	10	29	7	13
Irrigated (%)*	5	21	31	21	11	4	9	12	10	8
Land area (10⁶ha)	2 390	1 167	478	615	993	845	2 227	473	2 052	1 839
Climate	tropical; arid, humid	subtropical, tropical; arid	tropical, subtropical; humid, arid	tropical; humid	subtropical, temperate oceanic, continental; humid	tropical, temperate, oceanic subtropical; arid, humid	polar, continental, temperate oceanic; humid, arid	temperate oceanic, some subtropical; humid, arid	tropical, subtropical; mostly humid	continental subtropical, polar, temp. oceanic, mid, arid
Pop. (10⁶)	566	287	1145	451	1333	27	289	510	447	277
Ag. pop. (%)	62	32	63	49	59	17	13	8	27	3
Pop./ha cropland	3.6	3.4	5.4	5.7	12.6	0.5	1.3	3.7	2.9	1.2
Ag. prod. (10⁶t)										
Cereals	57	79	258	130	433	24	180	255	111	388
Roots and tubers	111	12.5	26	50	159	3	65	79	45	22
Pulses	5.7	4.1	14.4	2.5	6.3	2	6	7	5.8	2.2
S. cane and beet	60	39	297	181	103	32	62	144	494	56
Meat	6.7	5.5	5.7	6.4	39.6	4.5	17	42	20.5	33.5
1991 GNP/cap.**	350	1 940	320	930	590	13 780	2 700	15 300	2 390	22 100
Annual growth**	-1.2	-2.4	3.1	3.9	7.1	1.5	N.A.	2.2	-0.3	1.7
Ag. (% of GDP)**	>30 %	10-19%	>30%	20 to>30%	20-29%	<6%	10-29%	<6%	10-19%	<6%

* Agricultural land includes grazing and cropland, reported as a percentage of total land area. Cropland is reported as a percentage of agricultural land. Irrigated area is reported as a percentage of cropland.

** GNP is in 1991 USA dollars; annual growth, per cent per annum, is for the period 1980-1991. Source: Computed from FAO (1992); GNP per caput, GNP growth rates, and agriculture as a share of the economy are from *World Development Indicators* in World Bank (1993) and temperature and climate classes from Rötter et al. (1995). Note: East Asia GNP excludes Japan. Also, regional GNP data generally include only those countries for which data are given in Table 1 in *World Development Indicators*. Countries with more than 4 million population for which GNP data are not available include Vietnam, Democratic Republic of Korea, Afghanistan, Cuba, Iraq, Myanmar, Cambodia, Zaire, Somalia, Libya and Angola.

assess vulnerability. While the table is too aggregated to identify specifically vulnerable populations, it is indicative of where many of these people are likely to be.

Because of the wide range of uncertainty in precipitation, the only climatic dimension likely to enter significantly in an assessment of vulnerability is temperature. Cool regions are more likely to be limited by low temperatures and thus warming may prove beneficial – these areas may still suffer if precipitation changes are adverse. But, further warming is unlikely to benefit already warm regions. Thus, global warming appears somewhat stacked against the already warm areas. Coincidentally (or not), these regions tend to also be home to some of the world's poorest.

The focus on hunger and malnutrition as a first priority does not mean that other types of vulnerability are unimportant. Regional economic development, land degradation, or increased environmental stress resulting from agricultural production under a changed climate are important concerns as well.

ADAPTATION POTENTIAL AND POLICIES

The hierarchy of damage considerations as above – hunger, regional economic, farmer/farm sector, and yield vulnerability – helps to focus on adaptive strategies that reduce vulnerability. How can we avoid yield failures? If yields fail, what other crops can be grown? If farming becomes uneconomic, what other opportunities for employment exist? If the people of the region can no longer produce food, what other sources of food are available and how will they earn the income necessary to purchase food or what other means does the society in which they live have to provide food assistance?

Historically, farming systems have adapted to changing economic conditions, technology and resource availabilities and have kept pace with a growing population (Rosenberg, 1992; CAST, 1992). Evidence exists that agricultural innovation responds to economic incentives such as factor prices and can relocate geographically (Hayami and Ruttan, 1985; CAST, 1992). A number of studies indicate that adaptation and adjustment will be important to limit losses or to take advantage of improving climatic conditions (e.g., US National Academy of Sciences, 1991; Rosenberg, 1992; Rosenberg and Crosson, 1991; CAST, 1992; Mendelsohn *et al.*, 1994).

Despite the successful historical record, issues of future adaptation to climate change arise with regard to whether the rate of change of climate and required adaptation would add significantly to the disruption likely due to future changes in economic conditions, technology and resource availabilities (Gommes, 1993; Harvey, 1993; Kane and Reilly, 1993; Smit, 1993; Norse, 1994; Pittock, 1994; Reilly, 1994). If climate change is gradual, it may be a small factor that goes unnoticed by most farmers as they adjust to other more profound changes in agriculture stemming from new technology, increasing demand for food, and other environmental concerns such as pesticide use, water quality, and land preservation. However, some researchers see climate change as a significant addition to future stresses; where adapting to yet another stress such as climate change may be beyond the capability of the system. Part of the divergence in views may be due to different interpretations of adaptation which include: the prevention of loss, tolerating loss, or relocating to avoid loss (Smit, 1993). And, while the technological potential to adapt may exist, the socio-economic capability to adapt likely differs for different types of agricultural systems (Reilly and Hohmann, 1993).

THE TECHNOLOGICAL POTENTIAL TO ADAPT

Nearly all agricultural impact studies conducted over the past five years have considered some technological options for adapting to climate change. Among those that offer promise are:

- *Seasonal changes and sowing dates.* For frost-limited growing areas (i.e., temperate and cold areas), warming could extend the season, allowing planting of longer maturity annual varieties that achieve higher yields (e.g., Le Houerou, 1990; Rowntree, 1990a, b). For short-season crops such as wheat, rice, barley, oats, and many vegetable crops, extension of the growing season may allow more crops per year, autumn planting, or, where warming leads to regular summer highs above critical thresholds, a split season with a short summer fallow may be possible. For subtropical and tropical areas where growing season is limited by precipitation or where the cropping already occurs throughout the year, the ability to extend the growing season may be more limited and depends on how precipitation patterns change. A study for Thailand found yield losses in the warmer season partially offset by gains in the cooler season (Parry *et al.*, 1992).
- *Different crop variety or species.* For most major crops, varieties exist with a wide range of maturities and climatic tolerances. For example, Matthews *et al.* (1994a, b) identified wide genetic variability among rice varieties as a reasonably easy response to spikelet sterility in rice that occurred in simulations for South and Southeast Asia. Studies in Australia showed that responses to climate change are strongly cultivar dependent (Wang *et al.*, 1992). Longer-season cultivars were shown to provide a steadier yield under more variable conditions (Connor and Wang, 1993). In general, such changes may lead to higher yields or may only partly offset losses in yields or profitability. Crop diversification in Canada (Cohen *et al.*, 1992) and in China (Hulme *et al.*, 1992) has been identified as an adaptive response.
- *New crop varieties.* The genetic base is broad for most crops but limited for some (e.g., kiwi fruit). A study by Easterling *et al.* (1993) explored how hypothetical new varieties would respond to climate change (also reported in McKenney *et al.*, 1992). Heat, drought and pest resistance; salt tolerance, and general improvements in crop yield and quality would be beneficial (Smit, 1993). Genetic engineering and gene mapping offer the potential for introducing a wider range of traits. Difficulty in assuring traits are efficaciously expressed in the full plant, consumer concerns, profitability and regulatory hurdles have slowed the

introduction of genetically engineered varieties compared with early estimates (Reilly, 1989; Caswell *et al.*, 1994).

- *Water supply and irrigation systems.* Across studies, irrigated agriculture is in general less negatively affected than dryland agriculture but adding irrigation is costly and subject to the availability of water supplies. Climate change will also affect future water supplies. There is wide scope for enhancing irrigation efficiency through adoption of drip irrigation systems and other water-conserving technologies (FAO, 1989, 1991) but successful adoption will require substantial changes in how irrigation systems are managed and how water resources are priced. Because inadequate water systems are responsible for current problems of land degradation, and because competition for water is likely to increase, there likely will be a need for changes in the management and pricing of water regardless of whether and how climate changes (Vaux, 1990, 1991; World Bank, 1994). Tillage method and incorporation of crop residues are other means of increasing the useful water supply for cropping.
- *Other inputs and management adjustments.* Added nitrogen and other fertilizers would likely be necessary to take full advantage of the CO_2 effect. Where high levels of nitrogen are applied, nitrogen not used by the crop may be leached into the groundwater, runoff into surface water, or be released from the soil as nitrous oxide. Additional nitrogen in groundwater and surface water has been linked to health effects in humans and affects aquatic ecosystems. Studies have also considered a wider range of adjustments in tillage, grain drying and other field operations (Kaiser *et al.*, 1993; Smit, 1993).
- *Tillage.* Minimum and reduced tillage technologies in combination with planting of cover crops and green manure crops offers substantial possibilities to reverse existing soil organic matter, soil erosion, and nutrient loss and to combat potential further losses due to climate change (Rasmussen and Collins, 1991; Logan, 1991; Edwards *et al.*, 1992; Langdale *et al.*, 1992; Peterson *et al.*, 1993; Brinkman and Sombroek, this volume). Reduced and minimum tillage techniques have spread widely in some countries but are more limited in other regions. There is considerable current interest in transferring these techniques to other regions (Cameron and Oram, 1994).
- *Improved short-term climate prediction.* Linking agricultural management to seasonal climate predictions (currently largely based on the El Niño Southern Oscillation Phenomenon), where such predictions can be made with reliability, can allow management to adapt incrementally to climate change. Management/climate predictor links are an important and growing part of agricultural extension in both developed and developing countries (McKeon *et al.*, 1990, 1993; Nicholls and Wong, 1990).

THE SOCIO-ECONOMIC CAPABILITY TO ADAPT

While identifying many specific technological adaptation options, Smit (1993) concluded that necessary research on their cost and ease of adoption had not yet been conducted.

One measure of the potential for adaptation is to consider the historical record on past speeds of adoption of new technologies (Table 10.5). Adoption of new or different technologies depends on many factors: economic incentives; varying resource and climatic conditions; the existence of other technologies (transportation systems and markets); the availability of information; and the remaining economic life of equipment and structures (e.g., dams and water supply systems).

Specific technologies can only provide a successful adaptive response if they are adopted in appropriate situations. A variety of issues has been considered, including land-use planning, watershed management, disaster vulnerability assessment, consideration of port and rail adequacy, trade policy, and the various programmes countries use to encourage or control production, limit food prices, and manage resource inputs to agriculture (CAST, 1992; US OTA, 1993; Smit, 1993; Reilly *et al.*, 1994; Singh, 1994). For example, studies suggest that current agricultural institutions and policies in the USA may discourage farm management adaptation strategies such as altering crop mix, by supporting prices of crops not well-suited to a changing climate, providing disaster payments when crops fail, or prohibiting imports through import quotas (Lewandrowski and Brazee, 1993).

Existing gaps between best yields and the average farm yields remain unexplained but many are due in part to socio-economic considerations (Oram and Hojjati, 1995; Bumb, 1995); this adds considerable uncertainty to estimates of the potential for adaptation particularly in developing countries. For example, Baethgen (1994) found that a better selection of wheat variety combined with improved fertilizer regime

Table 10.5. Speed of adoption for some major adaptation measures

Adaptation	Adjustment time (yrs)	Reference
Variety adoption	3-14	Dalrymple, 1986; Griliches, 1957; Plucknett *et al.*, 1987; CIMMYT, 1991
Dams and irrigation	50-100	James and Lee, 1971; Howe, 1971
Variety development	8-15	Plucknett *et al.*, 1987; Knudson, 1988
Tillage systems	10-12	Hill *et al.*, 1994; Dickey *et al.*, 1987; Schertz, 1988
New crop adoption: soybeans	15-30	FAO, *Agrostat* – various years
Opening new lands	3-10	Medvedev, 1987; Plusquellec, 1990
Irrigation equipment	20-25	Turner and Anderson, 1980
Transportation system	3-5	(A. Talvitie, World Bank, pers. comm., 1994)
Fertilizer adoption	10	Pieri, 1992; Thompson and Wan, 1992

could double yields achieved at a site in Uruguay to 6 t/ha under the current climate with current management practices. Under the UKMO climate scenario, yields fell to 5 t/ha, still well above 2.5-3.0 t/ha currently achieved by farmers in the area. On the other hand, Singh (1994) concluded that the normal need to plan for storms and extreme weather events in Pacific island nations creates significant resiliency. Whether technologies meet the self-described needs of peasant farmers is critical in their adaptation (Cáceres, 1993). Other studies document how individuals cope with environmental disasters, identifying how strongly political, economic and ethnic factors interact to facilitate or prevent coping in cases ranging from the Dust Bowl disaster in the USA to floods in Bangladesh to famines in the Sudan, Ethiopia and Mozambique (McGregor, 1994). These considerations indicate the need for local capability to develop and evaluate potential adaptations that fit changing conditions (COSEPUP, 1992). Important strategies for improving the ability of agriculture to respond to diverse demands and pressures, drawn from past efforts to transfer technology and provide assistance for agricultural development, include:

- Improved training and general education of populations dependent on agriculture, particularly in countries where education of rural workers is currently limited. Agronomic experts can provide guidance on possible strategies and technologies that may be effective. Farmers must evaluate and compare these options to find those appropriate to their needs and the circumstances of their farm.
- Identification of the present vulnerabilities of agricultural systems, causes of resource degradation, and existing systems that are resilient and sustainable. Strategies that are effective in dealing with current climate variability and resource degradation are also likely to increase resilience and adaptability to future climate change.
- Agricultural research centres and experiment stations can examine the 'robustness' of present farming systems (i.e. their resilience to extremes of heat, cold, frost, water shortage, pest damage and other factors) and also test the robustness of new farming strategies as they are developed to meet changes in climate, technology, prices, costs and other factors.
- Interactive communication that brings research results to farmers and farmers' problems, perspectives and successes to researchers is an essential part of the agricultural research system.
- Agricultural research provides a foundation for adaptation. Genetic variability for most major crops is wide relative to projected climate change. Preservation and effective use of this genetic material would provide the basis for new variety development. Continually changing climate is likely to increase the value of networks of experiment stations that can share genetic material and research results.
- Food programmes and other social security programmes would provide insurance against local supply changes. International famine and hunger programmes need to be considered with respect to their adequacy.
- Transportation, distribution and market integration provide the infrastructure to

supply food during crop shortfalls that might be induced in some regions because of climate variability or worsening of agricultural conditions.
- Existing policies may limit efficient response to climate change. Changes in policies such as crop subsidy schemes, land tenure systems, water pricing and allocation, and international trade barriers could increase the adaptive capability of agriculture.

Many of the above strategies will be beneficial regardless of how or whether climate changes. Goals and objectives among countries and farmers vary considerably. Current climate conditions and likely future climates also vary. Building the capability to detect change and evaluate possible responses is fundamental to successful adaptation. Thus, even without having clear predictions of climate change, is it possible to identify some strategies that reduce potential vulnerability.

REFERENCES

Adams, R.M., McCarl, B.A., Dudek, D.J. and Glyer, J.D. 1988. Implications of global climate change for western agriculture. *Western J. Agric. Economics* **13**: 348-356.

Baethgen, W.E. 1994. Impact of climate change on barley in Uruguay: yield changes and analysis of nitrogen management systems. In: *Implications of Climate Change for International Agriculture: Crop Modelling Study*. C. Rosenzweig and A. Iglesias (eds.). US Environmental Protection Agency, Washington DC. pp. 1-13.

Bohl, H.G., Downing, T.E. and Watts, M.J. 1994. Climate change and social vulnerability: toward a sociology and geography of food insecurity. *Global Environ. Change* **4**(1): 37-48.

Brklacich, M. and Smit, B. 1992. Implications of changes in climatic averages and variability on food production opportunities in Ontario, Canada. *Climatic Change* **20**: 1-21.

Brklacich, M., Stewart, R., Kirkwood, V. and Muma, R. 1994. Effects of global climate change on wheat yields in the Canadian prairie. In: *Implications of Climate Change for International Agriculture: Crop Modelling Study*. C. Rosenzweig and A. Iglesias (eds.). US Environmental Protection Agency, Washington DC. pp. 1-23.

Bumb, B. 1995: Growth potential of existing technology is insufficiently tapped. In: *Population and Food in the Early 21st Century: Meeting Future Food Demand of an Increasing World Population*. N. Islam (ed.). Occasional Paper. International Food Policy Research Institute (IFPRI), Washington DC. pp. 191-205.

Cáceres, D.M. 1993. *Peasant Strategies and Models of Technological Change: A Case Study from Central Argentina*. M.Phil. Thesis. University of Manchester, Manchester.

Cameron, D. and Oram, P. 1994. *Minimum and Reduced Tillage: Its Use in North America and Western Europe and its Potential Application in Eastern Europe,*

Russia, and Central Asia. International Food Policy Research Institute, Washington DC. 121 p.

Carter, T.R., Porter, J.R. and Parry, M.L. 1991. Climatic warming and crop potential in Europe: prospects and uncertainties. *Global Environ. Change* **1**: 291-312.

CAST [Council for Agricultural Science and Technology]. 1992. *Preparing U.S. Agriculture for Global Climate Change*. Task Force Report No. 119, CAST, Ames, Iowa. 96 p.

Caswell, M.F., Fuglie, K.O and Klotz, C.A. 1994. *Agricultural Biotechnology: An Economic Perspective*. AER No. 687. US Department of Agriculture, Washington DC. 52 p.

CIMMYT. 1991. *Annual Report: Improving the Productivity of Maize and Wheat in Developing Countries: An Assessment of Impact*. Centro Internacional de Mejoramiento de Maiz y Trigo, Mexico City.

Cohen, S., Wheaton, E. and Masterton, J. 1992. Impacts of climatic change scenarios in the prairie provinces: A case study from Canada. *SRC Publication No. E-2900-4-D-92*. Saskatchewan Research Council, Saskatoon, Canada.

Connor, D.J. and Wang, Y.P. 1993. Climatic change and the Australian wheat crop. In: *Proceedings of the Third Symposium on the Impact of Climatic Change on Agricultural Production in the Pacific Rim*. S. Geng (ed.). Central Weather Bureau, Ministry of Transport and Communications, Republic of China.

COSEPUP [Committee on Science, Engineering and Public Policy]. 1992. *Policy Implications of Global Warming*. National Academy Press, Washington DC.

Dalrymple, D.G. 1986. *Development and Spread of High-Yielding Rice Varieties in Developing Countries*. 7th edn. US AID, Washington DC.

Darwin, R., Tsigas, M., Lewandrowski, J. and Raneses, A. 1995. *World Agriculture and Climate Change: Economic Adaptation*. Report No. AER-709. Economic Research Service, Washington DC.

Dickey, E.C., Jasa, P.J., Dolesh, B.J., Brown, L.A. and Rockwell, S.K. 1987. Conservation tillage: perceived and actual use. *J. Soil Water Conserv.* (Nov.-Dec.): 431-434.

Downing, T.E. 1992. *Climate Change and Vulnerable Places: Global Food Security and Country Studies in Zimbabwe, Kenya, Senegal, and Chile*. Research Report No. 1, Environmental Change Unit, University of Oxford. 54 p.

Easterling, W.E. III, Crosson, P.R., Rosenberg, N.J., McKenney, M., Katz, L.A. and Lemon, K. 1993. Agricultural impacts of and responses to climate change in the Missouri-Iowa-Nebraska-Kansas (MINK) region. *Climatic Change* **24**: 23-61.

Edwards, J.H., Wood, C.W., Thurow, D.L. and Ruff, M.E. 1992. Tillage and crop rotation effects on fertility status of a hapludult soil. *Soil Science Society of America Journal* **56**: 1577-1582.

Fischer, G., Frohberg, K., Parry, M.L. and Rosenzweig, C. 1994. Climate change and world food supply, demand and trade. *Global Environ. Change.* **4**(1): 7-23.

FAO. 1961-1990. *Agrostat 1961-1990*. FAO, Rome.

FAO. 1989. Guidelines for designing and evaluating surface irrigation systems. *Irrigation and Drainage Paper 45*. FAO, Rome.

FAO. 1991. Water harvesting. *AGL Miscellaneous Paper 17*. FAO, Rome.

FAO. 1992. *Agrostat*. FAO Statistics Division, Rome.

Gommes, R. 1993. Current climate and population constraints on world agriculture. In: *Agricultural Dimensions of Global Climate Change*. H.M. Kaiser and T.E. Drennen (eds.). St. Lucie Press, Delray Beach, Florida. pp. 67-86.

Griliches, Z. 1957. Hybrid corn: an exploration in the economics of technological change. *Econometrica*. **25**: 501-522.

Harvey, L.D.D. 1993. Comments on 'an empirical study of the economic effects of climate change on world agriculture'. *Climatic Change* **21**: 273-275.

Hayami, Y. and Ruttan, V.W. 1985. Disequilibrium in world agriculture. *Agricultural Development: An International Perspective*. Johns Hopkins Univ. Press, Baltimore. pp. 367-415.

Hill, P.R., Griffith, D.R., Steinhardt, G.C. and Parsons, S.D. 1994. *The Evolution and History of No-Till Farming in the Midwest*. Purdue University, West Lafayette, IN.

Howe, C. 1971. Benefit-cost analysis for water system planning. *Water Resources Monograph 2*. American Geophysical Union, Washington DC.

Hulme, M., Wigley, T., Jiang, T., Zhao, Z., Wang, F., Ding, Y., Leemans, R. and Markham, A. 1992. *Climate Change due to the Greenhouse Effect and its Implications for China*. CRU/WWF/SMA, World Wide Fund for Nature, Gland, Switzerland.

James, L.D. and Lee, R.R. 1971. *Economics of Water Resources Planning*. McGraw-Hill, New York.

Kaiser, H.M., Riha, S.J., Wilks, D.S., Rossiter, D.G. and Sampath, R. 1993. A farm-level analysis of economic and agronomic impacts of gradual global warming. *Amer. J. Agr. Econ.* **75**: 387-398.

Kane, S. and Reilly, J. 1993. Reply to comment by L.D. Danny Harvey on 'an empirical study of the economic effects of climate change on world agriculture'. *Climatic Change* **21**: 277-279.

Kane, S., Reilly, J. and Tobey, J. 1992. An empirical study of the economic effects of climate change on world agriculture. *Clim. Change* **21**: 17-35.

Knudson, M. 1988. *The Research and Development of Competing Biological Innovations: The Case of Semi- and Hybrid Wheats*. Ph.D. dissertation, University of Minnesota, St. Paul.

Langdale, G.W., West, L.T. and Bruce, R.R. 1992. Restoration of eroded soil with conservation tillage. *Soil Technology* **5**: 81-90.

Le Houerou, H.N. 1990. Global change: vegetation, ecosystems and land use in the Mediterranean basin by the twenty-first century. *Israel J. Botany* **39**: 481-508.

Leemans, R. and Solomon, A.M. 1993. Modelling the potential in yield and distribution of the earth's crops under a warmed climate. *Clim. Res.* **3**: 79-96.

Lewandrowski, J.K. and Brazee, R.J. 1993. Farm programs and climate change. *Climatic Change* **23**: 1-20.

Logan, T.J. 1991. Tillage systems and soil properties in North America. *Tillage Research* **20**: 241-270.

Matthews, R.B., Kropff, M.J. and Bachelet, D. 1994a. Climate change and rice production in Asia. *Entwicklung und Ländlicherraum* **1**: 16-19.

Matthews, R.B., Kropff, M.J., Bachelet, D. and van Laar, H.H. 1994b. The impact of global climate change on rice production in Asia: A simulation study. *Report No. ERL-COR-821*. US Environmental Protection Agency, Environmental Research Laboratory, Corvallis.

McGregor, J. 1994. Climate change and involuntary migration. *Food Policy*, **19**(2): 121-132.

McKenney, M.S., Easterling, W.E. and Rosenberg, N.J. 1992. Simulation of crop productivity and responses to climate change in the year 2030: The role of future technologies, adjustments and adaptations. *Agricultural and Forest Meteorology* **59**: 103-127.

McKeon, G.M., Day, K.A., Howden, S.M., Mott, J.J., Orr, D.M., Scattini, W.J. and Weston, E.J. 1990. Management for pastoral production in northern Australian savannas. *Journal of Biogeography* **17**: 355-372.

McKeon, G.M., Howden, S.M., Abel, N.O.J. and King, J.M. 1993. Climate change: adapting tropical and subtropical grasslands. In: *Proceedings XVII International Grassland Congress*. pp. 1181-1190.

Medvedev, Z.A. 1987. *Soviet Agriculture*. W.W. Norton, New York.

Mendelsohn, R., Nordhaus, W. and Shaw, D. 1994. The impact of climate on agriculture: a Ricardian approach. *Amer. Econ. Rev.* **84**: 753-771.

Moore, M.R. 1991. The Bureau of Reclamation's new mandate for irrigation water conservation: purposes and policy alternatives. *Water Resources Res.* **27**: 145-155.

Nicholls, N. and Wong, K.K. 1990. Dependence of rainfall variability on mean rainfall, latitude and the Southern Oscillation. *Journal of Climatology* **3**: 163-170.

Norse, D. 1994. Multiple threats to regional food production: environment, economy, population? *Food Policy* **19**(2): 133-148.

Oram, P.A. and Hojjati, B. 1995. The growth potential of existing agricultural technology. In: *Population and Food in the Early 21st Century: Meeting Future Food Demand of an Increasing World Population*. N. Islam (ed.). Occasional Paper. International Food Policy Reserach Institute (IFPRI), Washington DC. pp. 167-189.

Parry, M.L., Carter, T.R. and Konijn, N.T. (eds.). 1988a. *The Impact of Climate Variations on Agriculture: Volume 1: Assessments in Cool Temperate and Cold Regions*. Kluwer Academic, Dordrecht. 876 p.

Parry, M.L., Carter, T.R. and Konijn, N.T. (eds.). 1988b. *The Impact of Climatic Variations on Agriculture: Vol. 2 Assessment in Semi-arid Regions*. Kluwer Academic, Dordrecht.

Parry, M.L., Duinker, P.N., Morison, J.I.L., Porter, J.H., Reilly, J. and Wright, L.J. 1990. Agriculture and forestry. In: *Climate Change: The IPCC Impacts Assessment*. W.J. McG. Tegart, G.W. Sheldon and D.C. Griffiths (eds.). Australian Government Printing Office, Canberra. pp. 2-1 - 2-45.

Parry, M.L., Blantran de Rozari, M., Chong, A.L. and Panich, S. (eds.). 1992. *The

Potential Socio-Economic Effects of Climate Change in South-East Asia. United Nations Environment Programme, Nairobi.

Peterson, G.A., Westfall, D.G. and Cole, C.V. 1993. Agroecosystem approach to soil and crop management research. *Soil Science Society of America Journal* **57**: 1354-1360.

Pieri, C. 1992. *Fertility of Soils: The Future for Farmers in the West African Savannah.* Springer-Verlag, Berlin.

Pittock, A.B. 1994. Climate and food supply. *Nature* **371**: 25.

Plucknett, D.L., Smith, N.J.H., Williams, J.T. and Anishetty, N.M. 1987. *Gene Banks and the World's Food.* Princeton University Press, Princeton.

Plusquellec, H. 1990. The Gezira Irrigation Scheme in Sudan: objectives, design, and performance. *Tech. Paper No. 120.* World Bank, Washington DC.

Rasmussen, P.E. and Collins, H.P. 1991. Long-term impacts of tillage, fertiliser and crop residue on soil organic matter in temperate semiarid regions. *Advances in Agronomy* **45**: 93-134.

Reilly, J. 1989. *Consumer Effects of Biotechnology.* AIB No. 581. US Department of Agriculture, Washington DC. 11 p.

Reilly, J. 1994. Crops and climate change. *Nature* **367**: 118-119.

Reilly, J. and Hohmann, N. 1993. Climate change and agriculture: the role of international trade. *Amer. Econ. Rev.* **83**: 306-312.

Reilly, J., Hohmann, N. and Kane, S. 1993. *Climate Change and Agriculture: Global and Regional Effects Using an Economic Model of International Trade.* MIT-CEEPR 93-012WP. Massachusetts Institute for Technology, Center for Energy and Environmental Policy, Boston.

Reilly, J., Hohmann, N. and Kane, S. 1994. Climate change and agricultural trade: who benefits, who loses? *Global Envir. Change* **4**: 24-36.

Reilly, J., Baethgen, W., Chege, F.E., van de Geijn, S.C., Erda, L., Iglesias, A., Kenny, G., Patterson, D., Rogasik, J., Rötter, R., Rosenzweig, C., Sombroek, W. and Westbrook, J. 1996. Agriculture in a changing climate: Impacts and adaptation, In: *Changing Climate: Impacts and Response Strategies*, Report of Working Group II of the Intergovernmental Panel on Climate Change. Chapter 13. Cambridge University Press, Cambridge, UK (in press).

Rosenberg, N.J. 1992. Adaptation of agriculture to climate change. *Climatic Change* **21**: 385-405.

Rosenberg, N.J. 1993. *Towards an Integrated Assessment of Climate Change: The MINK Study.* Kluwer Academic Publishers, Boston. 173 p.

Rosenberg, N.J. and Crosson, P.R. 1991. *Processes for Identifying Regional Influences of and Responses to Increasing Atmospheric CO_2 and Climate Change: the MINK Project, An Overview.* DOE/RL/01830T-H5. Resources for the Future and US Department of Energy, Washington DC. 35 p.

Rosenzweig, C. 1985. Potential CO_2-induced climate effects on North American wheat-producing regions. *Climatic Change* **7**: 367-389.

Rosenzweig, C. and Parry, M.L. 1994. Potential impacts of climate change on world food supply. *Nature* **367**: 133-138.

Rosenzweig, C., Curry, B., Richie, J.T., Jones, J.W., Chou, T.Y., Goldberg, R. and Iglesias, A. 1994. The effects of potential climate change on simulated grain crops in the United States. In: *Implications of Climate Change for International Agriculture: Crop Modelling Study.* C. Rosenzweig and A. Iglesias (eds.). US Environmental Protection Agency, Washington DC. pp. 1-24.

Rötter, R., Stol, W., van de Geijn, S.C. and van Keulen, H. 1995. *World Agroclimates. 1. Current Temperature Zones and Drylands.* AB-DLO, Research Institute for Agrobiology and Soil Fertility, Wageningen. 55 p.

Rowntree, P.R. 1990a. Predicted climate changes under 'greenhouse-gas' warming. In: *Climatic Change and Plant Genetic Resources.* M. Jackson, B.V. Ford-Lloyd and M.L. Parry (eds.). Belhaven Press, London. pp. 18-33.

Rowntree, P.R. 1990b. Estimates of future climatic change over Britain. Part 2. Results. *Weather* **45**: 79-89.

Schertz, D.L. 1988. Conservation tillage: an analysis of acreage projections in the United States. *J. Soil Water Conser.* May-June: 35-42.

Seino, H. 1993. Impacts of climatic warming on Japanese agriculture. In: *The Potential Effects of Climate Change in Japan.* N. Shuzo, H. Hideo, H. Hirokazu, O. Toshiichi and M. Tsuneyuki (eds.). Environment Agency of Japan, Tokyo. pp. 15-35.

Sen, A. 1981. *Poverty and Famines: An Essay on Entitlement and Deprivation.* Oxford University Press, London.

Sen, A. 1993. The economics of life and death. *Scientific American* (May) **268** (5): 40-47.

Singh, U. 1994. Potential climate change impacts on the agricultural systems of the small island nations of the Pacific. Draft, Los Baños, Phillipines. IFDC-IRRI.

Smit, B. (ed.). 1993. *Adaptation to Climatic Variability and Change.* Occasional Paper No. 19. University of Guelph, Guelph, Canada. 53 p.

Thompson, T.P. and Wan, X. 1992. *The Socioeconomic Dimensions of Agricultural Production in Albania: A National Survey.* PN-ABQ-691. Intnl. Fert. Dev. Center, US AID, Washington DC.

Tobey, J., Reilly, J. and Kane, S. 1992. Economic implications of global climate change for world agriculture. *Journal of Agricultural and Resource Economics* **17**: 195-204.

Turner, H.A. and Anderson, C.L. 1980. *Planning for an Irrigation System.* American Association for Vocational Instructional Materials, Athens, GA.

Umali, D.L. 1993. Irrigation-induced salinity: a growing problem for development and the environment. *World Bank Tech. Paper No. 215.* World Bank, Washington DC.

US National Academy of Sciences [US NAS]. 1991. *Report on Adaptation to Climate Change.* Committee on Science, Engineering, and Public Policy, National Academy Press, Washington DC.

US OTA [US Congress, Office of Technology Assessment]. 1993. *Preparing for an Uncertain Climate. Vol. 1.* OTA-O-567. US Government Printing Office, Washington DC.

Vaux, H.J. 1990. The changing economics of agricultural water use. *Visions of the Future: Proceedings of the 3rd National Irrigation Symposium*. ASAE, St. Joseph, Michigan. pp. 8-12.

Vaux, H.J. 1991. Global climate change and California's water resources. In: *Global Climate Change and California*. J.B. Knox and A.F. Scheuring (eds.). University of California Press, Berkeley. pp. 69-96.

Wang, Y.P., Handoko, J.R. and Rimmington, G.M. 1992. Sensitivity of wheat growth to increased air temperature for different scenarios of ambient CO_2 concentration and rainfall in Victoria, Australia – a simulation study. *Climatic Research* **2**: 131-149.

Woodward, F.I. 1993. Leaf responses to the environment and extrapolation to larger scales. In: *Vegetation Dynamics and Global Change*. A. Solomon and H. Shugart (eds.). International Institute for Applied Systems Analysis, Chapman Hall, New York. pp. 71-100.

World Bank. 1993. *World Development Report*. Washington DC.

World Bank. 1994. *A Review of World Bank Experience in Irrigation*. Report **13676**. World Bank Operations Department, Washington DC.

11. Integrating Land-Use Change and Evaluating Feedbacks in Global Change Models: The IMAGE 2 Approach

RIK LEEMANS, GERT JAN VAN DEN BORN AND LEX BOUWMAN
Global Change Department, National Institute of Public Health and the Environment, RIVM, Bilthoven, The Netherlands

During the last decades concern about negative impacts of changing atmospheric conditions has increased. Initially most of this concern was focused on local and regional air pollution, but later global aspects were also considered. One of the major global concerns is the increase of 'greenhouse gases' (GHGs: H_2O, CO_2, CH_4, CO, CFCs, N_2O, tropospheric O_3 and its precursors) that influence the earth's radiative balance. The atmospheric concentrations of these gases have significantly increased since the start of the industrial revolution. Such increase could well lead to large changes in regional and global climate (Houghton et al., 1992).

The sources and sinks of the different GHGs are heterogeneous and difficult to evaluate comprehensively on a global scale. The concentrations of these gases are influenced by a series of chemical processes, many of which are intrinsically linked to each other (Prinn, 1994). Besides these important chemical processes in the atmosphere, the final concentrations are strongly determined by several oceanic and terrestrial processes. For example, oceans take up atmospheric CO_2 through diffusive and biological processes and the magnitude of both is determined by climatic properties, the oceanic energy balance and the ocean circulation (e.g., Sarmiento and Bender, 1994; Woods and Barkmann, 1993). Processes altering the terrestrial sources and sinks are important determinants of many different GHG fluxes between the atmosphere and biosphere (Leemans, 1996). Sources, sinks and fluxes of GHGs from the terrestrial biosphere are more difficult to quantify, because they are defined by local soil, vegetation and climatic properties and the actual land use. All these factors strongly differ geographically and interact with each other on specific time scales. Changes in these fluxes are currently dominated by human activities. This means that any comprehensive assessment of future atmospheric levels of GHGs requires incorporation of these heterogenous spatial and temporal properties, and of human activities affecting the terrestrial biosphere.

In this paper the focus is on processes in the terrestrial biosphere that determine the fluxes of GHGs. Firstly, the linkages and feedbacks between the different processes and compartments in the earth system are discussed with most emphasis on those processes which influence properties of the terrestrial biosphere. Then there is a short review of importance of land-use change, specifying its character and the problems involved in adequately evaluating its magnitude. Finally, the development of an earth system model (IMAGE 2; Alcamo, 1994) is presented that integrates climate change issues and evaluates their importance in a more comprehensive manner. Some applications of the model with respect to mitigation of GHGs will be presented and discussed.

IMPORTANCE OF FEEDBACKS AND LINKAGES

PHYSICAL AND BIOGEOCHEMICAL FEEDBACKS

Feedback processes influence the exchange rates of GHGs between the atmosphere, biosphere and oceans and modify the residence time of those gases in the different compartments. Feedback processes largely determine the dynamic response of the earth system. These processes include not only the natural geophysical feedbacks, such as changes in climate and ocean circulation, and biogeochemical feedbacks, such as changes in biological activity, atmospheric chemistry and oceanic CO_2 uptake, but also the important anthropogenic feedbacks such as changes in human activities, energy use and land use (see next section).

Geophysical feedbacks alter the radiative forcing characteristics of the atmosphere and include changes in characteristics and distribution of clouds, ice and snow. These feedbacks are incorporated in the global climate models and their strength determines the sensitivity of the climate system to changes in radiative forcing (e.g., Lashof, 1989; Houghton *et al.*, 1990, 1992). Biogeochemical feedback processes involve changes in the sources and sinks of greenhouse gases and changes in surface properties, such as albedo and transpiration. Direct and indirect effects are distinguished.

Direct effects are those processes that influence atmospheric and ocean chemistry (Lashof, 1989; Prinn, 1994) and biospheric processes. The most important feedback process is probably the enhancement of plant growth by increased levels of atmospheric CO_2. This so-called CO_2 fertilization effect has been observed in many greenhouse and ecosystem experiments (e.g., Strain and Cure, 1985, Drake, 1992). Increased CO_2 levels also improve the water-use efficiency (WUE) of plants and change plant growth under water-limited conditions (e.g., Morison, 1985).

The indirect biogeochemical feedbacks involve the consequences of climate change. A large number of those effects have been identified. Examples are changes in ocean circulation, shifts of vegetation zones, and changes in plant growth rates. A temperature increase influences photosynthesis and respiration in a complex manner (Larcher, 1980), enhancing plant growth in regions with temperature constraints. However, enhanced plant growth does not automatically lead to an increase in C storage within an ecosystem. Litter decay is simultaneously stimulated (Raich and Schlesinger, 1992).

Finally climate change alters, like an increased WUE, the distribution of vegetation and agricultural crops (Leemans and van den Born, 1994). Climate strongly defines the large-scale vegetation and crop distribution patterns and this correlation has be used to assess the global impacts of climate change (for a review see Leemans *et al.*, 1996).

Some of these responses are illustrated in Figures 11.1 and 11.2, where the BIOME model (Prentice *et al.*, 1992) is applied to assess changes in the distribution of potential vegetation, characteristic for prevailing stable climatic conditions. Human impacts and other (e.g., successional) influences are not taken into account. The effects of changes in WUE are simulated by shifting the drought tolerances in the BIOME model (Klein Goldewijk *et al.*, 1994). The maps (Figure 11.1) clearly show the large regional differences. Vegetation shifts due to a warmer climate can mainly be observed in the boreal zone, while the effects of increased WUE are mainly located in the tropics, where forests and savanna ecosystems expand. These changes are summarized in Figure 11.2 and could have a pronounced effect on the global C cycle. It was often stated that under a warmer climate the resulting equilibrium biosphere could store significantly more C (e.g., Smith *et al.*, 1992), but it becomes apparent that the transient vegetation response could well be different. Transient vegetation dynamics could temporally generate large CO_2 emissions through a rapid decline of ecosystems (Neilson, 1993; Smith and Shugart, 1993), while the successive recovery could last many decades.

This short review indicates that both direct and indirect effects must be considered when evaluating the impacts and consequences of increasing levels of GHGs on ecosystems and agrosystems. This is especially true when scenarios for future trends are developed and evaluated with integrated assessment models for the earth system. Such models too often lack appropriate parameterization of such feedback processes and can therefore generate misleading results. Despite the limitations of these models, they are still the only tool to assess the whole range of climate issues comprehensively, from sources, through processes to impacts and their interactions. In order to create adequate assessments, the developers of global earth system models and disciplinary experts must communicate and try to understand on one hand the modeller's need for simplification and generalization and, on the other hand, the expert's need for adequate observations and experimentation.

IMPORTANCE OF LAND-USE CHANGE

Sustaining an increasing population requires a continuous flow of agricultural and forest products, such as food, fodder, timber and fuelwood. These products all require land to be produced. Increasing this flow requires more productivity. Productivity can increase through intensification of agricultural practices, improved cropping systems and increase in agricultural areas. Such changes influence the biogeochemical properties of land and are therefore important determinants of the size of sources, sinks and fluxes of GHGs. Up to now only the obvious and large-scale conversions of land (deforestation, reclamation of wetlands, etc.) have been addressed with respect to their importance for global C fluxes to the atmosphere.

270

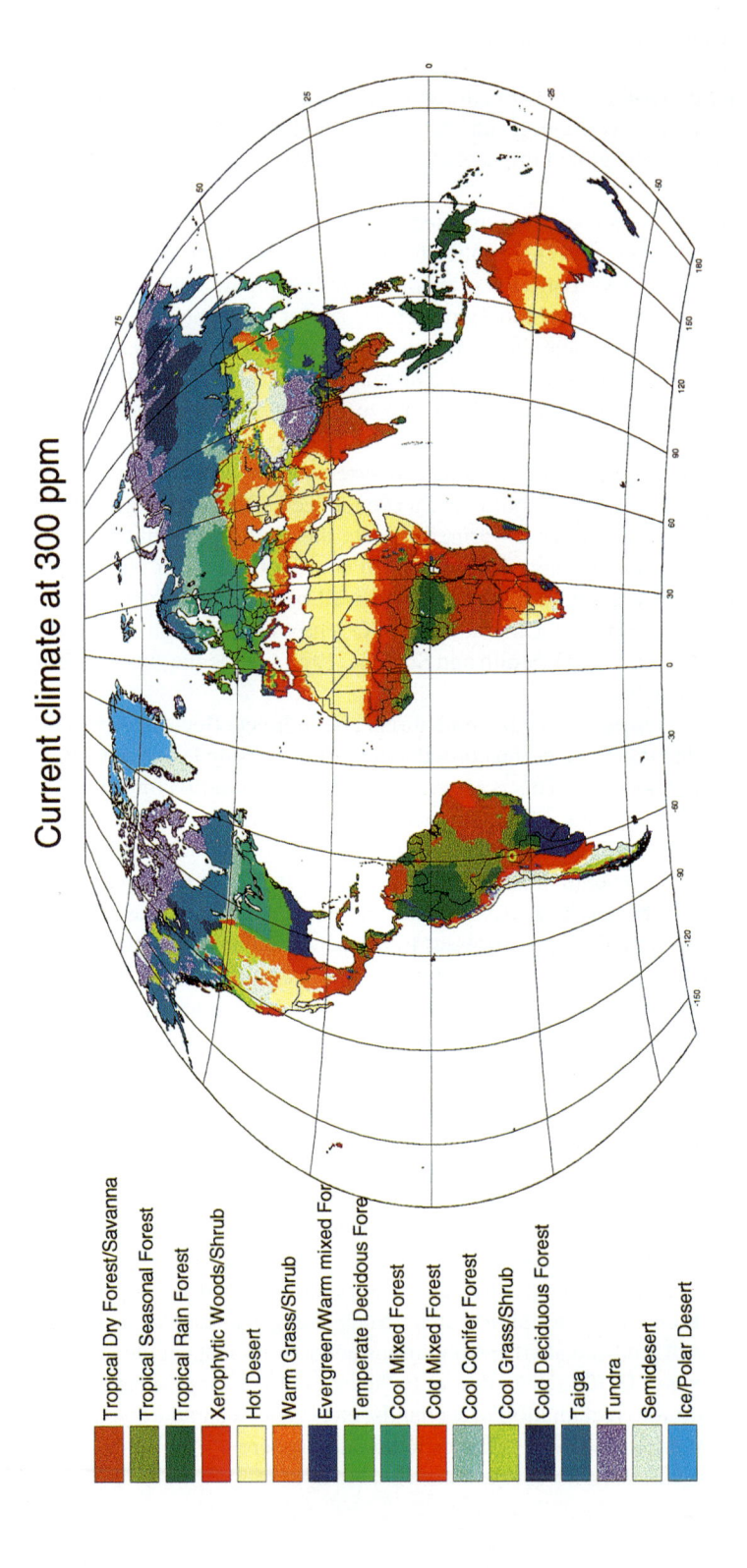

Figure 11.1. Global maps with the distribution of potential vegetation, based on the BIOME model (Prentice et al., 1992) for different climates and biogeochemical feedbacks. (1) The distribution under current climate

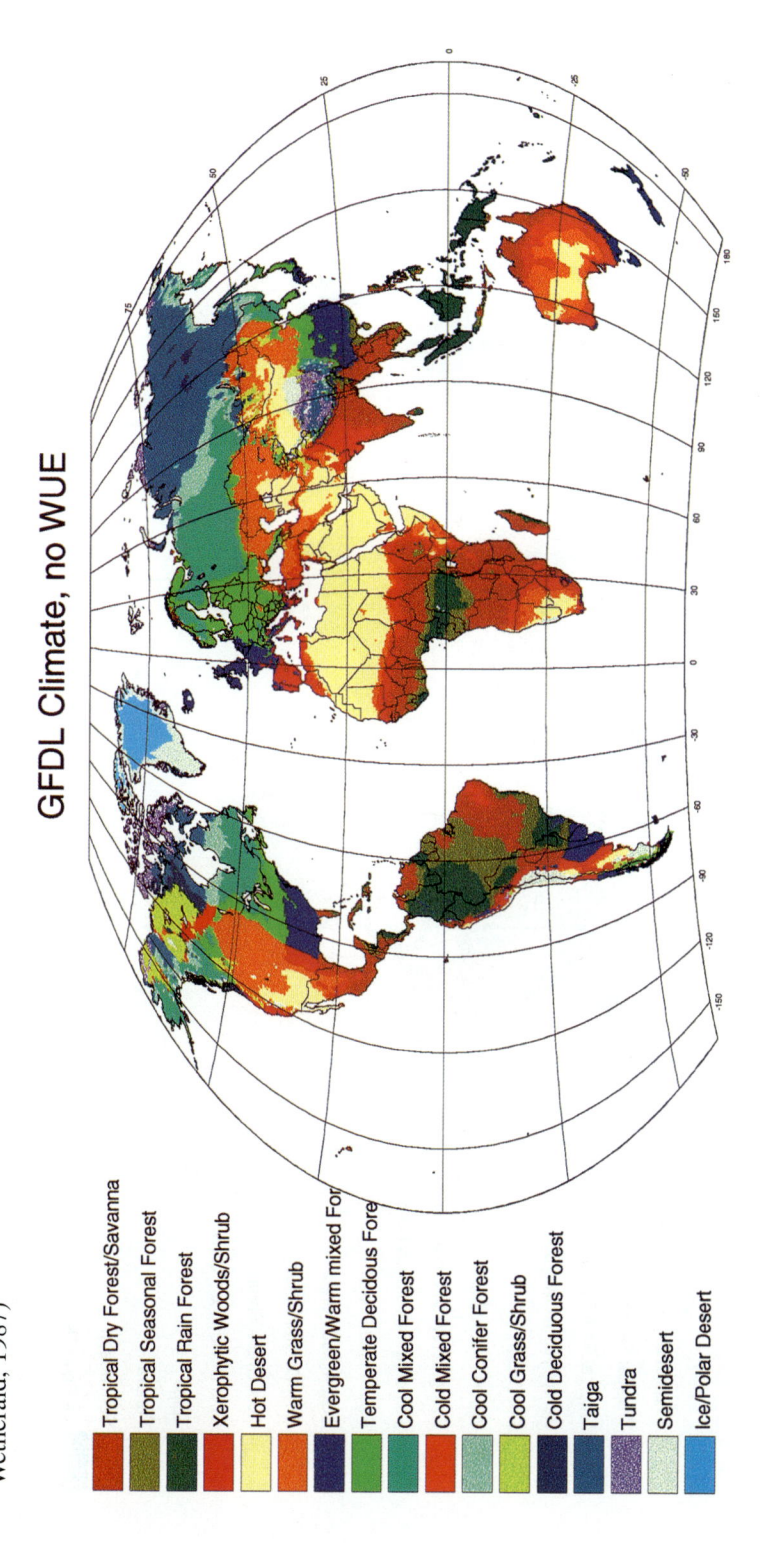

Figure 11.1. Global maps with the distribution of potential vegetation, based on the BIOME model (Prentice et al., 1992) for different climates and biogeochemical feedbacks. (2) Distribution under a doubled CO_2 climate according to the GFDL GCM (Manabe and Wetherald, 1987)

272

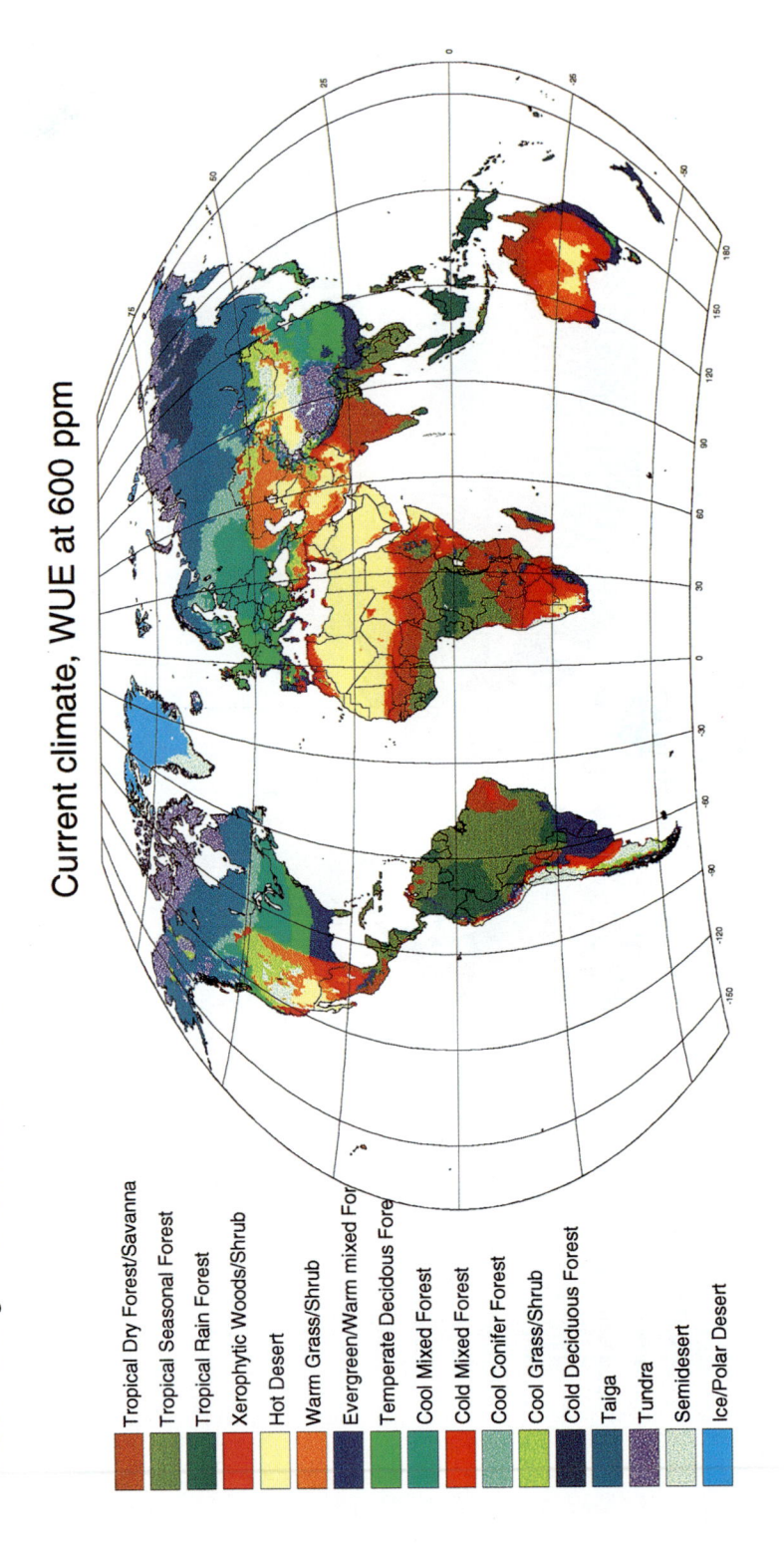

Figure 11.1. Global maps with the distribution of potential vegetation, based on the BIOME model (Prentice et al., 1992) for different climates and biogeochemical feedbacks. (3) Distribution under an enhanced WUE effect for doubled CO_2

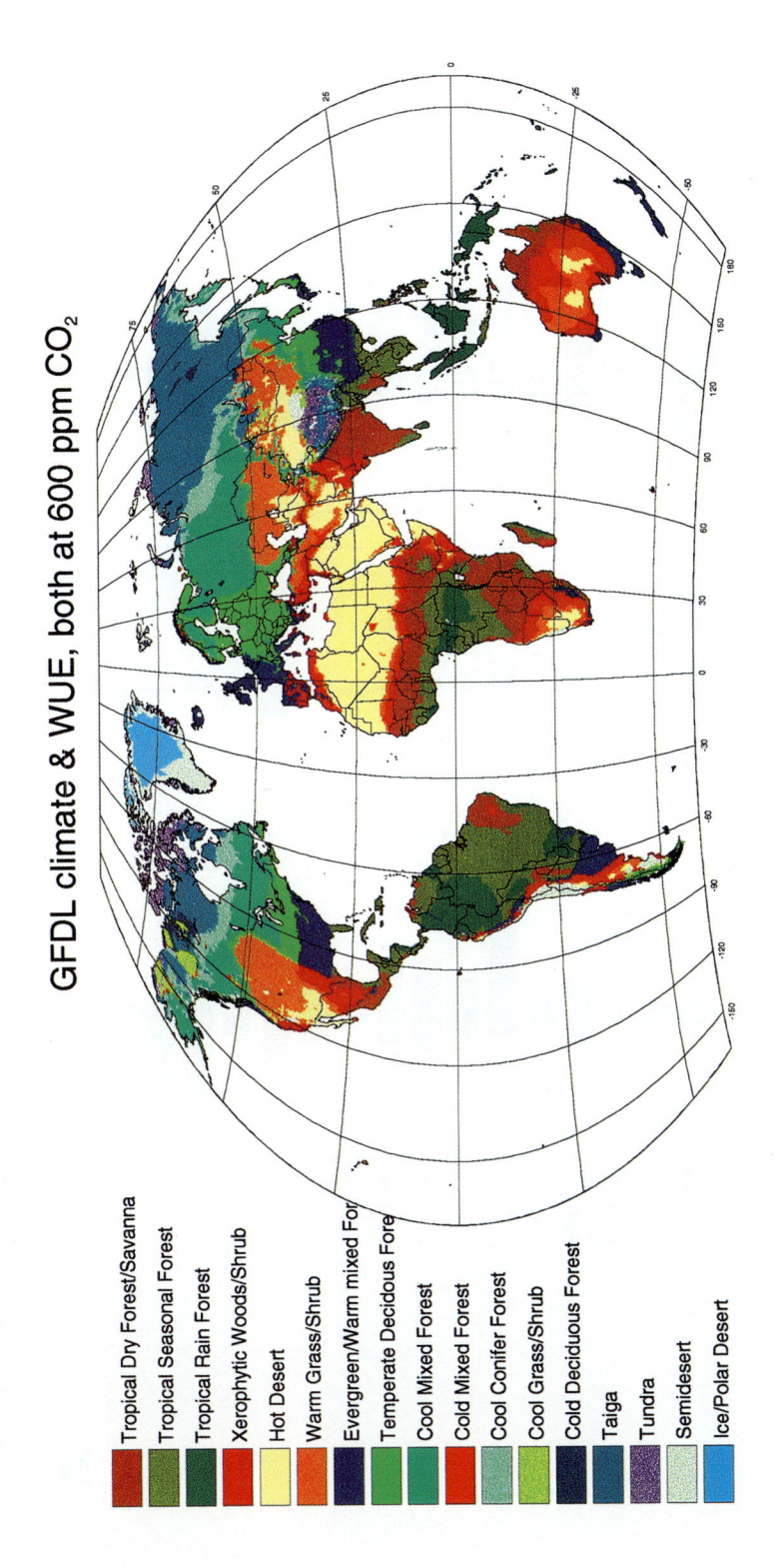

Figure 11.1. Global maps with the distribution of potential vegetation, based on the BIOME model (Prentice et al., 1992) for different climates and biogeochemical feedbacks. (4) Combined distribution of (2) and (3)

GFDL climate & WUE, both at 600 ppm CO_2

- Tropical Dry Forest/Savanna
- Tropical Seasonal Forest
- Tropical Rain Forest
- Xerophytic Woods/Shrub
- Hot Desert
- Warm Grass/Shrub
- Evergreen/Warm mixed For
- Temperate Decidous For
- Cool Mixed Forest
- Cold Mixed Forest
- Cool Conifer Forest
- Cool Grass/Shrub
- Cold Deciduous Forest
- Taiga
- Tundra
- Semidesert
- Ice/Polar Desert

Figure 11.2. Shifts in distribution of potential vegetation for doubled CO_2 conditions (GFDL: only climate change; WUE: enhanced WUE; GFDL and WUE: combined effect)

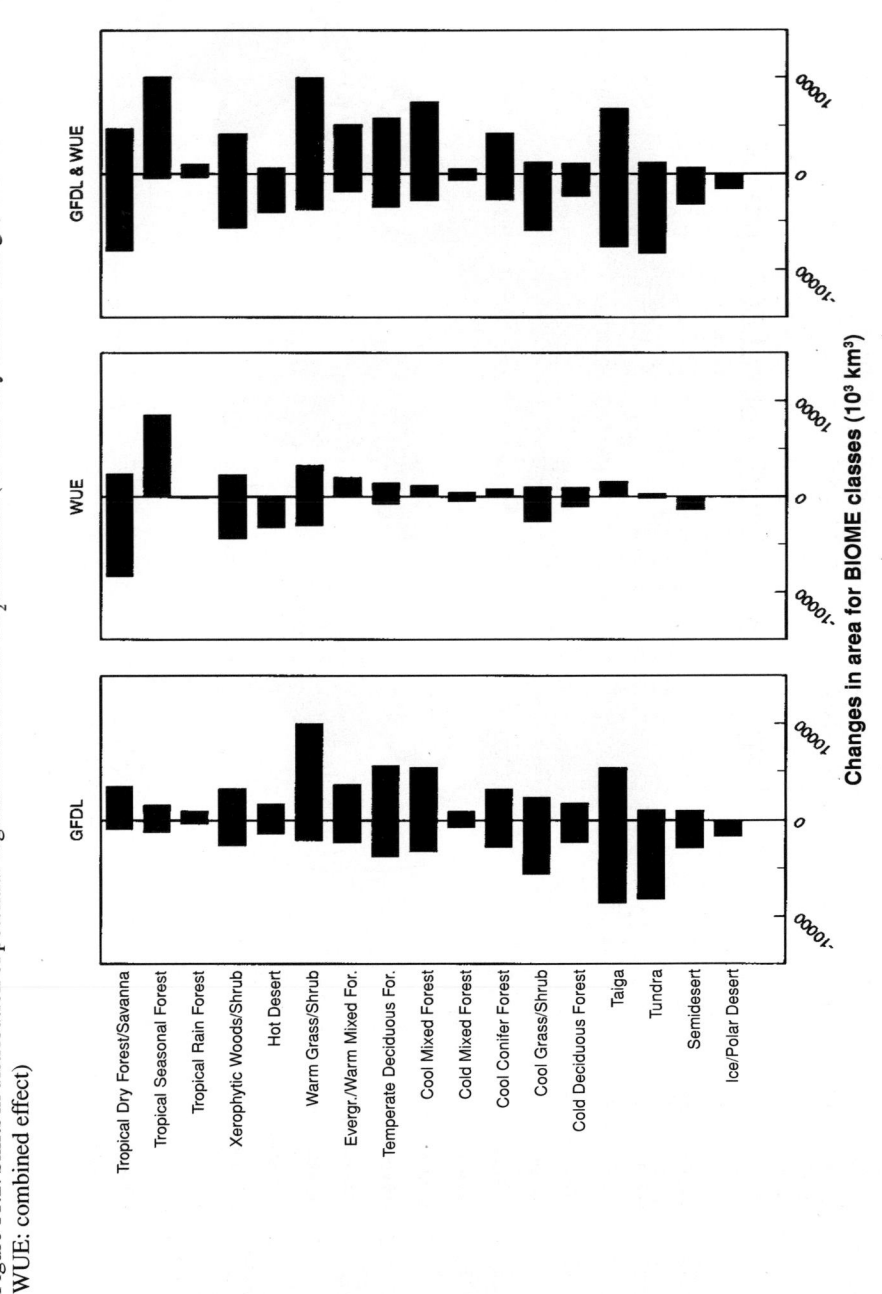

Tropical deforestation resulted in a large flux of CO_2 and other GHGs. Forests are felled, burned and removed to create pasture and croplands. These land-use practices are common under swidden agricultural systems. During recent decades, however, deforestation rates have been accelerating and the resulting flux of CO_2 accounted for 25% of the total increase of atmospheric CO_2 (Watson et al., 1992). After a while large parts of the reclaimed land were abandoned, often degraded, and secondary forest developed. Recently more attention has been paid to the balance between deforestation and forestation in the tropics (e.g., Skole and Tucker, 1993). This balance is important with respect to the total global fluxes. The resulting heterogenic spatial and temporal patterns make a precise assessment of the total flux from deforestation difficult.

Many evaluations of the total 'land-use' flux are calculated through the total global C budget. Data are used from the observed atmospheric concentrations, the assumed (=modelled) uptake by the oceans and the well-defined emissions from fossil-fuel combustion. The global budget is then balanced with a derived flux from the terrestrial biosphere. The magnitude of the terrestrial flux needed from terrestrial biosphere is generally smaller than the deduced emissions from tropical deforestation. The conclusion of such a deconvolution exercise is that a significant sink is needed to balance the global C cycle. This missing sink is assumed to be located in the terrestrial biosphere, because the processes determining all other fluxes appear to be well understood. In recent years many studies claimed to have identified at least part of this missing sink (e.g., Kauppi et al., 1992; Wofsy et al., 1993; Fisher et al., 1994), but the actual location and the processes involved are difficult to localize and observe. The missing sink emphasizes the importance of land use, land cover and land management in assessing sources, sinks and fluxes of GHGs (Leemans, 1996).

The above discussion has mainly focused on C, but similar conclusions can be drawn for other greenhouse gases, such as CH_4 and N_2O (Bouwman, 1995). For example, changes in land cover alter the uptake of CH_4 by soils, different agricultural practices lead to changed CH_4 emissions, N_2O emissions are influenced by the timing and amount of fertilizer applications. Such examples indicate that the spatial pattern of GHG emissions from the terrestrial biosphere is very heterogeneous and influenced by physical, biogeochemical, socio-economic and technical factors. Actual land use and its resulting land cover are important controls. Especially when mitigation policies are evaluated, globally aggregated assessments are no longer valid. State-of-the-art assessments should be dynamic, geographical and regionally explicit and include the most important aspects of the physical subsystem, the biogeochemical subsystem and land use and changes therein (Figure 11.3).

MODEL APPROACHES

Many approaches to model the consequences of global climate change have been developed. Most of these models only cover one or a few aspects of the whole chain of emissions, processes and impacts. Emission models largely focus on the energy

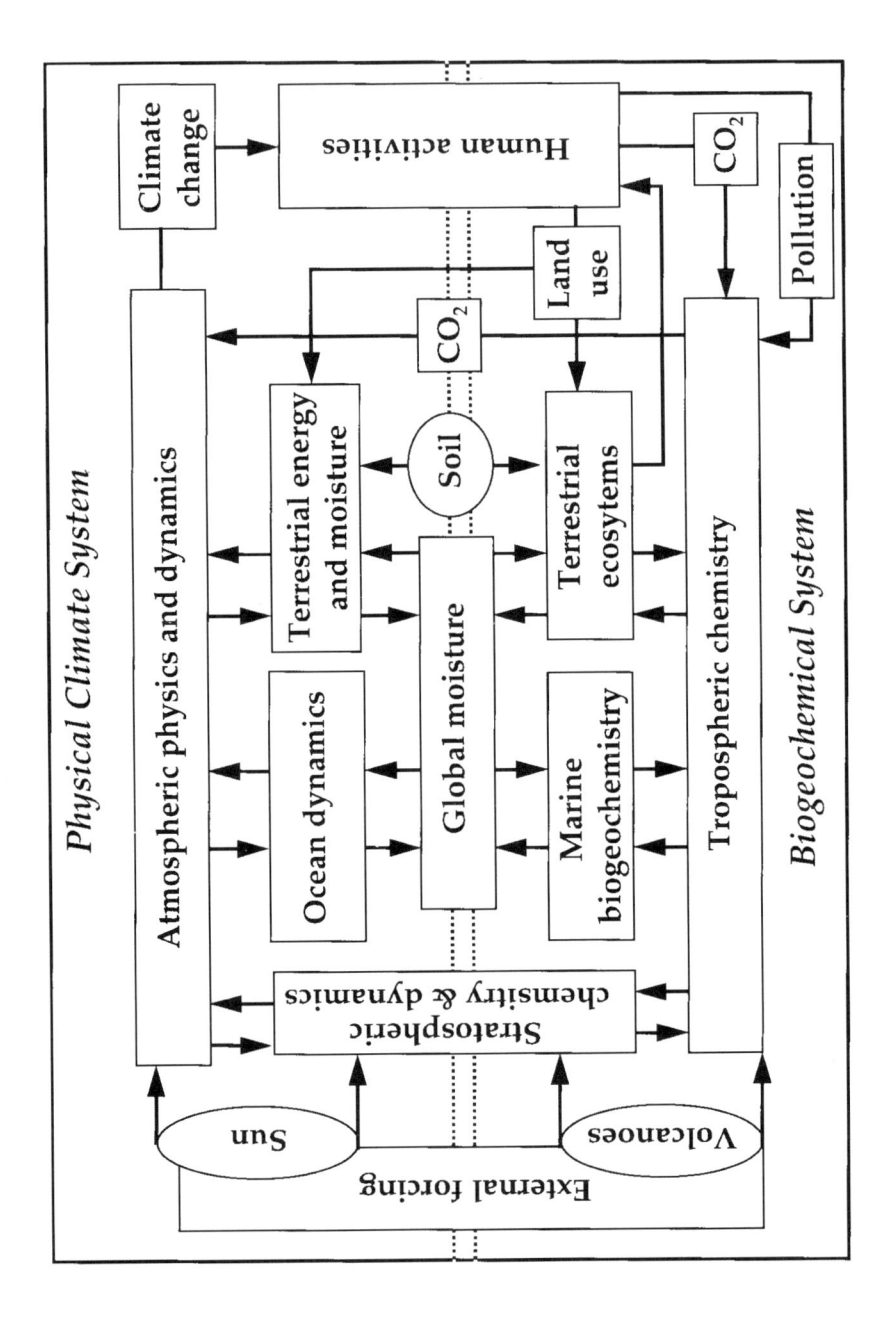

Figure 11.3. Flow diagram of the system earth (redrawn after IGBP, 1994)

sector and are therefore often derived from macro-economic or technology models (see reviews in Nakícenovíc *et al.*, 1994). Emissions from land use are not adequately covered in such emission models. More process-oriented models simulate aspects of atmospheric chemistry, climate and the biogeochemical cycles. The most advanced models are General Circulation Models (GCMs) that simulate the temporal and spatial characteristics of the climate system, and C-cycle models, that simulate the global C budget. The most diverse group of models addresses impacts. Many different sectors have their own model approaches which all simulate one or more aspects of changing atmospheric composition and/or climate change. In particular, agricultural, ecological and hydrological impact models have been developed (e.g. Parry *et al.*, 1990; Cramer and Solomon, 1993; Kwadijk, 1993). Efforts are now under way to harmonize the large diversity in impact assessment approaches and probably an improved, standardized methodology will be agreed upon (Carter *et al.*, 1994).

Recently an approach has emerged that integrates different sources of GHG emissions with the processes in the atmosphere and, if evaluated as necessary, with different impacts. Among the first of such models was IMAGE (Integrated Model to Assess the Greenhouse Effect; Rotmans *et al.*, 1990). This model simulated the dynamics of different GHG emission sources and computed dynamically the final atmospheric GHG concentrations considering the influence of atmospheric chemistry and the C cycle. These concentrations were fed into a radiative forcing model that computed a globally average annual climatic change. The only impact considered was sea-level rise. The model was applied to develop the first global IPCC scenarios (Leggett *et al.*, 1992).

The innovative approach of this IMAGE model has now several competitors. The successor of IMAGE was ESCAPE, an integrated model that was developed for policy evaluation in the EU (Rotmans *et al.*, 1994). The structure of IMAGE has been improved by including different impact modules in a geographically explicit way. The ESCAPE framework has been the basis for two further model developments. Hulme *et al.* (1994) concluded that the complete ESCAPE framework was too complex and developed a derived, highly parameterized global model that is suitable to determine the sensitivity of the climate system and rapidly evaluate the consequences of different policy options. However, feedbacks and linkages between subsystems are not adequately covered in this global model.

Currently several groups are involved in developing such integrated assessment models with a more adequate treatment of feedbacks and linkages, further integrating the structure of the ESCAPE framework. Some of them focus on only a single region, such as the Asian-Pacific region (the AIM model, Morita *et al.*, 1994) or North America (GCAM, Edmonds *et al.*, 1994), but include high-resolution impact modules and linkages between them, while the trends of other regions in GHG emissions are prescribed and the global biogeochemical and physical processes are simulated in a more aggregated manner. Other global models are being developed by several institutes (Dowlatabadi and Morgan, 1993). One of the few fully implemented and documented integrated global models is IMAGE 2 (Alcamo, 1994). Its innovation includes a dynamic simulation of land cover change, driven by socio-economic, atmospheric and climatic factors. The IMAGE 2 framework is especially developed to link all relevant processes comprehensively, while simultaneously retaining regional and local characteristics.

THE IMAGE 2 APPROACH

The overall objective of the IMAGE 2 model is to simulate, on the basis of political and socio-economic scenarios, plausible future trends of GHG concentrations in the atmosphere and to determine their impacts on physical, biogeochemical and human systems. To accomplish this, important scientific and policy aspects of global climate change must be linked. The time horizon of the simulations should be several decades in order to be able unambiguously to discriminate the effectiveness of policy measures. This objective leads to several scientific goals, such as to provide insights into the relative importance of different linkages and feedbacks in the earth system, to provide estimates of sources of uncertainty and to help identify gaps in our knowledge and data in order to help set the agenda for climate change research. The related policy-oriented goals are to provide a quantitative basis for analysing the societal cost and benefits of various measures to address climate change.

Here only the purpose of each IMAGE submodel is summarized. For a full description of all models and their linkages, reference is made to the papers in Alcamo

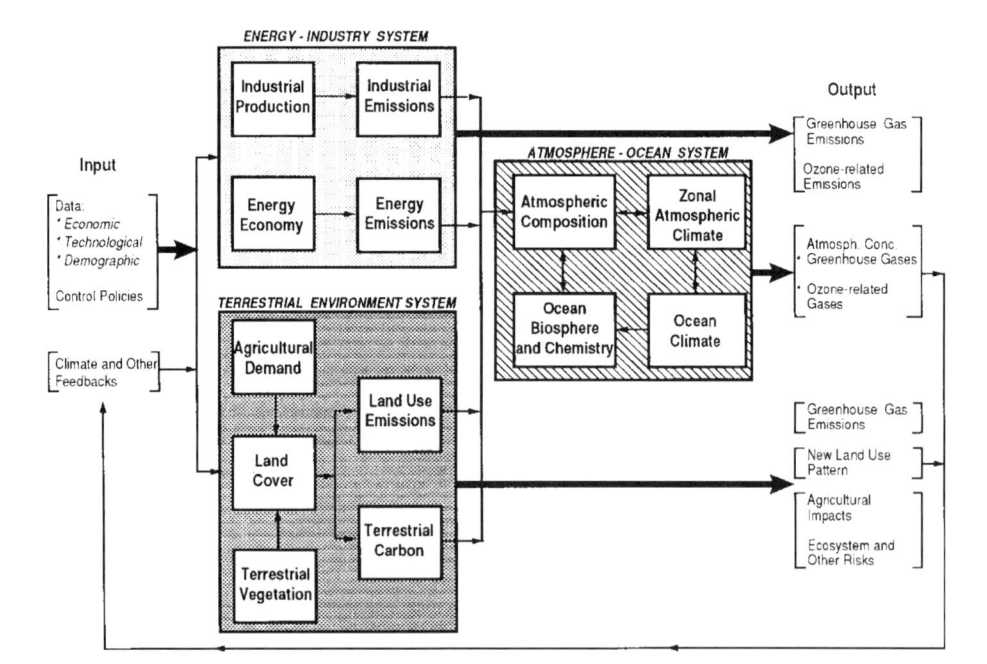

Figure 11.4. Framework of models and linkages in IMAGE 2 (after Alcamo, 1994)

(1994). The IMAGE model consists of three fully linked submodels (Figure 11.4): the Energy-Industry System (EIS), the Terrestrial Environment System (TES) and the Atmosphere-Ocean System (AOS). The input data of the socio-economic models are mainly prescribed economic, technologic and demographic trends combined with different control policies, while the environmental models are mainly driven by climatic, vegetation, crop and soil properties.

The models making up EIS compute the emissions of GHGs in 13 larger, somewhat homogeneous regions (Table 11.1) as a function of energy consumption and industrial production. End-use energy consumption is computed from various economic driving forces. Sector-specific emission factors are obtained from literature and calibrated against data for global and regional emissions. The models are designed such that the effectiveness of improved energy efficiency and technological development on future emissions in each regions can be evaluated and that the set of models can be used to assess the consequences of different policies and socio-economic trends on future emissions.

TES simulates the changes in global land cover based on climatic and socio-economic factors. The roles of land cover and management are used to compute the fluxes of CO_2 and other greenhouse gases from the terrestrial biosphere to the atmosphere. The structure of TES is discussed in more detail below.

The AOS computes the build-up of GHGs in the atmosphere and the resulting climate change averaged over latitudinal bands. AOS thus computes transient changes

Table 11.1. Main characteristics of the regions represented in the IMAGE model and some scenario assumptions. The figures given here are taken from the conventional wisdom scenario, which is derived from the IPCC baseline scenario (IS92a; Leggett et al., 1992). (The yield increase index is relative to 1990)

Region	Population ($\times 10^6$) 1990	2100	Economic growth (%) 1990-2025	2025-2100	Yield increase (%) 2025	2100
Canada	27	27	2.06	1.31	1.83	1.91
USA	250	295	2.09	1.25	1.62	1.66
Latin America	448	877	1.85	2.20	2.14	2.95
Africa	642	2 875	1.57	2.39	1.50	1.73
Western Europe	378	388	2.06	1.31	1.45	1.57
Eastern Europe	123	148	1.87	1.18	1.42	1.56
CIS	289	347	1.87	1.18	1.98	2.31
Near East	203	937	1.36	1.98	1.41	1.67
India and South Asia	1 171	2 644	2.97	2.84	2.34	2.98
China and centrally planned countries	1248	1963	4.23	3.07	1.50	1.93
East Asia	371	837	2.97	2.84	2.00	2.81
Oceania	23	24	2.71	1.28	1.64	1.84
Japan	124	130	2.71	1.28	1.02	0.99
Global	5 297	11 492				

in climate resulting from changes in all GHG emissions. As a starting point the atmospheric concentration of CO_2 is altered through CO_2 uptake by the oceans. Then, temporal trends of the average tropospheric concentrations of GHGs are computed accounting for the chemical atmospheric reactions involving O_3, OH and CO with other gases and sulphate aerosols. The final atmospheric levels of the GHGs (including H_2O) are used to compute the earth's energy balance. The cooling effect of backscattering of solar radiation on sulphate aerosols is included (Taylor and Penner, 1994). Surface heat exchange with land and oceans, and changes in albedo from land cover change, are considered in the final calculation of temperature change. Other climatic characteristics (e.g., precipitation) are crudely coupled by scaling the latitudinal temperature change towards the results from GCMs.

TES and EIS are linked to each other by the demand for biofuels (fuelwood and biomass). This linkage allows IMAGE 2 to evaluate the land cover consequences for future demands for biofuels. EIS and TES are linked in several ways with AOS. The most important linkages are through the emissions of GHGs and changed albedo of land cover patterns. AOS provides EIS and TES with a changed climate that is directly used to determine the energy demands for heating and cooling, compute potential vegetation patterns, crop distribution and yields, and modify the rates of N_2O emissions. The final concentrations of CO_2 also directly influence plant productivity. These dynamic linkages allow IMAGE 2 to be directly applied to assess the sensitivity of the system for different feedback processes.

All submodels have their specific domain with their own spatial and temporal resolution, scale and dimensionality. Despite this heterogeneity, we have tried to integrate all submodels through their obvious linkages. All socio-economic models (EIS and the agricultural demand and land-use models) distinguish 13 major regions. These models are thus regionally explicit and can be calibrated and validated with a large series of different tabular databases for countries and regions. All environmental models for the terrestrial biosphere are implemented on a 0.5×0.5° grid. Each grid cell is characterized by its climate (Leemans and Cramer, 1991), soil (Zobler, 1986), and land cover (Olson *et al.*, 1985). Spatial heterogeneity in environmental characteristics and local environmental influences on biogeochemical processes are thus adequately covered. Innovative rule–based systems are used to integrate all these dimensions comprehensively.

Modelling land cover change

Modelling land cover change is difficult. Although many models simulate some effect of land-use change, such changes involve often only prescribed tropical deforestation. Contraction, expansion and shifts of agricultural lands, intensification, and different crops and forest management practices are rarely included. Such changes have to be included in global emission models because their cumulative effects are significant. We have developed a land cover change model that is forced by a heterogeneous set of physical, biogeochemical and socio-economic variables. Although, many other models and studies (e.g., FAO, 1993a; Rosenzweig and Parry, 1994) stress the importance of macro-economic driving forces, we have not emphasized those.

Our approach is based on a set of well-established ecological and agricultural models. Potential vegetation patterns are modelled with the BIOME model (Prentice et al., 1992). This model uses climatic variables to delimit the distribution of different Plant Functional Types (PFTs, cf. deciduous broad-leaved trees, evergreen needle-leaved trees, C_4 grasses). Each PFT is characterized by its specific climatic constraints defined by the accumulated heat during the growing season, drought, and frost tolerances. Biomes are assemblies of these PFTs (Figure 11.1). BIOME allows us to evaluate shifts in vegetation due to climatic change and changes in water-use efficiency (Figures 11.1 and 11.2) and the consequences of these shifts on total C sequestration.

Crop distributions are modelled using the agro-ecological zone approach (Brinkman, 1987; Leemans and Solomon, 1993). Crops require a minimum growing period (the period with adequate moisture and heat supply) to grow and mature. This growing period, together with several other climatic parameters (see Leemans and Solomon, 1993), defines the potential rainfed distribution of crops. Yield is computed with a simple model (FAO, 1987) that distinguishes different crop types (C_3 or C_4, tropical or temperate, legumes or non-legumes) and estimates biomass production for each crop under the prevailing climatic conditions during the growing period. The economic yield is a fraction of this production and is adjusted for soil quality. The result of these two models is thus a potential vegetation and crop yield for each 0.5x0.5° gridcell.

Transformation of land cover is governed by the local, regional and global demand for agricultural commodities (grain, meat, wood and biofuels). This demand is satisfied by specific land use in croplands, pasture lands, rangelands and (managed) forests. Land use is strongly related to the potential productivities and technically achievable yields for each gridcell. We developed an agricultural demand model to estimate the societal demand for agricultural products. These calculations are performed for each world region and use the same socio-economic input as EIS (Table 11.1).

Regional agricultural demand is calculated from the per caput human consumption for different crops and meat products based on a assumed elasticity relationship between consumption and per caput income. These elasticity coefficients are estimated from 1970-1990 data of FAO (1991), and are the main adjustable parameters in this demand model. Total human (as opposed to animal) demands for these products are then computed for a given income and population scenario. A similar procedure is used to compute the total regional meat demand. This demand is then converted into numbers of livestock, and their feed requirements (concentrates from crops and residues and roughage from range and pasture lands). These feed requirements are added to the total regional human crop demand. This demand is then adjusted for trade between regions. The roughage demand is converted into a demand for grassland by multiplying productivity per animal and the average required rangelands per animal.

The total regional demand for agricultural products is reconciled with the potential local distribution and yields of crops, computed earlier. Land cover is initialized for 1970 conditions using an aggregated version of the database of Olson et al. (1985). The agricultural crop demands are assumed to be satisfied in the single class 'agricultural land', while the demand for rangelands can be satisfied by several types of grasslands, depending on what occurs locally. The basic idea of the dynamic land cover change model is to change land cover until the total regional demands are satisfied. The model thus generates future land cover logically consistent with demand and production potential. This is probably not realistic but it is congruous with available global data sets and some basic driving forces (Turner et al., 1990). Improvement of the analyses, incorporating different land-use systems, political structures and cultural characteristics, is far beyond the current requirements and analysis in IMAGE 2. However, we are eager to learn from a better understanding from emerging projects (e.g. Turner et al., 1993) and experiences with other models (e.g., Fischer et al., 1988).

Actual changes of future land cover patterns are difficult to estimate, because the actual driving forces are heterogeneous, differ per region and are in general poorly understood (Turner et al., 1993). Our approach is therefore to prescribe a set of transparent logical rules to match regional land demand with its potential. These rules, together with a management factor (which takes into account the yield increase by fertilizer application and technological improvement, cf. Table 11.1), define future patterns. For example, there are two weighted rules that guide the assignment of new agricultural areas: (1) new areas should occur adjacent, if possible, to existing agricultural land, because of the availability of infrastructure, transport and population; (2) the new land should develop in those areas of the highest production potential. Similar sets of rules are specified to satisfy rangelands and fuelwood demand, but the demand for cropland is satisfied first. Land not under agricultural management is assumed to succeed towards the potential vegetation as determined by the BIOME model. This land cover model results in a dynamically changing land cover on the terrestrial grid. The land cover patterns are consistent with a changed climate and are used to drive the terrestrial C cycle. One major improvement of this approach over earlier models is that an increase of agricultural area does not *per se* lead to deforestation, but it is a function of the agro-ecological suitability of all types of land. Agriculture could thus expand into grasslands as well as forests.

Carbon fluxes

The C-cycle model estimate the C sources, sinks and fluxes resulting from natural processes and land cover change. The C-cycle model computes, for each land cover type, Net Ecosystem Productivity (NEP), which is the difference between the Net Primary Productivity (NPP) and soil respiration. NPP is a characteristic of each land cover type, but its actual value is corrected according to local (temperature and moisture availability) and global (atmospheric CO_2 concentration) conditions and the time passed since the last land cover change. NPP and soil respiration are modified under the influence of climate change, assuming characteristic response functions for photosynthesis, plant respiration and soil respiration (e.g., Larcher, 1980). CO_2 fertilization is a function of available PFTs in each land cover type (such as C_3, C_4, annual, perennials or trees), temperature, moisture, soil fertility and altitude. The simulated NEP is consistent with the simulated climatic conditions.

NPP is partitioned into land cover specific biomass components (leaves, branches, stems and roots), each with a specific longevity, and non-living matter (litter, humus and charcoal), each with a specific turnover time. We have initialized the model starting from 1900. Most IMAGE 2 simulations start therefore (in 1990) with NEP conditions that are close to equilibrium with prevailing climatic and atmospheric conditions. This near-equilibrium condition shift when the environment changes or land is converted into another type. If natural vegetation is converted, a fraction of the biomass is burned on site, immediately releasing CO_2, while the remainder is allocated to the non-living pools. Such change results in an additional long-term low CO_2 flux from these sites. If agricultural land is abandoned, we assume that the potential natural vegetation returns. However, such succession takes time and we assume that NPP slowly builds up following a logistic function.

Land-use emissions

The Land Use Emissions model relates global land use to the flux of emissions of CH_4, N_2O, CO, NO_x, and Volatile Organic Carbons (VOCs, excluding CH_4) (cf. Table 11.2). In addition, the model estimates the emissions resulting from biotic processes unrelated to human activity, such as N_2O emissions from soils in unmanaged forests,

Table 11. 2. Sources described in the IMAGE 2 Land Use Emissions Model, gas species emitted and the type of calculation and presentation. GE = geographically explicit; R = regional total; G = global total. Adapted from Kreileman and Bouwman (1994)

Source	Species	Type of calculation
Wetland rice fields	CH_4	GE
Natural wetlands	CH_4	GE
Landfills	CH_4	R
Domestic sewage treatment	CH_4	R
Animals	CH_4	R
Animal waste	CH_4, N_2O	R
Termites	CH_4	G
Methane hydrates	CH_4	G
Aquatic sources	CH_4, N_2O	G
Biomass burning	CH_4, CO, NO_x, N_2O, VOC	
-Deforestation		GE
-Savanna burning		GE
-Agricultural waste burning		GE
Natural soils	N_2O	GE
Agricultural fields	N_2O	GE
Deforestation	N_2O	GE

and trace gas emissions from aquatic systems. The calculations are important in determining land use and cover related GHG emissions and can be used to evaluate strategies for reducing these emissions.

This model is based on current estimates for the various sources and species. Given the limitations of the available data, the model presents grid-based estimates for a number of different sources, including CH_4 emission from rice, wetlands, emissions of CH_4, CO, NO_x, N_2O and VOC from deforestation, savanna burning and agricultural waste burning and N_2O from natural soils, N fertilizer and deforestation (Table 11.3; Bouwman, 1995). For some of the sources (landfills, domestic sewage treatment, termites, CH_4 hydrates and aquatic sources) geographically explicit calculations are not yet possible because of data limitations (Table 11.2).

APPLICATIONS OF IMAGE 2

The IMAGE 2 model has been tested and calibrated against data from 1900 to 1970 and can for this period reproduce trends in regional energy consumption and energy related emissions, land cover, terrestrial flux of CO_2 and emissions of other GHGs into the atmosphere. The time horizon for the actual simulations extends from 1970 to 2100 and each simulation is characterized by a specific set of input options that describe policy and socio-economic trends (Table 11.1). Here we present a scenario which is an adaptation of the intermediate IPCC scenario (IS92a, Leggett et al., 1992). The scenario makes 'conventional' assumptions about future demographic, economic and technological driving forces (Table 11.1; Alcamo, 1994). There are no climate-related policies, except those that were already agreed upon (e.g., Montreal Treaty on CFC emissions). This scenario further assumes an increase in the energy use efficiency and a regionally specific vehicle utilization. This baseline scenario can be used as a baseline against which to compare other scenarios.

Land-use change and emissions

We compared the emissions calculated for the years 1970, 1990 and 2050 for CH_4 and N_2O with different reference estimates for selected land-use-related sources (Table 11.3; Watson et al., 1992). From these reports either the methodology has been adopted, giving by definition similar results, or emission factors have been adapted to arrive at equal global totals for 1990. In some cases our results slightly differ from those of the reference studies, for example for CH_4 emissions from animal waste. Major differences are found for all emissions from deforestation. The global deforestation rate for the baseline scenario is 155 000 km²/year for the period 1970 to 1980 and 140 000 km²/year for 1980 to 1990. Official inventories of deforestation by FAO (1993b) show that rates may have been respectively 115 000 km²/year and 155 000 km²/year. Reference estimates for biomass burning are based on these FAO deforestation estimates for the early 1980s and result therefore in lower emissions than our baseline scenario (Table 11.2).

Our calculations are sensitive to assumptions of demographic, technological and

Table 11.3. Estimated global emissions for 1970, 1990 and 2050, and reference estimates for 1990 and 2050. Reference estimates for 1990 are from Watson et al. (1992) unless indicated otherwise. Reference estimates for 2050 are from Pepper et al. (1992), scenario IS92a. Emissions are expressed as Tg CH_4/yr and Tg N_2O-N/yr. The range of uncertainty of the various estimates is not indicated here, but can be found in Watson et al. (1992). Adapted from Kreileman and Bouwman (1994)

Source	Emission (Tg/yr)				
	1970 IMAGE	1990 IMAGE	1990 REF.	2050 IMAGE	2050 REF.
a. Sources of CH_4					
Wetland rice fields	53	59	60	52	87
Animals	66	79	80	161	173
Animal waste	12	14	14[a]	28	54
Biomass burning					34
Deforestation[b,c]	16	14	6[d]	12	14
Savanna burning	17	16	13[d]	6	–
Agricultural waste burning	7	8	9[d]	–	
b. Sources of N_2O					
Fertilizer induced	0.4	0.9	2.2[e]	2.0	4.2[e]
Deforestation[c]	0.2	0.2	0.4[f]	0.1	1.1
Animal waste	0.5	0.6	0.4[g]	1.5	

[a] Gibbs and Woodbury (1993).
[b] Estimate for deforestation excludes shifting cultivation, which contributes about 10 Tg CH_4/yr (Crutzen and Andreae, 1990).
[c] For 1970 no values are available, since this calculation involves the history of deforestation over a period of some years. For deforestation effects on soil N_2O emission as well as for direct biomass burning the year 1975 is presented.
[d] Crutzen and Andreae (1990).
[e] Pepper et al. (1992). This estimate of total N_2O loss, including fertilizer-induced and background losses from arable land. From this number the amount of circa 1 Tg N_2O-N/yr needs to be subtracted to arrive at the 1990 fertilizer-induced loss.
[f] Bouwman (1995).
[g] Khalil and Rasmussen (1992).

economic development and the productivity of land (crop yields and animal productivity). Slight changes in the balance between agricultural demand and production could cause large differences in land use and associated emissions. There is a disagreement between the baseline scenario and the data from Watson et al. (1992) for future agricultural emissions, as shown for rice (Table 11.3). The assumed increase in rice yield (t/ha) in the baseline scenario for South and East Asia and the Centrally Planned Asian countries is 1.1% per year for the period 1990-2010, or a 25% increase over these 20 years. The assumed annual yield increases by Watson et al. (1992) are respectively 1.2% and 0.5%. Both estimates are far less optimistic than those of FAO (1993b) that assume a global rice yield increase of 1.5% per year for the same period leading to a total yield increase of about 35%. This illustrates the uncertainty in the forecasts of rice productivity and area and the consequences for the associated CH_4

emissions (Table 11.3). In general, our baseline also assumes far lower animal productivities than FAO (1993b).

The increase in crop productivity is coupled to increases in the use of agricultural inputs, represented in the model by N fertilizer and a management factor (Table 11.1). The annual increase in global N-fertilizer use in our baseline scenario is about 2% for the period 1990-2025 and 0.3% for 2025 to 2050. These assumptions lead to a doubling of N-fertilizer use between 1990 and 2025. The associated N_2O losses are directly related to fertilizer use. It is not unrealistic to assume that fertilizer use will continue to increase along with the projected increases in crop productivity. However, the projected increase in both the baseline and Watson et al. (1992) for the coming decades is somewhat slower than for the period 1970-1990.

The baseline scenario generates globally a large increase in the extent of agricultural land (Table 11.5). Although the global population more than doubles, the increase in agricultural area is only around 15%. This is comparable to the observed increase during the last half century (Plucknett, 1994). This relatively small increase in extent is accompanied by a large increase of agricultural productivity on existing agricultural land.

Although the global increase is modest, there are large regional differences. Agriculturally highly developed regions (especially North America and Europe) increase their productivity significantly, while the population is largely stable. Part of the increase is exported, but there is still a large amount of land that is abandoned. We assume that this land returns slowly to its potential vegetation (often forest), which creates a significant C sink in these regions. The decrease in agricultural land starts immediately in the simulation. Latin America follows a similar path during the next century, but initially there are still significant amount of deforestation.

Other regions (especially Africa, India and China) display different patterns; the increase in population and income (and its linkage to changed dietary preferences) amplify agricultural demand. This demand cannot be satisfied by an increase in productivity and imports and leads to a large increase in agricultural land. By the end of next century almost all suitable agricultural land will be occupied. This will result in deforestation and consequently large GHG fluxes from these regions. The model thus simulates a shift of the major loci of deforestation from Latin America towards Africa and Southeast Asia. The strongly increased human pressures on land in these regions could have severe consequences for the development of sustainable land management, food security and biodiversity.

Forestation as a mitigation option

The model allows not only the evaluation of the consequences of different socio-economic trends, but can also be used to evaluate different mitigation options. There are many different mitigation options (cf. Nakícenovíc et al., 1994), but their effectiveness or suitability can differ per region. It is therefore required to develop a comprehensive capability to evaluate such options. IMAGE 2 provides some possibilities, from which we present two examples.

One of the most often presented mitigation options is forestation. This option increases the C sequestered in forests and remains effective if the total extent of forests continues to increase (NEP of old-growth forests are in balance with respect to the C cycle), or if the resulting forest products are used in a durable way (construction wood), replace other CO_2 sources (woody biomass could replace fossil fuel) or both (construction wood could replace cement).

The simulation of IMAGE 2 satisfies the demand for agricultural products first. This demand drives land cover patterns and the remainder is natural vegetation with its specific C dynamics and storage. Only those regions that climatologically can be forests are suitable for the forestation mitigation option, but a large portion of these regions is used for other purposes. Early estimates of forestation potential were therefore probably much too optimistic because, to offset large amounts of fossil-fuel-related CO_2 emissions, large areas have to be forested annually (Sedjo and Solomon, 1989). Forestation in shelterbelts, along fields and roads and in degraded lands is, although important for other reasons, alone not sufficient.

Table 11.4 lists the suitable and available extent for forestation under the baseline scenario. From these figures it is apparent that large regional differences occur. Forestation is a possible option in the high and mid latitudes of developed regions. In the high latitudes there is, in 2050, even a large potential increase due to the polewards shift of boreal and temperate forests due to climatic warming (Figure 11.1). This phenomenon is a negative feedback, which actually makes the forestation option more effective. In Latin America forestation possibilities also improve somewhat, but in many other regions only few opportunities are available. In these regions, C sequestration in shelterbelts and on degraded lands is probably the only valid option.

Biofuels as a mitigation option

Another mitigation option is the use of biomass for energy generation. An efficient use of renewable biofuels and biomass for generation of energy could decrease the dependence on fossil fuels. Such a programme is already being implemented in

Table 11.4. Land suitability and availability (not used for other purposes) for the forestation mitigation option, according to IMAGE 2 simulation for the conventional wisdom scenario

Region	Available in 1990 (10^6 km^2)	Suitable in 1990 (10^6 km^2)	Available in 2050 (10^6 km^2)	Suitable in 2050 (10^6 km^2)
Canada and USA	6.1	10.2	7.9	11.7
Latin America	6.2	11.1	6.5	14.3
Africa	2.5	4.5	0.2	6.3
Europe	1.3	3.9	1.4	4.2
CIS	8.5	11.5	9.6	14.1
India and South Asia	0.5	0.9	0.0	1.3
China	1.9	4.5	0.2	5.5
East Asia	1.3	3.2	0.5	3.6

Brazil, where ethanol from sugar cane is used to replace petrol. Biofuels applied for transportation are derived from oil crops or from fermentation of sugar or starch. The techniques involved are not yet competitive with energy derived from fossil oils. However, new technologies could make very efficient use of biomass as an energy source (Johansson *et al.*, 1993). For example, the foreseen development of fuel cells, in which biomass is converted into a hydrogen-rich gas, promises the possibility of small- and large-scale electricity production with an efficiency ranging from 50 to 80% with few environmental impacts. Many future scenarios assume therefore large amounts of energy derived from renewable biomass. Especially in the second half of the next century when fuel-cell systems should be widely operational, biomass could be the major source of renewable energy (Johansson *et al.*, 1993; Kassler, 1994).

We make several assumptions for the use of biomass as an energy source in the baseline scenario. The use of such biomass increases from 5 EJ now (10^{18} J: current global energy use is 300 EJ) to 75 EJ in 2050 and 210 EJ in 2100. These figures are based on reasonable mixes of energy carriers, the annual growth in energy demand induced by the increasing global population and its income, a decline in oil and natural gas resources and the preference for renewable energy sources over fossil fuels. We assume for the baseline that the required biomass is simply available from crop and forest residues and waste. This amount of biomass thus does not require new cropland. However, the biofuels used in the transportation sector (ethanol) are still derived from sugar cane and maize but this relatively small additional demand has been included in the simulated extent of agricultural land.

We derived two different scenarios from the baseline scenario. These scenarios only differ with respect to their source of biomass-energy level. In the second scenario, biomass crops, we assumed that such large amounts of biomass are not simply available and that a large portion should be grown specifically in the form of short-rotation forests and other biomass crops, such as *Miscanthes* spp. These crops are effective sources, because they generate a high yield of 40-60 t/ha/year and have a wide potential distribution. We assumed that 40% of the total biomass-related energy demand is satisfied by such crops. The third scenario, no-biomass fuel, is the opposite approach. Renewable energy from biomass is not feasible and the total modern biomass fuels are replace by fossil oils.

Table 11.5 presents some of the results of these three scenarios. The baseline and the biofuel-crop scenario result in lower atmospheric concentrations of CO_2, but significantly higher CH_4 concentrations, than the no-biofuel scenario. From such results we must conclude that the effectiveness of mitigation options should be evaluated simultaneously for all GHGs. The final concentration of a single GHG is not a good indicator because reducing one species could enhance another. CH_4 has a much larger global warming potential. Despite the differences in atmospheric concentrations between the scenarios, the global temperature increase is therefore not significantly different between the scenarios. Most differences are buffered through changes in land cover and the accompanying climatic and biogeochemical feedbacks.

The differences in the simulated extent of agricultural lands are large (Table 11.5). The large amount of additional demand for biomass crops result in a 30 or 65%

Table 11.5. Agricultural extent under different scenario options for the generation of energy. The difference between the scenarios is the source of a large fraction of energy carrier (1. All biomass comes from residuals. 2. Biofuel crops. 60% of the biomass comes from biofuel crops. 3. Biomass is replaced by oil)

Property	Scenario 1 2050	Scenario 2 2050	Scenario 3 2050	Scenario 1 2100	Scenario 2 2100	Scenario 3 2100
Atmospheric CO_2 concentration (ppmv)	522	534	539	777	821	857
Atmospheric CH_4 concentration (ppmv)	2.5	2.6	2.4	2.3	2.4	1.7
Average surface temperature (°C)	+1.2	+1.3	+1.2	+2.1	+2.3	+2.2
Change in agricultural area (26.7 10^6 km^2)	+9%	+30%	+9%	+14%	+65%	+15%
Change in forest area (47.2 10^6 km^2)	−26%	−32%	−26%	−27%	−31%	−27%

increase in, respectively, 2050 and 2100. This increase leads to both a larger deforestation and expansion into more arid regions, such as steppes and savannas. Here the importance for a comprehensive evaluation with respect to land use becomes apparent. The increasing land requirements in the biofuel-crop scenario result in a smaller total sink in the terrestrial biosphere. This has consequences for the simulated atmospheric CO_2 concentration and is the main reason that this scenario yields higher concentrations than the baseline, where it was assumed that biomass was freely available. The consequences of biomass energy scenarios must be carefully evaluated and land-use issues should not be neglected in such assessment.

CONCLUDING REMARKS

The added value of integrated assessment models is thus their ability to generate insights that cannot be easily derived from individual natural or social science component models that have been developed in the past. The current integrated assessment models with respect to global climate change all have a mix of policy and scientific goals. This has lead to a large diversity of different, but complementary approaches. Only a few models include a simulation of land use and land cover change. This omission limits a comprehensive evaluation of feedbacks and effective linkages among economic, energy, agriculture and forestry sectors and withholds adequate evaluation of many C cycle or land cover related mitigation options. The above examples illustrate the usefulness of these integrated models in creating a greater understanding of the linkages and relationships among the climate system, the biosphere and human activities. Models such as IMAGE 2 (Alcamo, 1994) and AIM (Morita *et al.*, 1994) will be enhanced continuously. More consistent scenario developments, better algorithms and better initialization and validation data will improve future applications. A major conclusion from such analysis should be that from the integration of a series of disciplinary expertise and insights emerges a better understanding, which could further disciplinary and multidisciplinary research.

Without such understanding, integrated global change assessments will continue to emphasize energy-industry-transport emissions, a globally aggregated C cycle with limited land-use change, and the climate system, while natural ecosystems and agrosystems will only be addressed in impact studies. Excluding them from integrated assessments of climate change will lead to non-sustainable climate policies. Fortunately, major international research programmes (Diversitas, WCRP, IGBP and HDP) are now addressing land use and land cover change more systematically, which means that we now enter an exciting new era in global change research.

ACKNOWLEDGEMENTS

The preparation of this paper was funded by the Dutch Ministry of Housing, Planning and the Environment under contract MAP410 to RIVM and the National Research Programme 'Global Air Pollution and Climate Change'. The Terrestrial Environment System of the IMAGE 2 model is an official contribution to IGBP-GCTE core research.

REFERENCES

Alcamo, J. (ed.). 1994. *IMAGE 2.0: Integrated Modeling of Global Climate Change.* Kluwer Academic Publishers, Dordrecht.

Bouwman, A.F. 1995. *Compilation of a Global Inventory of Emissions of Nitrous Oxide.* Ph.D. Thesis. Agricultural University, Wageningen.

Brinkman, R. 1987. Agro-ecological characterization, classification and mapping. Different approaches by the international agricultural research centres. In: *Agricultural Environments. Characterization, Classification and Mapping.* A.H. Bunting (ed.). CAB International, Wallingford, UK. pp. 31-42.

Carter, T.R., Parry, M.L., Harasawa, H. and Nishioka, S. 1994. IPCC technical guidelines for assessing impacts of climate change. *IPCC Special Report CGER-1015-'94.* Intergovernmental Panel on Climate Change, WMO and UNEP, Geneva.

Cramer, W. and Solomon, A.M. 1993. Climatic classification and future global redistribution of agricultural land. *Clim. Res.* **3**: 97-110.

Crutzen, P.J. and Andreae, M.O. 1990. Biomass burning in the tropics: impact on atmospheric and biochemical cycles. *Science* **250**: 1669-1678.

Dowlatabadi, H. and Morgan, M.G. 1993. Integrated assessment of climate change. *Science* **259**: 1813 and 1932.

Drake, B.G. 1992. The impact of rising CO_2 on ecosystem production. *Water Air Soil Pol.* **64**: 25-44.

Edmonds, J., Wise, M. and MacCracken, C. 1994. Advanced energy technologies

and climate change: an analysis using the global change assessment model (GCAM). *Report PNL-22*. Pacific Northwest Laboratory, Washington DC.
FAO. 1987. Report on the agro-ecological zones project. Vol. 3. Methodology and results for South and Central America. *World Soil Resources Report* **48/3**. FAO, Rome.
FAO. 1991. Agrostat PC. Land Use. *Computerized Information Series* **1/7**. FAO, Rome.
FAO. 1993a. Agriculture: Towards 2010. *Conference Report* **C93/24**. FAO, Rome.
FAO. 1993b. Summary of the final report of the forest resources assessment 1990 project for the tropical world, 11th session of the Committee on Forestry. *Report D1/U9798*. FAO, Rome.
Fischer, G., Froberg, K., Keyzer, M.A. and Pahrik, K.S. 1988. *Linked National Models: A Tool for International Food Food Policy Analysis*. Kluwer, Dordrecht.
Fisher, M.J., Rao, I.M., Ayarza, M.A., Lascano, C.E., Sanz, J.I., Thomas, R.J. and Vera, R.R. 1994. Carbon storage by introduced deep-rooted grasses in the South American savannas. *Nature* **371**: 236-238.
Gibbs, M.J. and Woodbury, J.W. 1993. Methane emissions from livestock manure. In: *Proceedings of the International Workshop on Methane and Nitrous Oxide. Methods for National Inventories and Options for Control*. A. Van Amstel (ed.). National Institute of Public Health and Environmental Protection, Bilthoven. pp. 81-91.
Houghton, J.T., Jenkins, G.J. and Ephraums, J.J. (eds.). 1990. *Climate Change: The IPCC Scientific Assessment*. Cambridge University Press, Cambridge.
Houghton, J.T., Callander, B.A. and Varney, S.K. (eds.). 1992. *Climate Change 1992: The Supplementary Report to the IPCC Scientific Assessment*. Cambridge University Press, Cambridge.
Hulme, M., Raper, S.C.B. and Wigley, T.M.L. 1994. An integrated framework to address climate change (ESCAPE) and further developments of the global and regional climate modules (MAGICC). In: *Integrative Assessment of Mitigation, Impacts and Adaptation to Climate Change*. N. Nakicenovic (ed.). IIASA, Laxenburg. pp. 112-134.
IGBP. 1994. Global modelling and data activities 1994-1998. *IGBP Report* **30**. International Geosphere-Biosphere Programme, Stockholm.
Johansson, T.B., Kelly, H., Reddy, A.K.N. and Williams, R.H. (eds.). 1993. *Renewable Energy: Sources for Fuels and Electricity*. Island Press, Washington DC.
Kassler, P. 1994. *Energy for Development*. Shell Selected Paper, Shell, London.
Kauppi, P., Mielikäinen, K. and Kuusela, K. 1992. Biomass and carbon budget of European forests, 1971 to 1990. *Science* **256**: 70-74.
Khalil, M.A.K. and Rasmussen, R.A. 1992. The global sources of nitrous oxide. *J. Geophys. Res.* **97**: 14651-14660.
Klein Goldewijk, K., van Minnen, J.G., Kreileman, G.J.J., Vloedbeld, M. and Leemans, R. 1994. Simulating the carbon flux between the terrestrial environment and the atmosphere. *Wat. Air Soil Pollut.* **76**: 199-230.
Kreileman, G.J.J. and Bouwman, A.F. 1994. Computing land use emissions of greenhouse gases. *Wat. Air Soil Pollut.* **76**: 231-258.

Kwadijk, J. 1993. The impact of climate change on the discharge of the River Rhine. *Nederl. Geogr. Stud.* **171**: 1-201.

Larcher, W. 1980. *Physiological Plant Ecology.* Springer-Verlag, Berlin.

Lashof, D.A. 1989. The dynamics greenhouse: Feedback proceses that may influence future concentrations of atmospheric trace gases and climatic change. *Clim. Change* **14**: 213-214.

Leemans, R. 1996. Incorporating land-use change in Earth system models illustrated by IMAGE 2. In: *Global Change and Terrestrial Ecosystems.* B. Walker and W. Steffen (eds.). Cambridge University Press, Cambridge (in press).

Leemans, R. and Cramer, W. 1991. The IIASA database for mean monthly values of temperature, precipitation and cloudiness on a global terrestrial grid. *Research Report RR-91-18.* International Institute for Applied Systems Analysis, Laxenburg.

Leemans, R. and Solomon, A.M. 1993. The potential response and redistribution of crops under a doubled CO_2 climate. *Clim. Res.* **3**: 79-96.

Leemans, R. and van den Born, G.J. 1994. Determining the potential global distribution of natural vegetation, crops and agricultural productivity. *Wat. Air Soil Pollut.* **76**: 133-161.

Leemans, R., Cramer, W. and van Minnen, J.G. 1996. Prediction of global biome distribution using bioclimatic equilibrium models. In: *Effects of Global Change on Coniferous Forests and Grassland.* J.M. Melillo and A. Breymeyer (eds.). John Wiley, New York (in press).

Leggett, J., Pepper, W.J. and Swart, R.J. 1992. Emissions scenarios for the IPCC: an update. In: *Climate Change 1992. The Supplementary Report to the IPCC Scientific Assessment.* J.T. Houghton, B.A. Callander and S.K. Varney (eds.). Cambridge University Press, Cambridge. pp. 71-95.

Manabe, S. and Wetherald, R.T. 1987. Large-scale changes in soil wetness induced by an increase in carbon dioxide. *J. Atmosph. Sci.* **44**: 1211-1235.

Morison, J.I.L. 1985. Sensitivity of stomata and water use efficiency to high CO_2. *Plant Cell Environ.* **8**: 467-474.

Morita, T., Matsuoka, Y., Kainuma, M., Kai, K., Harasawa, H. and Dong-Kun, L. 1994. Asian-Pacific Integrated Model to assess policy options for stabilizing global climate. *AIM Report 1.0.* National Institute for Environmental Studies, Tsukuba, Japan.

Nakícenovíc, N., Nordhaus, W.D., Richels, R. and Toth, F.L. (eds.). 1994. *Integrative Assessment of Mitigation, Impact, and Adaptation to Climate Change.* International Institute for Applied System Analysis, Laxenburg.

Neilson, R.P. 1993. Vegetation redistribution: a possible biosphere source of CO_2 during climatic change. *Wat. Air Soil Pollut.* **70**: 659-673.

Olson, J., Watts, J.A. and Allison, L.J. 1985. *Major World Ecosystem Complexes Ranked by Carbon in Live Vegetation: A Database.* Report NDP-017. Oak Ridge National Laboratory, Oak Ridge, Tennessee.

Parry, M.L., Porter, J.H. and Carter, T.M. 1990. Climatic change and its implications for agriculture. *Outlook on Agriculture* **19**: 9-15.

Pepper, W., Leggett, J., Swart, R., Watson, J., Edmonds, J. and Mintzer, I. 1992.

Emission scenarios for the IPCC: an update. Assumptions, methodology and results. *Report IPCC Working Group I*. Geneva.
Plucknett, D.L. 1994. Science and agricultural transformation. *IFPRI Lecture Series*. International Food Policy Research Institute, Washington DC.
Prentice, I.C., Cramer, W., Harrison, S.P., Leemans, R., Monserud, R.A. and Solomon, A.M. 1992. A global biome model based on plant physiology and dominance, soil properties and climate. *J. Biogeogr.* **19**: 117-134.
Prinn, R.G. 1994. The interactive atmosphere: global atmospheric-biospheric chemistry. *Ambio* **23**: 50-61.
Raich, J.W. and Schlesinger, W.H. 1992. The global carbon dioxide flux in soil respiration and its relationship to vegetation and climate. *Tellus* **44B**: 81-99.
Rosenzweig, C. and Parry, M.L. 1994. Potential impact of climate change on world food supply. *Nature* **367**: 133-138.
Rotmans, J., de Boois, H. and Swart, R.J. 1990. An integrated model for the assessment of the greenhouse effect: the Dutch approach. *Clim. Change* **16**: 331-356.
Rotmans, J., Hulme, M. and Downing, T.E. 1994. Climate change implications for Europe: An application of the ESCAPE model. *Glob. Env. Change* **4**: 97-124.
Sarmiento, J.L. and Bender, M. 1994. Carbon biogeochemistry and climate change. *Photosynth. Res.* **39**: 209-234.
Sedjo, R.A. and Solomon, A.M. 1989. Climate and forests. In: *Greenhouse Warming: Abatement and Adaptation*. N.J. Rosenberg, W.E. Easterling III, P.R. Crosson and J. Darmstadter (eds.). Resources for the Future, Washington DC. pp 105-119.
Skole, D. and Tucker, C. 1993. Tropical deforestation and habitat fragmentation in the Amazon: Satellite data from 1978 to 1988. *Science* **260**: 1905-1910.
Smith, T.M. and Shugart, H.H. 1993. The transient response of terrestrial carbon storage to a perturbed climate. *Nature* **361**: 523-526.
Smith, T.M., Leemans, R. and Shugart, H.H. 1992. Sensitivity of terrestrial carbon storage to CO_2 induced climate change: Comparison of four scenarios based on general circulation models. *Clim. Change* **21**: 367-384.
Strain, B.R. and Cure, J.D. 1985. *Direct Effects of Increasing Carbon Dioxide on Vegetation*. Report DOE/ER-0238. US Department of Energy, Washinghton DC.
Taylor, K.E. and Penner, J.E. 1994. Response of the climate system to atmospheric aerosols and greenhouse gases. *Nature* **369**: 734-737.
Turner, B.L. II, Clark, W.C., Kates, R.W., Richards, J.F., Mathews, J.T. and Meyer, W.B. (eds.). 1990. *The Earth as Transformed by Human Action: Global and Regional Changes in the Biosphere over the Past 300 Years*. Cambridge University Press, Cambridge.
Turner, B.L., Moss, R.H. and Skole, D.L. 1993. *Relating Land Use and Global Change: A Proposal for an IGBP-HDP Core Project*. IGBP Report No.24 and HDP Report No. 5. International Geosphere-Biosphere Programme and the Human Dimensions of Global Environmental Change Programme, Stockholm.
Watson, R.T., Meira Filho, L.G., Sanheuza, E. and Janetos, A. 1992. Greenhouse gases: sources and sinks. In: *Climate Change 1992. The Supplementary Report to*

the IPCC Scientific Assessment. J.T. Houghton, B.A. Callander and S.K. Varney (eds.). Cambridge University Press, Cambridge. pp. 27-46.

Wofsy, S.C., Goulden, M.L., Munger, J.W., Fan, S.M., Bakwin, P.S., Daube, B.C., Bassow, S.L. and Bazzaz, F.A. 1993. Net exchange of CO_2 in a mid-latitude forest. *Science* **260**: 1314-1317.

Woods, J. and Barkmann, W. 1993. The plankton multiplier – positive feedback in the greenhouse. *J. Plankton Res.* **15**: 1053-1074.

Zobler, L. 1986. *A World Soil File for Global Climate Modeling.* Technical Memorandum. NASA, New York.

12. Global Change Impacts on Agriculture, Forestry and Soils: The Programme of the Global Change and Terrestrial Ecosystems Core Project of IGBP

BERNARD TINKER together with JAN GOUDRIAAN, PAUL TENG, MIKE SWIFT, SUNE LINDER, JOHN INGRAM AND SIEBE VAN DE GEIJN
Department of Plant Sciences, University of Oxford, UK

The concept of global change is now well understood and to a large extent accepted amongst scientists, agencies and the informed public. It is usually also accepted that the driving forces for changes are: (i) change in atmospheric composition; (ii) climate change (which arises from the first); and (iii) land-use change (driven by both socio-economic factors and by climate change). A number of associated topics, such as ozone/UV-B, acid rain and other forms of pollution are sometimes included, but are not central to the discussion in this paper. Of these factors, change in atmospheric composition is well documented, and a reasonably good prediction of its future progress can be given. Climate change may be occurring, but it is still possible that what is being observed is a normal fluctuation (IPCC, 1992). Land-use change cannot yet, therefore, be associated with climate change in the global sense, but it is occurring very rapidly because of socio-economic driving forces.

The *potential* impacts of these three factors are agreed to be massive, and could well be catastrophic in some cases, while beneficial in others. However, the progress and impact of climate change is difficult to predict, atmospheric composition may be controlled to an as yet unknown extent by human action, and the progress of land-use change depends upon demographic, social and economic factors which cannot be predicted with accuracy over long periods. The details of the impacts are therefore extremely uncertain, especially for rainfall.

The International Geosphere-Biosphere Programme (IGBP) was set up, under the auspices of the International Council of Scientific Unions, to conduct research into these matters. Of the Core Projects of the IGBP, Global Change and Terrestrial Ecosystems (GCTE) (Steffen *et al.*, 1992) has the most obvious and direct involvement with the land surface. Within GCTE, Focus 3 has the specific responsibility for agriculture, forestry and soils (Table 12.1). GCTE has responsibility for work on both

Table 12.1. Structure of GCTE Focus 3

Activity 3.1:	**Effects of Global Change on Key Agricultural Systems**
Task 3.1.1:	Experiments on Key Crops with Changed Atmospheric Composition and Climate, on Different Soils
Task 3.1.2:	Modelling Growth of Key Crops Under Changed Atmospheric Composition and Climate
Task 3.1.3:	Global Change Impact on Pastures and Rangelands and Associated Animal Production
Activity 3.2:	**Changes in Pests, Diseases and Weeds**
Task 3.2.1:	Global Monitoring Network and Data Sets for Pests, Diseases and Weeds
Task 3.2.2:	Distributions, Dynamics and Abundance of Pests and Diseases Under Global Change
Task 3.2.3:	Weed Distribution, Dynamics and Abundance Under Global Change
Activity 3.3:	**Effects of Global Change on Soils**
Task 3.3.1:	Global Change Impact on Soil Organic Matter
Task 3.3.2:	Soil Degradation Under Global Change
Task 3.3.3:	Global Change and Soil Biology
Activity 3.4:	**Effects of Global Change on Multi-species Agro-ecosystems**
Task 3.4.1:	Experimental Studies on the Relationship between Plant Species Number and Function in Agricultural Systems
Task 3.4.2:	Modelling Complex Agricultural Systems
Task 3.4.3:	Long-term Agricultural Experimental Network
Activity 3.5:	**Effects of Global Change on Managed Forests**
Task 3.5.1:	Experimental and Observational Studies of Managed Forests
Task 3.5.2:	Modelling Global Change Impact on Function, Structure and Productive Capacity of Managed Forests

impacts of global change, and also on feedbacks which further control the extent of global change. For example, the change and extent of different types of vegetation has important implications for the regional climate. GCTE's main responsibility is, however, for studies of impacts.

At a series of international 'open' workshops, GCTE Focus 3 has considered carefully what research can and should be done, in a situation where the climatic predictions are so uncertain. Given this uncertainty on a worldwide scale, let alone on a site-specific basis, the research priorities have aimed at improved system understanding leading to more robust predictive models. Because of the global scale of the problem it has been essential to base the research programmes on international collaboration, so that the results have generic value. National programmes will subsequently fill out the more local details. The work undertaken should therefore be of a precautionary and preparative type, so that it simplifies and clarifies the tasks which must be undertaken as and when it becomes clear that climate change is occurring, and the extent and direction of the changes. This is one reason why systems analysis and the preparation of better predictive models occur so frequently in our programme.

It seems likely that when climate change becomes apparent, there will be a period of increasing turbulence within agriculture. Farmers and advisers are familiar with

THE GLOBAL CHANGE AND TERRESTRIAL ECOSYSTEM PROJECT   297

the possibility of weather variation in a single year or over a short run of years, as in many regional or local droughts. Two questions which they are always called upon to decide are, firstly whether these new conditions will hold in the following year, because they have to take immediate decisions on cropping; and secondly whether this change may last for several years, because they then may have to adjust their farming system. A long-term trend, possibly with greater annual variability, is however something with which farmers are not accustomed to cope.

Much will depend upon the degree of reliability of local climate predictions. If climate cannot in fact be predicted in detail at the local scale, the essential characteristic must be flexibility and rapid response. This may involve change of cultivar, change of agronomic variable such as sowing date, change to a different crop, or change to a completely different agricultural system, e.g., from arable to grassland. In some cases, it may mean abandonment of farming, as happened so often during the drought of the 1930s in the American Midwest (Worster, 1979). If the predictability of climate change on a local basis becomes better, then farmers can plan these changes in a more rational way. In either case, the best tool for guiding their response is likely to be a set of the best possible predictive crop models, tested over widely different climates, photoperiods and soil types.

IGBP has no specific funding, other than a limited amount of money for its own internal administrative operation. Its work proceeds in two main ways: (i) by the launching of major collaborative projects which directly address specific aspects of the GCTE Operational Plan (Steffen *et al.*, 1992), and which gain national or international funding; (ii) by the coordination and gathering together of many projects (usually nationally funded) which have similar aims. The latter strategy aims to build a global picture by synthesizing what is usually locally based research. The total programme cannot therefore be a tightly structured and organized one, such as is possible where there is a large funding stream to determine the research to be done. We are to some extent guided by Lord Rutherford's statement 'We have no money, so we have to think!'. However, that thinking has produced a number of important initiatives, some already in being, some still at the planning stage, which we explain below.

OVERVIEW OF THE GCTE AND FOCUS 3 STRUCTURES

The GCTE Core Project encompasses a very wide range of terrestrial biomes and scientific disciplines. Given this wide scope, the early planning phase of GCTE recognized the need for a highly structured programme, to ensure both that appropriate collaboration at the international scale is established and that important areas (either geographically or thematically) were not overlooked or duplicated. GCTE has thus been designed around four major Foci: ecosystem function; ecosystem structure and composition; global change impact on agriculture, forestry and soils; and ecological complexity.

GCTE is hierarchical in design, each *Focus* being divided into *Activities*, each of which are in turn subdivided into *Tasks*. This arrangement allows Tasks of a similar nature to be clustered and necessary collaboration at the Task level to be maximized. There are, however, numerous instances where a particular scientific topic spans Activities or indeed Foci. To minimize duplication of effort while maximizing the benefits of interdisciplinary study, inter-Foci and/or inter-Activity aspects have been identified and are strengthened as much as possible. Good examples of such collaboration are, respectively, the system response to elevated CO_2, and the pests and weeds component within the GCTE Crop Networks; both are described later in this paper.

Within Focus 3, four major types of production system have been identified: monocrop agriculture, pasture and grazing systems, multi-species cropping (including agroforestry and rotational systems), and forestry. These have been grouped into three Activities:

1. *Key agricultural systems.* This includes a representative selection of monoculture food crops, together with improved pastures and rangelands; the emphasis in this Activity is on harvestable product, i.e., the harvested portion of the crop or animal.
2. *Multi-species agro-ecosystems.* These are the norm in much of the world, be they spatially or temporally mixed. Modelling spatially complex agricultural systems is in its infancy, relative to modelling monocrop production or even rotation farming.
3. *Managed forests.* These encompass the spectrum from intensively managed plantation crops through more natural forests which are either selectively logged or put to some other use with socio-economic implications.

Two further Activities cross-cut these three major areas: one addresses the effect of global change on pests, diseases and weeds and the other addresses the effect of global change on soils. The full list of the Focus 3 Activities and their associated Tasks is given in Table 12.1, and their rationale and work plans are discussed below. (As GCTE planning has progressed, new Activities have been assigned the next sequential number. While this avoids the possibility of confusion brought about by re-numbering Activities and Tasks, it must be stressed that the number does not imply importance.)

The need for within-GCTE collaboration has already been highlighted. There is however an equal need to develop appropriate collaboration with other international programmes, be they other IGBP Core Projects (notably IGAC[1], BAHC[2] and LUCC[3]) or non-IGBP initiatives; GCTE's role is often one of international coordination and engendering collaboration, linking new initiatives and synthesizing results within a common, internationally agreed framework. This aspect will be especially necessary for linking to the human dimension, particularly apposite when dealing with ecosystems from which a 'harvest' is taken; global change impact on agriculture and forestry is more than just impact on biology – it is impact on food, wealth, development and people.

---

[1] International Global Atmospheric Chemistry project
[2] Biospheric Aspects of the Hydrological Cycle project
[3] Land Use and Land Cover Change project

EFFECTS OF GLOBAL CHANGE ON MAJOR FOOD CROPS (A COMPONENT OF ACTIVITY 3.1)

Anticipated changes in global rainfall and temperature patterns together with the established increase in atmospheric CO_2 will affect the production of crops throughout the world. Of the many crops which might form the subject of a research programme in global change, GCTE selected a shortlist of six for initial studies: wheat, rice, potatoes, cassava, maize and groundnuts. The crops were selected both in recognition of their importance in global food production, and as representatives of a diverse range of crop 'functional types'. GCTE has also recently included sorghum, a major crop of marginal tropical regions.

GCTE is launching a series of Crop Networks, one for each of these crops. These worldwide research networks are designed to promote integrated experimental and modelling research to determine the interactive effect of global change factors on these key agronomic species. Each will be led by a small working group of modellers, experimentalists, and pests and disease specialists. The Networks will build on the many crop experiments already established and on models already developed. Existing models, however, are generally constructed and validated using data from one locality, so the identification of modelling approaches that yield generic capability is needed. The worldwide network approach will greatly facilitate the required cross-comparison of models with appropriate data sets, and both modellers and experimentalists are essential to the given Network's success. The GCTE research will lead to the development of crop models that are robust under a wide range of changed environmental conditions.

The Crop Networks are designed to maximize the understanding of global change impact on the cropping system. This means on the crops *per se*, *and* on the crops' pests, weeds and soils (key aspects which are discussed elsewhere in this paper) – i.e., on the whole agro-ecosystem. This approach requires the consideration of not only all components of the system, but also the interactive effects of global change drivers. This integrative approach makes GCTE uniquely innovative. Furthermore, by taking the systems approach to crop modelling, these Networks cover many aspects of GCTE, and provide a forum for the global change community to undertake and discuss collaborative modelling research.

The Networks also strongly promote international collaboration in experimentation. One of the key experimental techniques, of equal interest to Focus 1 and Focus 3, is Free Air CO_2 Enrichment (FACE); FACE systems are now established in crops (wheat and cotton) and in pastures. (Methodological problems are encountered when using the technique in higher-stature systems, i.e., forests, but as the technique is so valuable, these are being tackled by various groups around the world.) The real, and unique value of FACE is that the site's microclimate is not disturbed; in addition, pests and diseases are not excluded from, or trapped in, the system. Another main advantage is size; being some 300-400 m^2 in area, subtreatments (e.g., nitrogen or water) can be incorporated and, even so, the experiment is not seriously disrupted by the destructive sampling of soils and vegetation from all plots. The question of whether

Open-Top Chambers (OTCs) can be 'calibrated' against FACE is important, as OTCs are being widely used. The issue is currently being researched by GCTE collaborators.

GCTE Crop Networks started with wheat, because models are already well developed and its physiology is well understood. An international workshop (Ingram, 1992) launched the GCTE Wheat Network, and initiated the evaluation of current wheat production models using a minimum set of standard model initialization parameters. A subsequent workshop (Goudriaan *et al.*, 1993) undertook a highly refined analysis of the Network's wheat production models, using a small number of agreed data sets. A summary of this most recent workshop's design and results is given below, to illustrate the effectiveness of working in this truly global, collaborative way.

The workshop (and preparative exercises) covered three main areas, and relied upon a series of pre-circulated datasets:

THE HIGHLY REFINED SENSITIVITY ANALYSIS (EXERCISE 1)

This required models to be run with common baseline data, but with artificially perturbed temperature (from −6 °C to +6 °C, at 1 °C increments) and CO_2 concentration (seven levels). Two sets of 30-year mean meteorological data were distributed to modellers in advance of the workshop, one set being for a winter wheat growing season (European data), the other being for spring wheat (North American data). Modellers were asked to fit genetic coefficients which resulted in standard phenological development dates for the current (zero °C offset) case. Output data included, *inter alia*, phenological development, leaf area index and grain yield.

Contrary to expectations, the variance of the models' results was unexpectedly large, even though the dates of emergence, of anthesis and of maturity had been prescribed (see Figure 12.1). Models, however, contain many feedbacks, making it difficult to trace the precise causes of deviation. (A good example is the major feedback route occurring through growth of leaf area – once leaf area in early growth is overestimated, its high value will tend to be amplified though a higher rate of photosynthesis.) For this reason, it was decided to 'force' the models in line by imposing the same prescribed time course of leaf area index on all models. This artificial move enabled exclusion of morphological model differences (specific leaf area, leaf area formation, tillering, leaf appearance), leaving only functional model differences (light interception, photosynthesis and respiration). The results of this forcing exercise are presented in Figure 12.2, from which it can be seen that the variance was only marginally reduced.

The models included in this exercise differ greatly in complexity, ranging from one with only 90 lines of code in GWBASIC to one requiring a Cray computer to run. There was no apparent relationship between model output values and model complexity.

Figure 12.1. Scatter plot of simulated yield against simulated biomass for the North American site □ and for the European site ■

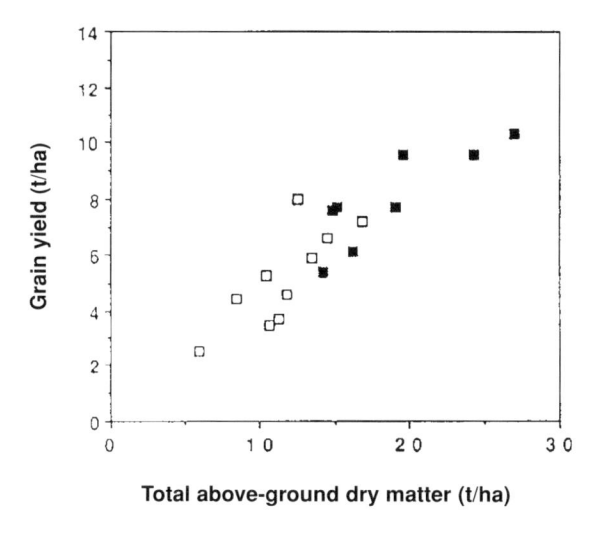

Figure 12.2. Scatter plot of simulated yield, as in Figure 12.1, but with the same imposed time course for leaf area index in all models

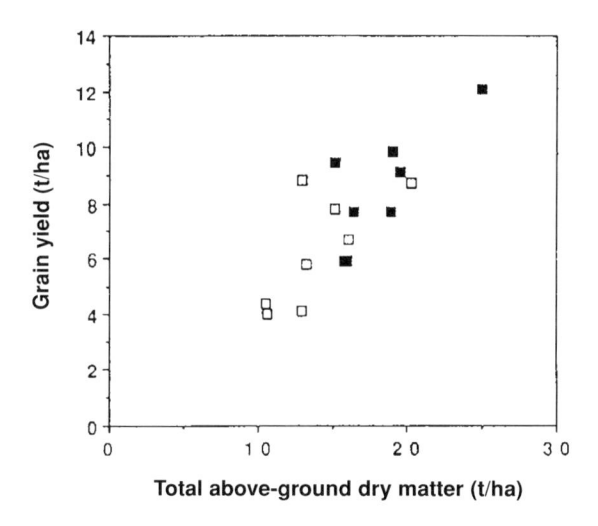

REAL DATA SETS AND MODEL ASSESSMENT (EXERCISE 2)

Together with the 30-year mean meteorological data for Exercise 1, modellers were sent meteorological, soil, crop management and phenology data for crops at the European (winter wheat) and North American (spring wheat) sites. Modellers were asked to bring their output data, including grain yield, for the cases provided. This exercise was designed to highlight the strengths and limitations of each model, and was *not* intended to be a 'horse race'; the models differ in objective and design, in operational scale and in geographic, climatic and phenological aspects.

The full results from Exercises 1 and 2 will be reported and discussed in two multi-authored scientific papers. One will discuss the sensitivity analysis, and the other will analyse the precise backgrounds of the model differences. It must be stressed, however, that the workshop did not aim to tell which model does the best job; rather, it set out to improve understanding of why the models behave as they did, and which components of the models were critical.

COMBINING NETWORK DATA SETS WITH NETWORK MODELS (EXERCISE 3)

This exercise aimed to promote the Network's data/model dialogue by using the Network's diverse experimental data sets for further model validation and comparison. The methodology needed to run each model for each available data set was established. The intention of the exercise was to set the scene for Network efforts that will identify which combinations of model and data set are most likely to improve understanding and to develop interesting research questions.

This 'workshop series' approach has proved a very productive way of working, and the networks for the other crops will be similarly designed. In addition to workshops, Network activities include the distribution of collated metadata sets to all formal collaborators and other interested parties. These metadata *describe* the nature of the Network's data sets and models; they do not list the data or model code *per se*. These metadata compilations are published in GCTE Report No. 2, and will be regularly updated. Formal Network collaborators share actual data and model code according to an agreed code of conduct; the IBSNAT data format was chosen and is now the GCTE standard for all the GCTE Crop Networks, although the conversion of all data sets to this standard is not yet complete; a GCTE Networks' Officer position has been funded by the Dutch Global Change Programme (NOP) to assist with this important task.

The GCTE Rice Network was launched with an international workshop held at IRRI in March 1994 (Ingram, 1994a), and network activities will be planned in detail at a workshop in 1995, co-funded by FAO. Similar Networks for potato and cassava are in advanced planning, and initial Network membership will be established in 1995.

These GCTE Networks represent a major step forward in international, collaborative agronomic science. They add value to the large investments in crop modelling that are being made in many countries, and being dynamic in both scientific content and participation, and are very much open to further scientists.

IMPROVED PASTURES AND RANGELANDS (A COMPONENT OF ACTIVITY 3.1)

Uncertainties surrounding the possible effect of global change on food production are not limited to crops. Therefore, to complement the Focus 3 work in this area, an important Task in Activity 3.1 (Task 3.1.3) deals with the potential impact of global change on livestock production. Global change will have far-reaching consequences for dairy, meat and wool production, though mainly via impacts on grass and range productivity; direct impacts on the animal itself are not expected to be great. This concept underpins GCTE's work plan in this area, as manifested in the Task's title: 'To predict the effects of global change on pasture and range composition and production, and the consequent effects on livestock production'.

Grasslands and rangelands form a continuum. At one end lie improved pastures, i.e., areas where the original vegetation has been removed and grassland is maintained by regular grazing and/or mowing, and fertilizer and/or herbicide applications. This definition obviously encompasses intensively managed grasslands (e.g., in the Netherlands), but it also includes the areas of sown grassland which are used less intensively for livestock production. Improved pastures can therefore range from simple monocultures to complex mixtures of species.

At the other end of the continuum lie rangelands. These occur in areas where either climate or soil factors are so limiting that intensive primary production is not viable. While rangelands may contain introduced species, this end of the continuum is characterized by systems where (i) the pasture has a substantial component of native species; (ii) the pasture is composed of more than three or four species; and (iii) there is little or no addition of fertilizer or irrigation.

Improved pastures are invariably restricted to commercial operations, whereas rangelands are used by commercial and subsistence farmers. Vegetation productivity and animal growth models will be formulated for each type and use, and ultimately a generic model will be developed that could be applied to intermediate types.

The improved pasture and rangelands components share two general objectives. These are:
(i) to predict the effects of changes in atmospheric composition and climate on forage production, and the consequential impact on livestock production at the patch and landscape scales; and
(ii) to predict the feedback effects to the atmosphere-climate system of changes in management induced by changes in atmospheric composition and climate.

Improved pastures and rangelands of course have specific individual objectives in addition to these. Systems with significant management intervention via, for instance, forage and animal germplasm improvement or irrigation, will provide opportunities for adaptation to meet changing environmental conditions; research is needed to prepare adaptive strategies to cope with the interactive effects of changes in such conditions. On the other hand, in more extensive systems (where adaptation is hard to implement due to the scale of operation), the research strategy is one of improving the ability to predict the effects of interactive changes in intrinsic and extrinsic forces.

The GCTE approach is to build on points of commonalty between the types of systems, rather than highlight their differences. This helps link to other areas of GCTE interested in, for instance, rangelands – Focus 2 includes an Activity dealing with patch-scale modelling, an important aspect when developing decision support systems for range managers.

GCTE also has an important role in linking natural and social sciences in this area; the rangelands component of this Task is currently leading Focus 3's links to the Human Dimensions Programme of global environmental change (HDP), although this will also become a major aspect of the developing research programme of Activity 3.4 'Effects of global change on multi-species agroecosystems' (see below). The essential point of contact is the proposed joint IGBP-HDP Core Project, Land Use/Cover Change (LUCC). The Task ultimately aims to predict the interactions between the biophysical driving factors and the extrinsic driving factors, the focus of LUCC. The socio-economic understanding of contrasting systems, on a region-by-region basis, will have immediate regional value for land-use planning. (Further details about this GCTE research programme can be found in Stafford Smith and Campbell, 1994.)

EFFECTS OF GLOBAL CHANGE ON MULTI-SPECIES CROPPING SYSTEMS (ACTIVITY 3.4)

GCTE Focus 3 launched its operational phase with research on monocropped systems, and substantial progress has been reported above. GCTE is now turning its attention to cropping systems designed around mixtures of species, and detailed planning is well under way, launched with an international workshop in Kenya in 1994.

Global change will impact on all agriculture systems to some extent. The effects of global change may however first be observed in multi-species agro-ecosystems, because the competitive advantage of the different species in the system may change. The resulting overall effect may be more pronounced than in monocropped systems. There are also two further, powerful reasons for highlighting multi-species agricultural systems in a programme of research on global change. The first is socio-economic; a major proportion of the world's population depend on such systems for their food and livelihood. The second is ecological; study of such systems can give insight into the relationship between complexity, stability, diversity and an ecosystem's capacity to respond to change.

In addition to spatially mixed systems, temporally mixed (i.e., rotation) agriculture offers numerous, well-established benefits, and it has long been a principal agricultural management technique. As with other aspects of this 'multi-species' component of GCTE research, the consequences of global change on rotational systems will be undertaken in collaboration with other GCTE Foci, Activities and Tasks, and with other international programmes; the residual effects on soil organic matter and nutrient dynamics, or the pest, disease and weeds management are obvious examples where maximum collaboration is needed.

Standard classifications recognize about six or eight major types of cultivation system. They vary greatly in the diversity of the plant and livestock components from shifting and recurrent fallow-based cultivation through permanent mixed and rotational crop cultivation to annual and perennial monocrops. Livestock production similarly ranges from pastoral nomadism utilizing natural savannas to intensive ranching on species-regulated grasslands. Furthermore, livestock and arable farming may also be combined in a range of mixed farming systems (Frissel, 1978; Grigg, 1974; Ruthenbag, 1980).

The major advances in world sufficiency in food production have come from intensive monocultures of improved varieties of a small range of cereals. These crops are often grown in rotation, and receive high inputs of manufactured fertilizers and energy-subsidized management of soil, water, pests and weeds. This type of agriculture is characteristic of the industrialized regions of the northern hemisphere and has also made considerable impact in a number of countries of the developing world by means of the so-called 'Green Revolution'. Most of the world's farmers, however, still depend for their food and income on multi-species systems of one type or other. In comparison with monocultures these systems are less researched, more complex to model and there are more options to consider. Their behaviour and productivity under global change is therefore less easy to predict.

Nonetheless, in recent years cropping systems of 'intermediate' complexity, such as those based on intercropping of a few plant species including those incorporating trees, have replaced intensive monocultures as the target for much of modern agricultural research, particularly in the tropical regions. This change of strategy is largely based on the hypothesis that multi-species systems are more sustainable and more environmentally conservative than monocrops. Research with intercropping (e.g., Francis, 1986) or alley cropping (e.g., Kang *et al.*, 1990) has shown that these two- or three-species systems may indeed gain comparable or even higher returns as intensive monocrops, although results are not conclusive. There have also been many claims for the greater *yield stability* of intercropping systems (e.g., Rao and Willey, 1980) although Vandermeer and Schultz (1990) could find no theoretical justification as to why this should necessarily be so.

The GCTE Activity dealing with multi-species systems under global change will build as much as possible on this existing research. There is, however, a clear need to increase basic understanding of the interactions occurring in temporally and spatially mixed crops, which will of course be of great benefit in the short term for improving yield and yield stability, and for longer-term global change studies.

The Activity has three major components, to be managed as individual, but mutually dependent Tasks:

EXPERIMENTAL STUDIES ON THE RELATIONSHIP BETWEEN PLANT-BASED DIVERSITY AND COMPLEXITY AND SYSTEM SUSTAINABILITY (TASK 3.4.1)

As outlined above there is a considerable but inconclusive body of evidence that cropping systems with more than one species of plant (e.g., intercropping or agroforestry) have both greater productivity and greater resilience in the face of disturbance than mono-cultural systems. These effects may be enhanced when the system is structurally and chemically complex as well as species diverse. This could be due, for example, to the plants forming complex microenvironments through multiple canopy layers, and providing a wide range of allelochemicals and other substances which influence the function of other members of the community. Research will rigorously evaluate the evidence for this hypothesis, particularly in the context of global change. The research programme will be developed in close collaboration with GCTE's Focus 4, 'Ecological Complexity'.

The influence of the plant system on the herbivore and decomposer subsystems will be important components of the studies. The below-ground community is essential to the maintenance of ecosystem function through its role in soil fertility by means of nutrient cycles regulation, control of soil organic matter dynamics and modification of soil structure. This is of particular significance to agricultural systems where the soil subsystem is constantly disturbed by management practices. The focus in this research will be on the role in system productivity and sustainability of key groups of the fauna and microflora such as earthworms, termites, nematodes, nitrogen-fixers and mycorrhiza. This is an area where improved system understanding will be particularly beneficial for improving current productivity and reducing yield variation; it is also the essential prerequisite to building improved predictive models, robust to global change.

MODELLING COMPLEX AGRICULTURAL SYSTEMS (TASK 3.4.2)

The time scale of change relevant to sustainability and global change is much longer than that of a realistic experimental programme and also extends beyond the usual planning horizon of the farmer. As stated elsewhere in this paper, the most rigorous way of extrapolating predictions of change beyond the short term is by means of simulation models. Whilst there is now a *comparatively* effective (but by no means adequate) suite of such models for individual crops such as wheat, maize or rice, the modelling of more complex systems such as intercrops and agroforestry (with the inclusion of competitive and/or synergistic effects), is in its infancy. More rapid advancement will come from an interactive effort between modellers tackling the issues of simulating complex cropping systems than if they operate in isolation. Further strength would be gained moreover from linkages with the monocrop and ecological modelling activities also being promoted in GCTE.

A further dimension to the modelling activities comes from the need to assess the economic and social implications of effects on agro-ecosystem performance, including modifications in agricultural practice, resulting from the impact of global change. GCTE scientists in Activity 3.4 will collaborate with social scientists from the Human Dimensions Programme and the CGIAR to develop economic-ecological models and/or decision support systems that link the biological and socio-economic dimensions of agroecosystem change.

LONG-TERM AGRICULTURAL EXPERIMENTS AND DATABASES AS A RESOURCE FOR GLOBAL CHANGE RESEARCH (TASK 3.4.3)

The only unequivocal way of determining the relationship between external factors and system response over time is by means of long-term experiments. The data from such monitoring can also be used to validate the predictions of simulation models. A large range of such experiments were initiated at various times over the last century, many of them in developing countries. Only a minority are still in existence and the data from many have never been published in the open literature. The need for publication of data, rehabilitation of selected experiments and establishment of experiments in new sites was discussed at a conference at Rothamsted Experimental Station, UK (Swift *et al.*, 1995). This discussion provided a strong foundation for developing this Task; this will be done in collaboration with the Task dealing with soil organic matter (see below), where long-term data sets are crucial for model development.

EFFECTS OF GLOBAL CHANGE ON MANAGED FORESTS (ACTIVITY 3.5)

Forests cover more than one-third of the land surface of the earth and are almost equally divided between the temperate and the tropical regions. While the intensity of management and use of the forests varies both within and between regions, forest and other wooded land constitutes the largest component of current and future land use in terms of area, and plays an important role in the global carbon balance. Global change is having, and will continue to have, a major impact on forest cover.

From a socio-economic point of view the objective of securing long-term wood supply to timber and pulp industries is an obvious imperative. Biomass production in less intensively managed forest ecosystems may, however, be equally important since it provides fuel, fodder and other utilities for a large proportion of the world's population. Hence, the term 'managed forests' is used here in a broad sense to include forest ecosystems managed and used for purposes other than industrial production. Managed forests thus cover the range from plantations (where management includes all silvicultural activities), through natural forest managed (in the real sense of the word) for wood production, to natural forests simply exploited for timber and other products.

Forests are long-lived communities; the rapidly changing atmospheric CO_2 concentrations and the rising temperatures associated with climate change are likely to have significant impacts not only on the forests of the future but also on forests already in the ground. The current changes in forest cover brought about by the socio-economic drivers of global change will be compounded by changes in these biophysical drivers. To provide the information needed to predict the interactive effects of global change on forests, and the knowledge required to provide the basis for appropriate management programmes, we need focused, coordinated forest research programmes throughout the world.

RESEARCH ON FOREST ECOSYSTEMS WITHIN GCTE

Studies of forest ecosystems occur in all four GCTE Foci, in other IGBP Core Projects and in many other international and national research programmes. There is, however, one Activity specially designated to study the 'Impacts of global change on managed forests', which lies within Focus 3. The primary aim of this Activity is to understand and evaluate potential effects of global change on future structure, biomass production, and yield of managed forests; and hence to identify resource management strategies for sustainable forestry under changed climatic conditions. This will require the prediction of both short-term forest responses to altered climate, disturbance and silvicultural practices, and effects on long-term sustainable site productivity and biodiversity; clearly close cooperation among all GCTE Foci is needed. The role of managed forests as carbon sinks or sources should be evaluated as well as the possibility of increasing carbon sequestering in forest ecosystems by means of silvicultural practices. This long-term emphasis will require conceptual advances in capacity to characterize ecosystem sustainability and resilience in response to altered rates of input, loss and cycling of carbon, water and mineral nutrients.

The GCTE Managed Forests Implementation Plan has been recently published jointly with the International Union of Forestry Research Organisations (IUFRO) (Landsberg et al., 1995). This document, which functions in support of the GCTE Operational Plan, provides a framework for planning and implementing scientific research concerned with forests. Emphasis has been placed on ensuring consistency in terms of concepts, procedures and data recording. These will lead to results that can be compared and used to test and validate models, and assess the impacts of global change on forest growth and production.

The research strategy underlying this plan is to sample a representative range of forest types, growing under different conditions in different parts of the world. Within these the establishment of identical, or very similar, experiments will allow evaluation of the variation of, and constraints on, the productivity and biological diversity of forests. Networks analogous to those coordinated in other Focus 3 Activities (e.g., the GCTE Wheat Network, discussed above) will be established. As more projects addressing suitably similar goals, or using similar techniques, 'come on-line', they will be encouraged to join the most appropriate network or other GCTE structure. This will maximize collaborative effort, and help synthesize results from around the world.

To ensure rapid progress in information and understanding, an agreed, standardized primary data recording system will be developed, which would be available for exchange and comparison. The use of Geographical Information Systems (GIS) with models, to deal with the heterogeneity of large areas of forest, and as a framework for the recording and analysis of remotely sensed information, is strongly advocated.

This Activity is designed around two closely related Tasks, one dealing with experimental and observational studies (Task 3.5.1), and the other with modelling (Task 3.5.2). While the latter Task will not be geographically specific, the former will be split between boreal and temperate, and tropical and semi-arid regions. A further Task specifically addressing plantation forestry is being considered, and would probably be split into tropical and temperate components.

STUDY SITES AND TRANSECTS

For assessing the impact of environmental conditions, and hence global change, on the growth and performance of forests it is important that study sites be located (i) in areas where significant change is expected; (ii) where there is reason to believe that the forests are likely to be susceptible to change; and (iii) where the forest ecosystems are economically important. Based on these criteria, the following systems have initial priority:
- boreal forests, which contain much of the world's available softwood and where considerable warming has been predicted, though physiological responses to CO_2 enhancement may be limited by low temperature and infertile soils;
- northern hemisphere temperate coniferous and mixed coniferous-deciduous forests which form the basis for most of the present timber and pulp industry;
- softwood and eucalypt plantations in semi-arid and subtropical regions where changes in precipitation may have considerable effects;
- tropical forests, where warm conditions may lead to a large CO_2 fertilization effect.

EXPERIMENTAL TREATMENTS, MEASUREMENTS AND OBSERVATIONS

Appropriate models should provide a framework for organizing and focusing measurements and experiments, and modelling should therefore be done *a priori* rather than *a posteriori*. It is not the primary role of models to 'integrate' the observations made in experiments; parameters can generally be adjusted to obtain a good 'fit' to observation without learning about the system or the model. Models also provide the only means of evaluating the likely consequences of various management actions, particularly in relation to the uncertainties arising from global change. It should be noted that regression-based models (which include most of the models developed as part of traditional forest research and management) are unable to predict the effects on growth of changing climatic conditions; they are derived from historical observations, without a mechanistic basis. There is therefore an urgent need to link process-based models to tree population models and hence to produce estimates of the harvestable product, which is the information that forest managers need.

Experimental work should be aimed not only at providing empirical information, but also at understanding the physiological control mechanisms and key processes underlying forest response to environmental factors, at both the tree and stand levels. The basic minimum requirement, at each experimental or observational site, is a series of baseline measurements and observations (Level I), preferably carried out over a long period. Where resources exist these baseline observations should be supplemented by more detailed studies on community dynamics (Level II) and, at some sites, measurements of the physiological processes governing forest growth and productivity (Level III).

Level I includes the normal measurements that can be made with minimal infrastructure but which provide the essential baseline information about the state of forests and their long-term growth patterns. Such measurements accumulated over long periods at many well-characterized sites will provide an extremely valuable database from which considerable information about forest growth and performance, and the effects of weather and climate, can be obtained.

Level II measurements are aimed at providing information about the factors that cause changes in species composition. Such studies are important for all natural stands since species composition, and hence biodiversity, are likely to become significantly affected by many aspects of global change.

Level III measurements are concerned with the physiological processes underlying and driving the growth and production of trees. These studies should be performed on specific sites, when experimental treatments increase water availability and mineral nutrients similar to field experiments under way in Australia, Sweden and the United States. Where possible, experiments should also incorporate elevated CO_2 and soil warming treatments. The sites should incorporate the same experimental treatments, and should follow the same protocols, as far as possible.

DATABASES AND GEOGRAPHICAL INFORMATION SYSTEMS (GISs)

The long-term objective must be to develop databases for each forest type which provide accurate (geo)references for site location, long-term climatic data, properly documented site descriptions and standardized records of research data and observations. Such records, over time and for a number of sites/transects, will allow the refinement and testing of a complete range of models as well as providing invaluable empirical information about the productivity and growth patterns of the world's forests. This will be a major GCTE product, also invaluable for monitoring forest cover change.

The technology has great potential: the expenditure of considerable resources on accurate documentation (in GIS) of forest biomass, species composition and a range of other parameters, is justified; in fact it is arguable that this should have a higher priority for funding than a great deal of other research, since without good information about forest resources, and their state, detailed information about processes is of limited value for policy and management purposes. Given these data in GIS, in compatible formats, for many regions of the world, it will be possible to compare productivity and evaluate the consequences of global change at sites around the world.

CHANGES IN PESTS, DISEASES AND WEEDS (ACTIVITY 3.2)

One of the first noticeable effects of global change may be changes in agronomic pests and weeds ('pests' here refers to insect pests and microbial pathogens). This is because global change will potentially affect the pest/weed-host relationship in one (or more) of three ways: by affecting the pest/weed population; by affecting the host population; and by affecting the pest/weed-host interaction. It is hypothesized that the net effect will manifest in one of several ways: (i) pests/weeds currently of minor significance may become key thereby causing serious losses; (ii) the distribution and intensity of current key pests/weeds may be affected, leading to changed effects on yield and also on mitigation techniques such as pesticides and integrated pest management; and (iii) the competitive abilities in weed-plant interactions may be affected through changes in ecophysiology. The goal of this Activity is therefore to determine, through worldwide networks for research, data sharing and modelling, the potential impact of global change on pest and weed distribution and dynamics; and from this, the socio-economic cost of such impact. A conceptual framework for generating specific outputs (assessments of impact) is presented in Figure 12.3.

All pest effects should be determined in the context of the crop model(s) that are being used to estimate global change effects on crops under pest-free environments. This Activity is therefore being developed hand-in-hand with GCTE's Activity 3.1 (Effects of global change on key agricultural systems). The data sets used for crop modelling and pest modelling will ultimately be fully integrated, although the interpretation of pest effects should be done only when the effects of global change on crops are established. This will allow tangible targets and outputs to be generated for an assessment of impact. Table 12.2 lists crop-pest combinations which will receive initial emphasis in the GCTE research programme.

GLOBAL MONITORING NETWORK AND DATA SETS FOR PESTS, DISEASES AND WEEDS (TASK 3.2.1)

This Task aims to establish appropriate data sets in support of the experimental and modelling work of Tasks 3.2.2 and 3.2.3, which are discussed below. GCTE will not attempt to include *all* pests, diseases and weeds (clearly a vast undertaking), but will concentrate initially on those pests, diseases and weeds of the six priority crops in Activity 3.1. This work will be linked to several other initiatives on global data networks, such as (i) a US initiative to develop a global database on crop losses; (ii) a USDA-ARS initiative to develop a global database on pest distribution and intensi-

312 GLOBAL CLIMATE CHANGE AND AGRICULTURAL PRODUCTION

Figure 12.3. Conceptual framework for generating specific outputs from GCTE Activity 3.2

Patch or site outputs

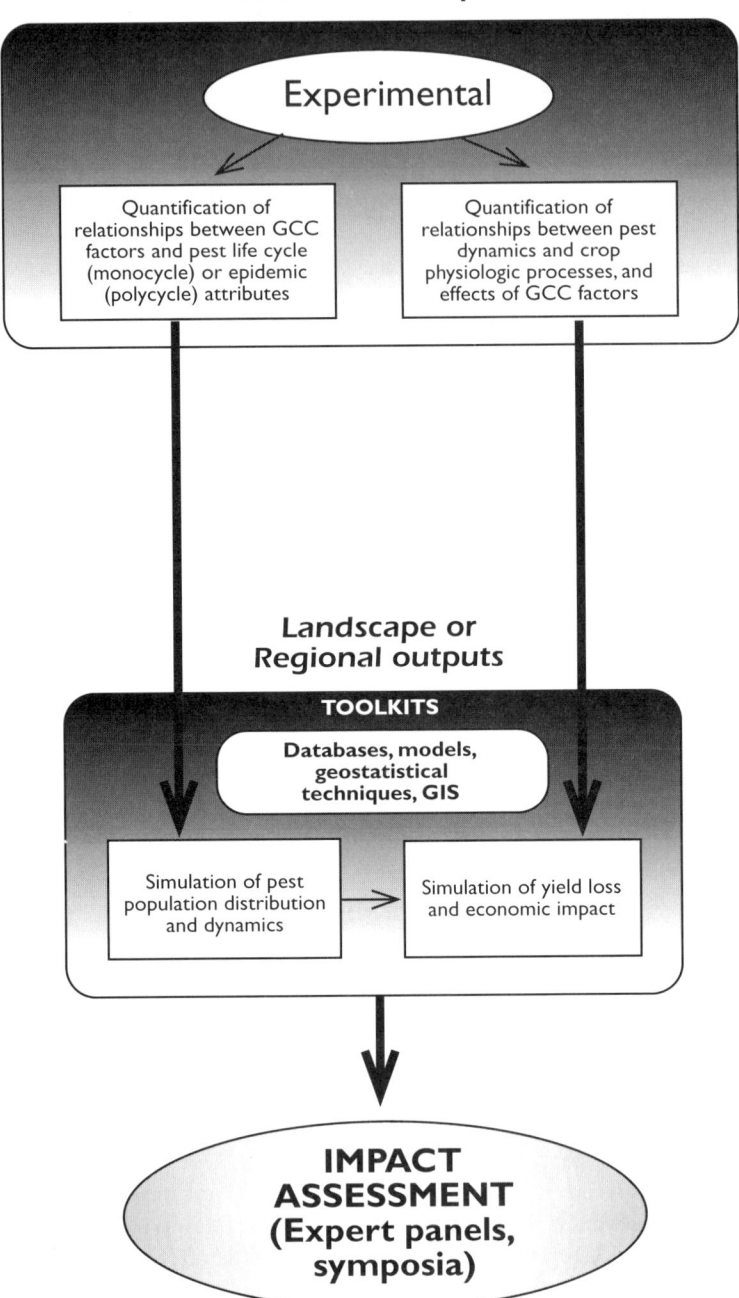

Experimental

- Quantification of relationships between GCC factors and pest life cycle (monocycle) or epidemic (polycycle) attributes
- Quantification of relationships between pest dynamics and crop physiologic processes, and effects of GCC factors

Landscape or Regional outputs

TOOLKITS

Databases, models, geostatistical techniques, GIS

- Simulation of pest population distribution and dynamics → Simulation of yield loss and economic impact

IMPACT ASSESSMENT (Expert panels, symposia)

THE GLOBAL CHANGE AND TERRESTRIAL ECOSYSTEM PROJECT

Table 12.2 Crop-pest combinations for assessment of global change impact

Agronomic species	Pest	Domain
Wheat	Aphids	Europe, North America
	Leaf rust	Europe, Americas, Oceania
	Septoria tritici	Europe, N America
	(To be refined after consultation with CIMMYT)	
Rice	Brown planthopper	Asia
	Blast	Asia
	Weeds	Asia
	(To be refined after consultation with IRRI)	
Cassava	Mealy bug	West Africa
	Mosaic virus	West Africa
	(To be refined after consultation with IITA)	
Maize	To be determined after consultation with CIMMYT	
Groundnut	Leaf spots	N America, West Africa
	(To be refined after consultation with IITA)	
Potato	Colorado P. Beetle	N America
	Late blight	Americas, Europe
	(To be refined after consultation with CIP)	

Note: The term 'Pest' is used here to include insects, pathogens and weeds.

ties; (iii) the former IBSNAT (International Benchmark Sites Network for Agrotechnology Transfer) Project Office at the University of Hawaii, with its minimum data sets (MDS) for crop and pest models; and (iv) the University of Hannover's global database on pest-induced crop losses for the major food crops.

DISTRIBUTIONS, DYNAMICS AND ABUNDANCE OF PESTS AND DISEASES UNDER GLOBAL CHANGE (TASK 3.2.2)

There is already much literature on global change-pest interactions, although many of these are conflicting. Table 12.3 (from Manning and von Tiedemann, 1993) shows one interpretation of the literature. GCTE will initially aim towards making an objective assessment of such literature to arrive at a common, accepted statement on global change effects on selected pests. Pest-crop combinations to be initially researched are shown in Table 12.2. Research will not, however, be restrictive in the number of combinations and GCTE welcomes further suggestions. Experimental and modelling studies will be developed in close collaboration with the Crop Networks, and representatives from Activity 3.2 are included in the Crop Networks' Working Groups.

Table 12.3. Effect of selected global change parameters on pathogens (from Manning and von Tiedemann, 1993)

GC factor	Type of pathogen	No. of reports	Nature of impact
Elevated CO_2	Bacteria	0	–
	Necrotrophic fungi	9	4(+), 4(0), 1(-)
	Biotrophic fungi	7	6(+), 0(0), 1(-)
UV-B	Bacteria	0	–
	Necrotrophic fungi	13	9(+), 2(0), 2(-)
	Biotrophic fungi	3	3(-)
Ozone	Bacteria	4	0(+), 1(0), 3(-)
	Necrotrophic fungi	25	16(+), 6(0), 3(-)
	Biotrophic fungi	14	6(+), 2(0), 6(-)

WEED DISTRIBUTION, DYNAMICS AND ABUNDANCE UNDER GLOBAL CHANGE (TASK 3.2.3)

Changing environmental conditions will alter the competitive advantage of one species over another in a given system. This will potentially affect ecosystem structure and composition, and has major implications not only for agricultural systems (where farmers strive to control weeds), but also for less managed systems (where invasive species may significantly alter the species composition). While the latter aspect is primarily of interest to Focus 2, the former clearly concerns Focus 3.

Within Focus 3, a two-track approach is being planned. One component will concentrate on those weeds currently important for the relatively narrow range of crop species covered by Activity 3.1. The other will address weeds in a broader sense, and will try to determine what will make an otherwise benign species an economically important one under global change; obviously individual species will be hard to spot *a priori*, but the improved understanding of what triggers a shift in species composition (the Focus 2 aspect) will be a valuable contribution.

An international workshop in May 1995 (sponsored by the Australian Government) will consider the effects of global change on the nation's pests and diseases, and GCTE will sponsor a symposium on 'Assessment of Global Change Effects on Pests' during the XII International Plant Protection Congress, to be held in June 1995 in the Netherlands. Meanwhile, the current international effort on the Activity's topics are being documented, and will comprise three parts: (i) a list of long-term data sets for weather, global change factors, crops and pests (part of Task 3.2.1); (ii) a list of scientists working on global change-pest effects estimation for the six crops (part of Task 3.2.2); and (iii) a list of scientists working on global change-weeds effects (part of Task 3.2.3).

EFFECTS OF GLOBAL CHANGE ON SOILS (ACTIVITY 3.3)

Soil science forms an obvious cross-cutting issue in GCTE, and it will be dealt with in all the Foci, as is necessary for their research. However, a number of research issues are specific to soils as such, and for this reason Activity 3.3 deals with soil science.

Soil properties vary from the transitory and rapidly variable, such as nitrate content, to the virtually permanent, such as texture. The number of properties that will be altered *directly* by changes in temperature, rainfall or CO_2 concentration is fairly small, though a few soils are morphologically poised, and may alter rather rapidly (Sombroek, 1990). Changes in climate will affect soils in a terrain, in the sense that their erosion potential will alter with rainfall, plant cover and cultivation; and changes in atmospheric CO_2 concentration may lead to changed soil organic matter quantity and type, via its impact on vegetation. The topics listed as Tasks below are those where global change is likely to have a direct or indirect impact on soils, and where this impact has important practical consequences.

In almost any activity concerning soil science, there will be a need for spatially referenced information. This is not a direct task of GCTE, but a geo-referenced World Soil Database is being assembled by a collaborative project involving FAO, ISRIC and USDA, with coordination being provided by the IGBP Data and Information System (IGBP-DIS); GCTE strongly supports this initiative.

GLOBAL CHANGE IMPACT ON SOIL ORGANIC MATTER (TASK 3.3.1)

Soil organic matter (SOM) is probably the single most important soil variable, and its level and properties contribute to both structure and fertility. Increased temperature and altered water status may change organic matter, but these effects can largely be modelled already. The potential effects of enhanced CO_2 in changing the root/shoot ratio of plants, root exudation and the chemical composition of plant tissue requires much more research, but could be very important for SOM dynamics. SOM models will need to be altered accordingly, for different types of vegetation and climates. This Task therefore aims to determine the impacts of global change, as expressed at the plant physiological, vegetation and ecosystem levels, on SOM dynamics.

Most of these subjects are of course already being researched heavily, especially given that global change in the sense of land-use change is of course the most important agent of loss of SOM at present (see Task 3.3.2). GCTE will add to and help to coordinate this work. In particular, it will do its best to ensure that all major enhanced-CO_2 experiments – particularly in FACE experiments – include studies on soil organic matter. GCTE is establishing a worldwide network to compare existing models under widely varying conditions of soil climate and vegetation, and to validate the models against the world's major long-term trials data sets.

GCTE has formed particular links with two international programmes, 'Alternatives to Slash and Burn' and the Tropical Soil Biology and Fertility Programme (Swift, 1991), because both focus, in different ways, upon the vegetation-soil-biota-SOM system. Both these are based in the tropics, but there are also important questions about the behaviour of SOM and related materials in higher latitudes during climate and land-use change.

SOIL DEGRADATION UNDER GLOBAL CHANGE (TASK 3.3.2)

Soil degradation is a wide-ranging and emotive term. GCTE has to focus, and it is doing so, on enhanced water erosion in the humid tropics, and wind erosion in semi-arid regions. Land-use change, or land cover change caused by climatic change, are the likely causes. The Task's objective, therefore, is to develop the capability to predict soil degradation by erosion caused by interactive changes in land use and climate. GCTE has established a worldwide network of erosion scientists (Ingram, 1994b), including modellers, experimentalists and scientists coordinating long-term monitoring studies (essential for validation erosion models). A detailed meeting held in early 1995 (sponsored by US EPA and USDA) planned in detail a model comparison workshop to be held in 1995, supported by NATO; the meeting also planned a major field campaign to be launched in West Africa in 1996. GCTE will endeavour to undertake genuinely strategic work, and to avoid site-specific studies. To ensure that research is in progress in the most-at-risk areas, the work will be conducted closely with bodies already involved in this area, such as ISRIC and FAO.

GLOBAL CHANGE AND SOIL BIOLOGY (TASK 3.3.3)

Originally this Task related to the production of greenhouse gases only. It was gradually perceived that its scope should be widened, because the behaviour of soil biota will be so critical in many global change situations, such as litter breakdown or new soil-borne diseases. The general questions in ecology will also apply to soil biota – speed of migration, invasion of new areas, speed of adaptation to new climates, and the classification of functional types. Soil microfauna and symbiotic organisms may be particularly interesting. It is absolutely essential, however, that the GCTE emphasis shall be on the processes mediated by soil biota, rather than on counting of species.

Work will be conducted in close collaboration with GCTE Focus 4 (Ecological Complexity) and with the IGBP Core Project 'International Global Atmospheric Chemistry' (IGAC).

CONCLUSION

The work outlined here is only one part of GCTE, i.e., Focus 3. There are strong links and contacts between this work, essentially designed for production, and the work dealing with the less managed ecosystems of the world. Cross-linking activities, such as the effect of CO_2 on both wild and managed plants, or the importance of changes in soil and soil organic matter, are established throughout. Focus 2, dealing with structures of ecosystems, has great relevance for mixed ecosystems and landscapes, where some land is agricultural and some is wild. In general, it has much to contribute towards the understanding and prediction of land use. GCTE is fully aware that it needs collaboration with the socio-economic approach in the Human Dimensions Programme, which stands in parallel to the IGBP. Furthermore, it will need the involvement of scientists from all aspects of the global research community, be they from national systems (industrialized *and* developing nations), the international agricultural research centres, UN organizations and programmes, or other bodies.

Focus 3 is very committed to the idea that its work should be useful. With the possibility, indeed the likelihood, of global change occurring, it is sensible to carry out preparatory work for the amelioration which will have to be undertaken. This falls squarely under the tasks of the IPCC Working Groups 2 and 3. Focus 3 cannot give exact prescriptions, because the future is too uncertain, but we believe that we can clear some of the ground, so that better decisions can be taken when they are needed.

REFERENCES

Francis, C.A. 1986. *Multiple Cropping Systems*. Macmillan, New York.

Frissel, M.J. (ed.). 1978. Cycling of mineral nutrients in agricultural ecosystems. *Developments in Agricultural and Managed-Forest Ecology* **3**. 355 p.

Goudriaan, J., van de Geijn, S.C. and Ingram, J.S.I. 1993. *Report of the GCTE Wheat Modelling and Experimental Data Comparison Workshop, November 1993*. GCTE Focus 3 Office, Oxford.

Grigg, D.S. 1974. *The Agricultural Systems of the World: An Introductory Approach*. Cambridge University Press, Cambridge.

Ingram, J.S.I. 1992. *Report of the GCTE Focus 3 Meeting: Effects of Global Change on the Wheat Ecosystem, July 1992*. GCTE Focus 3 Office, Oxford.

Ingram, J.S.I. 1994a. *Report of the GCTE Rice Ecosystems Workshop, March 1994*. GCTE Focus 3 Office, Oxford.

Ingram, J.S.I. 1994b. *Report of the GCTE Workshop Soil Erosion under Global Change, Paris, March 1994*. GCTE Focus 3 Office, Oxford.

IPCC. 1992. *Climate Change 1992 – The Supplementary Report to the IPCC Scientific Assessment*. J.T. Houghton, B.A. Callander and S.K. Varney (eds.). Cambridge University Press, Cambridge.

Kang, B.T., Reynolds, L. and Attah-Krah, A.N. 1990. Alley farming. *Advances in Agronomy* **43**: 315-359.

Landsberg, J.J., Linder, S. and McMurtrie, R.E. 1995. GCTE Activity 3.5: Effects of Global Change on Managed Forests – Implementation Plan. *GCTE Report No 4/ IUFRO Occasional Paper No 1*.

Manning, W.J. and von Tiedemann, A. 1993. Global climate change: potential impacts on plant-pathogen interactions. Paper presented at symposium on Global Climate Change: Implications in Plant Pathology, Sixth International Congress of

Plant Pathology, 28 July-6 August 1993, Montréal, Canada. *Paper S 18.2, Abstracts of ICPP*. 20 p.

Rao, M.R. and Willey, R.W. 1980. Evaluation of yield stability in intercropping: studies with sorghum/pigeon pea. *Experimental Agriculture* **16**: 105-116.

Ruthenbag, H. 1980. *Farming Systems in the Tropics*. 3rd edn. Clarendon Press, Oxford. 424 p.

Sombroek, W.G. 1990. Soils on a warmer earth: tropical and subtropical regions. In: *Soils on a Warmer Earth*. H.W. Scharpenseel, M. Schomaker and A. Ayoub (eds.). Elsevier, Amsterdam. pp. 157-174.

Stafford Smith, M. and Campbell, B. 1994. GCTE Task 3.1.3 Implementation Plan. *GCTE Report 3*. GCTE, Canberra.

Steffen, W.L., Walker, B.H., Ingram, J.S.I. and Koch, G.W. (eds.). 1992. Global Change and Terrestrial Ecosystems. The Operational Plan. *IGBP Report 21*, 95 pp.

Swift, M.J. (ed.). 1991. Soil fertility and global change: the role of TSBF studies in the IGBP. *Biology International, Special Issue No. 25*. 24 p.

Swift, M.J., Seward, P.D., Frost, P.G.H., Qureshi, J.M. and Muchena, F.N. 1995. Long-term experiments in Africa: developing a database for sustainable land-use under global change. In: *Long-term Experiments in Agricultural and Ecological Science*. R.A. Leigh and A.E. Johnston (eds). CAB International, Wallingford. pp 229-251.

Vandermeer, J.H. and Schultz, B. 1990. Variability, stability and risk in intercropping: some theoretical considerations. In: *Agroecology: Researching the Ecological Basis for Sustainable Agriculture*. S.R. Gliessman (ed.). Springer-Verlag, New York. pp. 205-232.

Worster, D. (1979). *Dust Bowl: The Southern Plains in the 1930s*. Oxford University Press, New York.

13. Global Climatic Change and Agricultural Production: An Assessment of Current Knowledge and Critical Gaps

FAKHRI A. BAZZAZ
Department of Organismic and Evolutionary Biology, Harvard University, Cambridge, Massachusetts, USA

WIM G. SOMBROEK
Land and Water Development Division, and Interdepartmental Working Group on Climate Change, FAO, Rome, Italy

The predicted changes in climate, especially increased atmospheric CO_2, temperature and precipitation, associated with changes in nitrogen deposition, tropo- and stratospheric ozone levels, UV-B radiation, etc. can have great impacts on world agricultural production and supply patterns. In order for agricultural production to be sufficient to meet the demands of the ever-growing human population, the impact of the climate must be understood and integrated in any future planning. The Food and Agriculture Organization of the United Nations is much concerned with this issue. The Organization formed an Interdepartmental Working Group on Climate Change and charged it with coordinating FAO activities in the critical area.

A Consultation 'Global climatic change and agricultural production: direct effects of hydrological and plant physiological processes' was held from 7 to 10 December 1993 at FAO Headquarters in Rome, with the support of the United Nations Environment Programme in Nairobi.

The objectives of the Consultation were: (1) to analyse and assess the effects of higher atmospheric carbon dioxide (CO_2) levels, higher ultraviolet (UV-B) radiation, higher near-surface ozone concentrations, higher temperatures and changing precipitation/evapotranspiration ratios on plant growth and food production; (2) to provide an overview of the state of knowledge on individual and combined effects, including a description of the processes, and availability of data for specific crops; (3) to identify gaps in our present knowledge and to focus on critical research needs.

The meeting was attended by 40 external participants, the 12 members of FAO's Climate Change Group and other interested FAO staff. The discussions concentrated on crops, with limited consideration of natural ecosystems. Because the majority of the presentations and discussions emphasized agricultural systems, it was decided to focus the book on agriculture alone. However, we believe that there are strong interactions between agricultural and natural ecosystems and these natural ecosystems play a significant role in the global carbon cycle which can greatly impact agriculture. The two are separated for convenience only. No discussions took place on the effects of climate change on oceanic biotic production because of the absence of specialists in that field among the assembled groups.

The chapters contained in this book are recent revisions of the presentations given in the meeting and therefore reflect the fast-growing research on the impact of global climatic change on agro-ecosystems. The views expressed in various chapters of the book reflect those of authors who wrote the individual chapter and therefore do not reflect the positions of their respective institutions, nor do they necessarily represent the views of FAO.

Due to increasing consumption of fossil fuels such as oil, gas and coal, in order to satisfy human needs, large quantities of carbon dioxide are being emitted into the atmosphere (Boden et al., 1994). In addition, agricultural and industrial activities add considerable amounts of methane (CH_4), nitrous oxides (N_2O) and chlorofluorocarbons (CFCs) to the atmosphere as well. Collectively, all of these gases lead to what is called the enlarged greenhouse effect because all of them absorb infrared radiation (Houghton et al., 1990). It is estimated that about 50% of CO_2 emitted to the atmosphere remains in it and the other 50% is taken up by the ocean and terrestrial ecosystems.

General circulation models based on the equivalent of doubling CO_2 concentration have predicted a global increase in mean global temperature ranging from 1.5 to 4.5°C. Furthermore, all current models show that the increase will be unequally distributed globally. For example, it is predicted that temperature rise in higher latitudes will be much more than in equatorial regions. These models also predict a change in precipitation with some regions receiving more rain than the present and others receiving much less rain.

Agriculture is totally dependent on weather and climate. Despite much effort by climatologists, there is considerable uncertainty about the potential impact of climate change on this sector. Little is known as to how, when, where and to what extent climate change will occur; one incontestable fact is the rising concentration of carbon dioxide (CO_2) in the earth's atmosphere. An additional certainty is the soundness of the basic greenhouse theory: the composition of the gas mix in the atmosphere strongly affects the planet's temperature. If the model scenarios are realistic, correctly reflecting future realities, such an increase may have serious consequences for agriculture and, in particular, for the regional food security in some regions (Ruttan, 1994). This security is already strained by increased demand and intensification of resource use, by the fast-growing human population and by an increase in per caput consumption of agricultural products (Rosenzweig and Parry, 1994). There is little question that agriculture must keep pace with the burgeoning human population which is expected to reach 10 thousand million in the 21st century. Since the timing, spatial pattern and magnitude of climate change are uncertain, policy-makers face a dilemma as to what measures, if any, should be taken to face the predictions of climate change. At present, some of the proposed preventive and mitigation measures would have enormous economic and social costs, particularly in relation to energy use. But these costs could even be larger in the future, if uncertainties are not resolved and we would learn that changes in the climate would be larger and more devastating than initially thought.

Thus, agriculture faces a particularly difficult dilemma: should it begin adapting at high cost to uncertain climatic changes while seeking the resolution of the scientific issues concerning the magnitude of climate change and its impact on agriculture, or choose the 'business-as-usual' principle and run the risk of leaving future generations unprepared when changes materialize? Although clear-cut answers might not be available for at least the next decade, improving scientific knowledge on the agronomic and ecological effects of any climate change, both adverse and positive, and on the ability of humans and ecosystems to adapt, might reduce the uncertainty and help formulate better policy (Fajer and Bazzaz, 1992). Therefore, whatever policies are adopted should be subject to frequent reassessment and must be flexible enough to accommodate change dictated by the gaining of new knowledge.

The panel of experts assembled by FAO, made up of persons with a wide range of backgrounds in various aspects of agriculture and natural ecosystem science, reached consensus on the following issues.

PRESENT KNOWLEDGE, PROBLEMS AND UNCERTAINTIES

A. EFFECTS ON PLANT PHYSIOLOGY AND SOIL PROCESSES

CO_2 fertilization and anti-transpiration effects

As plant type and plant productivity are major determinants of food production, it is critical to understand and quantify the response of the most important crops to changing environmental conditions. Different crop species have different responses to increased atmospheric CO_2 concentrations and to the combined changes in other factors, such as temperature, precipitation, pollutants, ultraviolet radiation (UV-B), etc. For example, free-air CO_2 enrichment experiments (FACE) in Arizona, USA, show that cotton is highly responsive to elevated CO_2 whereas wheat is much less responsive (see chapter by Allen *et al.* in this volume).

CO_2 is a key factor in photosynthesis and in plant growth. After diffusion into the plant through stomata, it is transformed by photosynthesis into carbohydrates. A large number of water molecules are lost by transpiration through the stomata for every CO_2 molecule entering the leaf. In a CO_2-rich environment, the larger concentration gradient forces more CO_2 into the plant, while partial closure of the stomata will reduce water losses from the leaf. As stomatal opening decreases in a high CO_2 environment, water loss from the plant is also reduced, increasing water-use efficiency. For example, there are indications that doubling present CO_2 concentrations reduces stomatal conductance (opening) by 30-60% depending on the species. The reduction in water consumption is called the CO_2 anti-transpiration effect. The water-use efficiency (WUE) of the plant improves, since less water is used for equal or more CO_2

transformed into dry matter. At the same time, net photosynthesis might increase, because photorespiration, which reduces carbon gain, is less at high CO_2 concentrations. In optimal conditions of light, moisture and availability of nutrients, this fertilization effect could increase above- and below-ground biomass production by 10 to 40% depending on crop type and even to higher levels such as the case in cotton. Earlier and more rapid leaf production is expected, and the incremental increase in biomass can benefit the root system even more. The earlier establishment of ground cover, because of the early canopy development, may also limit water loss by direct soil evaporation. This response is even greater at high temperatures, since the optimum temperature for photosynthesis increases with high atmospheric CO_2. In contrast, combined low temperatures and elevated CO_2 concentration could reduce plant growth.

The importance of the anti-transpiration and fertilization effects varies with crop type. For example, at double atmospheric CO_2 the biomass production of C_3 plants, including major crops such as rice, wheat, potatoes, beans, soybean, sunflower, groundnut and cotton, can be expected to increase, on average, by some 30%, provided other factors are not limiting.

On the other hand, and independently of CO_2, the physiology of C_4 plants permits a generally higher photosynthetic capacity than in C_3 plants. This efficiency, however, is quickly saturated at increased CO_2 concentration. Therefore, in a CO_2-rich environment, the net improvement in photosynthesis of C_4 plants is proportionally small (about 10%, mostly in stem), although the WUE might significantly improve (by about 40%). This category includes crops of major importance for food production, such as maize, sorghum, sugar cane and millet, but also tropical grasses, pasture, forage and some weed species that are critical for agricultural production. CAM plants seem to be less sensitive to CO_2 enrichment as well (Poorter, 1993).

Effects on soil fertility

A sudden doubling of the atmospheric CO_2 concentration and associated higher temperatures – as in most experimental set-ups, both enclosed and under free-air enrichment – may result in soil degradation including nutrient depletion. A potential increase in initial soil fertility under elevated atmospheric CO_2 can be expected if the increase in CO_2 occurs gradually, as in practice and as is the case in crop models that take into account the gradual transient increases. Additional litter is likely to raise soil organic matter content unless litter chemistry drastically changes which would cause a decline in litter decomposition rate. However, the higher soil temperature can stimulate microbial respiration and the decomposition of the organic matter (i.e., mineralization) and cause the release of nutrients which become available for plant uptake through the root system unless microbial competition with plants for the available nutrients is intensified. Furthermore, since both the nutrient uptake efficiency and the structure (length and density) of root systems improve under elevated CO_2 concentrations, overall plant nutrient uptake can also increase. Furthermore, there are indications that the expanded root system can penetrate more deeply into the soil and reach extra sources of moisture and nutrients.

An additional driving force is an enhancement of the symbiotic association between root systems and the fungi and bacteria of rhizosphere under high CO_2 concentrations. If these concentrations do not greatly alter the existing gas composition in the rhizosphere, the colonization of the root system by mycorrhizae (fungi in symbiosis with roots, facilitating plant uptake of occluded phosphate) and nitrogen-fixing bacteria (bacteria with the ability to incorporate atmospheric nitrogen into nitrogenous compounds which can be utilized by living organisms) is likely to improve the nutrient uptake by the host plants. However, it is possible that this process will ultimately slow down if changes in litter chemistry, e.g., an increase in C/N ratios and an increase in tannins, occur. In such a case there will be a greater demand for the use of additional fertilizers.

Uncertainties

This promising picture of improved food production under higher atmospheric CO_2 is modified by other factors that can limit growth. First, most of our understanding of the positive effects on crops relies on short-term and controlled studies usually at the individual plant level. Despite some evidence from field experiments such as FACE, extrapolation and generalization to large-scale field conditions or to long-term global food production are still uncertain. Also, since crop responses to climate change are site-specific and species-dependent, the knowledge on one kind of grouping or plants (e.g., annuals) may have little relevance to other species or groupings (e.g., perennials). Second, under conditions of limited soil nutrients or solar radiation (e.g., through enhanced cloud cover), higher CO_2 may not improve overall yields; in much of the world such stress conditions are the rule rather than the exception. In fact, nutrients and other present-day limitations might be more critical to agricultural production than the potential impact of climate change especially if urbanization and increased demand for agricultural products push agriculture toward marginal land.

Experimentation and scientific explanation provide additional arguments for a balanced assessment of the CO_2 fertilization and anti-transpiration effects. There are feedback mechanisms that might lower the direct effects of higher CO_2 and temperatures. For example, because transpiration rate per unit leaf area goes down when stomatal conductance decreases, leaf temperature can rise, triggering a general increase in canopy temperature and potentially higher water use. There are indications that the foliage ages more rapidly and the critical period of seed filling is shortened at high canopy temperature. Such feedbacks can undercut the decrease in crop seasonal water use that might result from the improvement in water-use efficiency of plants.

There are physiological and biochemical mechanisms that can further limit the long-term benefits of CO_2. For example, after an initial improvement, the growth rate of many perennial species exposed to elevated CO_2 concentrations tends to fall to near that of unexposed plants. One reason is a possible biochemical feedback, such as the reduction in the enzyme rubisco due to the accumulation of carbohydrates in leaves, that slows down photosynthesis and reduces phosphorus availability to carry carbohydrates into the growing parts of plants. There is also insufficient knowledge of the exact causes and consequences of this downregulation in various crops. It appears that sink strength plays a major role in this process as downregulation seems to occur more strongly in crops which are sink limited than in crops which are sink unlimited. The direct physiological responses of plants (e.g., changed stomatal resistance, leaf area index) may be different at different growth stages to varying CO_2 concentrations above and beyond the finding that early stages of plant development are more responsive to elevated CO_2 relative to later stages.

Interactions with pollutants in the atmosphere are also important for agriculture. Tropospheric ozone (O_3) levels, sulphur dioxide (SO_2) concentrations and UV-B radiation are likely to increase in parallel with the rise in CO_2. Depending on the species, it is estimated that crop damage from peak O_3 exposure may result in significant reduction in yield in the proximity of the source (e.g., downwind of major urban areas). Because most studies are performed in closed systems, there is considerable uncertainty on the applicability of such results in open-field environments where significant spatial and temporal variability of O_3 concentration are to be expected. Moreover, the damage attributed to O_3 alone is confounded by the interaction among elevated CO_2, increase in temperature and changes in water availability. For example, under some conditions, yield losses attributable to water stress can outweigh those resulting from O_3 exposure. Yet, all indications suggest that the increase in tropospheric O_3 will have a negative effect on crops; what is unknown is the magnitude of this effect at the local scale or in combination with other factors.

Losses of stratospheric ozone are contributing to higher levels of UV-B radiation but, although some experimental work suggests reduced growth, there is little field evidence that this would significantly affect plant growth. Furthermore, it seems uncertain that the interaction between UV-B and ozone will be of importance to crop productivity.

B. EFFECTS ON HYDROLOGICAL PROCESSES

Many countries allocate a large percentage of their freshwater resource to agriculture. Since water is a critical factor for crop yields, it is obvious that the advantages of elevated atmospheric CO_2 cannot be realized if water is limiting. There is strong evidence, however, that responsiveness to CO_2 increases in dry years in some grasslands. Although some recent models show increased drought in some regions and increased rainfall in other regions, available information on the potential effect of climate change on the global water availability is conflicting and remains largely fragmented, except for the nearly gratuitous statement that the hydrological cycle and ultimately the overall water supply are likely to be affected.

Current models predict that there may be an increase in the Indian monsoon and a decrease in moisture availability in the Amazon region and parts of tropical Africa. Numerous Global Circulation Models (GCMs) have attempted to predict these changes in rainfall patterns based on increased global warming potential. In most cases the predictions compare poorly with the observed data. There is no solid evidence that recorded changes in precipitation in the last three decades are related to the increasing atmospheric CO_2. Three reasons are cited for the mismatch between real and simulated data obtained from GCMs. Firstly, precipitation is the least reliable model output, because of the temporal and spatial variability in rainfall and more importantly because physical processes governing short-term variations are not well understood. Secondly, runoff resulting from precipitation depends on a large number of parameters many of which are specific to each river basin. GCMs rarely account for these parameters; their predictions are therefore inaccurate as long as the horizontal processes (i.e., overland flow, channel routing, groundwater movement, etc.) governing local water balance are largely ignored. Finally, many GCMs generally do not take into account specific cloud effects on regional heat balance, the influence of oceans, or extreme events (e.g., drought, flooding, etc.), the frequency of which is predicted to increase with climatic change.

However, in spite of their inherent deficiencies, models are valuable analytical tools for the assessment of the potential impacts of climate change on the water resources systems. Often these models use hydrological time series (e.g., measurement of river flow) to associate past meteorological events (e.g., precipitation) with recorded changes in lake levels and river flow regimes. This is not justified in view of the non-linear relationship between rainfall and runoff; small changes in precipitation have a magnified effect on river flow. For example, when regional precipitation decreased ~20% around Lake Chad, flow in the local river system was reduced ~50%. Similarly, in the Volga basin where regional precipitation is already increasing, the discharges will substantially increase if GCM predictions are correct. In regions with significant changes in precipitation, monitoring of river flow might therefore be a good indicator of any 'climate change effect'. However, even when based on this indicator, the analysis of hydrological time series of major lakes and river systems (Lake Chad, River Nile, Caspian Sea, Great Lakes, etc.) fail so far to detect any trend likely to be caused by global warming. Factors other than climate change (e.g., population growth, industrial and agricultural development, new methods of irrigation) seem to have a far greater impact on water resources.

C. EFFECTS ON GLOBAL FOOD SECURITY

The important question remains: how do the above processes influence the global food production system? Models have attempted to simulate climate change impacts on the world food situation by pulling together population data, components of the earth-atmosphere-ocean system and scenarios of climate change as well as estimates of potential changes in yields of important crops. The purpose of these models is to predict changes in land productivity and the geographical shift in agricultural land use as a function of changes in climate and food demand. The main outputs of these models reveal some new predictions, and confirm previous ones, that a doubling of atmospheric CO_2 is likely to cause the following:

- agro-ecological zones would shift because of temperature increase and improved water-use efficiency, with significant regional differences;
- crop yields and winter grazing in mid- and high-latitude regions (i.e., mostly developed countries) would improve because of increased photosynthesis, longer growing periods and extension of frost-free growing regions, provided optimum growth conditions are maintained, e.g., by judicious fertilizer and biocide use on agricultural land;
- in most developing countries, crop productivity would diminish (some 10% reduction in cereals), which could raise agricultural prices on local and world markets and increase the need for cereal imports, although the global food supply/demand ratio might change only little;
- there can be much risk in tropical and subtropical regions, and the greatest risk to food security would be in Sub-Saharan Africa. The magnitude of the threat will also depend on the behaviour of non-agricultural sectors of the economy in the future.

There remains much uncertainty around these predictions. For example, soil conditions in part of the new lands becoming available through shifting climatic zones may be unsuitable for sustainable crop production. New crop varieties may have to be developed. Also, the predictions are confounded by the uncertainties about the role of volcanoes, oceans and terrestrial ecosystems in global carbon fluxes. For example, recent evidence from the Arctic tundra (Oechel et al., 1994) suggests that due to soil warming the system is already a net source for atmospheric carbon. Present and future alterations of land use (deforestation, extensification or intensification of land use, etc.) are extremely important determinants of the terrestrial carbon fluxes into the atmosphere. At the same time, since elevated CO_2 concentrations can increase net primary biomass production, some enhancement of the terrestrial sink (i.e., more carbon storage in vegetation and soils), is to be expected, at least in the short term.

The inherent weaknesses of the current GCMs further exacerbate these uncertainties. In spite of recent advances, only few models, the so-called Coupled General Circulation Models (CGCMs) consider the role of oceans and the more realistic transient scenario (i.e., progressive build-up of CO_2). Spatial and temporal resolutions are still poor at the regional level and of little use at the local level where farmers must act. Important parameters, such as cloudiness, atmospheric aerosols, ocean-atmosphere-terrestrial linkages, cost of technological and social adaptations, extreme climatic events, lateral water transfers, availability of water resources and soil nutrients, and multiple and inter-cropping systems must be included in the models.

IMPLICATIONS AND NEEDS

Our current scientific knowledge provides a good understanding of plant physiology, morphology and growth. For example, plant response to varying temperatures, changing precipitation, soil moisture and humidity are well documented. Likewise, there are some vegetation models that adequately correlate the present distribution pattern of vegetation and climate. Nevertheless, a high degree of uncertainty still clouds the potential impacts of climate change on both managed and natural ecosystems. The reasons for uncertainty have been discussed in the previous sections. They relate mainly to the paucity of relevant climatic data, the incomplete understanding of the processes underlying the global circulation models, the coarseness of their spatial resolutions, and the insufficient knowledge on the long-term, direct and indirect biological and physical responses of crops and other ecosystems to elevated CO_2 and other correlated factors. The working groups at the Expert Consultation identified several research areas which would reduce uncertainty, improve knowledge, increase preparedness in the face of climate change and provide better grounds for policies related to climate change:

- Firm quantitative assessment is needed of site-specific crop responses as a function of time and growth stage, in particular for perennial crops and crops of greatest importance to food production. Such assessment needs to be properly structured, and its parameters should be identified and tested for priority crops and regions. Open-field multi-variable experiments should consider important plant biological stresses along with changing atmospheric composition (CO_2, O_3, SO_2, N_2O and other non-CO_2 greenhouse gases), precipitation, and UV-B radiation.
- As some of the soils of the new lands made available through shifting of climatic zones may be unfavourable for crop production, it is important that they are categorized and mapped to avoid or lessen the chances of inappropriate land-use choices.
- The long-term effects of elevated CO_2 on the uptake of nutrients (e.g., nitrogen, phosphorus, etc.) are unknown and studies on this subject should start soon. Nitrogen balances should be linked to plant-soil-water balances, and the impact of elevated CO_2 on biological nitrogen fixation and phosphorus uptake need to be quantified for various crops under field conditions.
- Agronomists need to work closely with climatologists at a regional level to provide a sound basis for optimizing crop, soil and water management under the changing conditions. Special attention should be given to evaluating probabilities of extreme events (droughts, floods) and their effects on plant growth and yield. This should be accompanied by concerted projects on plant breeding by conventional techniques and by using genetic engineering and selection for stress-resistant genotypes.
- The known physical mechanisms regulating plant water use should form the basis for assessing water balances for drought- or flood-prone regions, in specific soil conditions, and for selected crops and changing climatic scenarios. Water management assumes even a greater importance under a changing climate. Water resources harnessing criteria (e.g., for irrigation structures) should achieve the necessary flexibility to accommodate future changes, particularly in vulnerable areas of the world. Long-term, continuous climate and hydrographic

observations are essential not only for detecting climate change signals but also for judicious water resources development.
- A concerted effort should be directed to increase the quality of global modelling projections. High-resolution databases are needed on land use and land cover, soil carbon, nutrients and mineralization rates, crop cultivars and their climatic requirements, N_2O emission rates from fertilizer applications related to local climate, soil and vegetation conditions. Remote sensing, field-level evaluation of soil and terrain conditions, monitoring of nutrient inputs and use are important tools for these tasks.
- Improving high-resolution databases will not only help generate better GCM projections, it will also contribute to the planning of overall development process in the developing countries. Better information improves knowledge and helps devise good policies and sound agricultural management practices. These in turn would increase the resilience of production systems to inter- and intra-seasonal climatic variations and to global climate change. Forums for communicating progress on data collection and availability should be strengthened.
- Better socio-economic data on household income and expenditure are also critically important. Much of this information does not exist or is not at a suitable resolution and consistency, and therefore is not directly comparable between or within regions.
- There is a lack of basic biological knowledge about how tree species and forest ecosystems are affected by climate change. The rate at which a particular species is expected to move into a newly available area needs to be studied, as well as the effects of elevated CO_2, alone and in conjunction with other global change parameters on successional, mature, degraded and recovering forests in different world biomes.
- Readily available assessments (e.g., FAO's *Agriculture Towards 2010*) should extend their perspectives by enhancing their treatment of indicators of sustainability, with implications of climate change and consequences for greenhouse gas emissions.

CONCLUSIONS

Several qualitative judgements can be drawn from the Expert Consultation and the various preceding (updated) chapters that resulted from it:
- The scientific uncertainty surrounding the issue of climate change will not be resolved soon. The time scales of climate change are usually so long that observational studies are usually too short to provide adequate answers. The uncertainty is exacerbated by limitations in modelling techniques, especially at the local scale, and by the lack of knowledge about the complex biophysical responses in field conditions to global change.
- Although the rising CO_2 concentration in the atmosphere is currently the primary cause of climate change, the correlated changes in environmental condi-

tions (temperature, precipitation, O_3, UV-B, humidity, etc.) are likely to be as important as CO_2 in determining the responses of managed ecosystems. To determine any needed changes in management practices of graziers or ranchers, farmers and foresters, both positive and negative responses need to be fully understood and tested in field conditions.
- In spite of many uncertainties, global warming, if it happens, can be a serious problem that could have great implications on agriculture and on natural ecosystems.
- Feedbacks among biophysical, economic, social and technological mechanisms are likely to accentuate the uneven distribution of climate change impacts between developed and developing countries. Although global food security might not be affected, the developing countries are presently the least able to make the necessary adjustments. The most vulnerable regions should therefore receive priority for determining the impact on food security, even though the effects of social, economic and technological constraints on food security are likely to match or exceed those associated with climate change.
- Past and present activities of the industrial countries are currently the major sources of CO_2: it is their responsibility to reduce emissions first and prepare for the likely consequences; imposing reduction targets on agriculture in developing countries is impractical and non-equitable.
- Whether or not climate change takes place, improving the resilience of food production and minimizing risks against weather variability are essential if agriculture is to meet the challenges of ensuring food security, raising rural employment in developing countries and protecting natural resources and the environment. This can serve at the same time ('no-regrets policies') to prevent or mitigate the negative impacts of climate change. For example, good land husbandry and better agronomic management adapted to variable conditions are appropriate both to face higher inter-annual weather variability and the more modest and gradual global climate change. Similarly, crop and livestock breeding for heat and drought stresses is a pressing need on its own, because of growing human populations, but can also serve as a potential response to global climatic changes.

Whether climate change impact scenarios will ultimately materialize depends on how precipitation patterns change and on the magnitude of temperature increase and its spatial and temporal distribution. Unfortunately, there is no scientific consensus as yet on the answers to these questions, nor is there certainty that the slight global temperature increase observed in the 20th century was caused by the greenhouse effect. Nevertheless, past climate fluctuations have provided circumstantial evidence that temperature variations are linked to greenhouse gases. It would be perilous for agriculture to ignore the potential impacts of any global warming on the basis that the likelihood of warming is uncertain or because scientific consensus is not yet achieved. The Expert Consultation and the present consolidated texts may provide a contribution in identifying uncertainties and pointing to the emerging issues and needs of agriculture in the face of a potential global climate change.

A good part of the needs for further research is being addressed within the International Geosphere-Biosphere Programme (IGBP) of the International Council of Scientific Unions. In particular its research programme on Global Change and Terrestrial Ecosystems (GCTE) is relevant, and therefore a description of it is added to this publication as a free-standing chapter (No. 12).

REFERENCES

Boden, T.A., Kaiser, D.P., Sepanski, R.J. and Stoss, F.W. (eds.). 1994. *Trends '93: A Compendium of Data on Global Change*. US Department of Energy, Carbon Dioxide Information Analysis Center, Oak Ridge National Laboratory, Oak Ridge, Tennessee.

Fajer, E.D. and Bazzaz, F.A. 1992. Is carbon dioxide a 'good' greenhouse gas? Effects of increasing carbon dioxide on ecological systems. *Global Environmental Change: Human and Policy Dimensions* **2**: 301-310.

FAO, 1995. *World Agriculture: Towards 2010, An FAO Study*. N. Alexandratos (ed.). John Wiley, Chichester, UK, and FAO, Rome. 488p.

Houghton, J.T., Jenkins, G.J. and Ephramus, J.J. (eds.). 1990. *Climate Change: The IPCC Scientific Assessment*. Cambridge University Press, Cambridge.

Oechel, W.C., Cowles, S., Gurlke, N., Hastings, S.J., Lawrence, B., Prudhomme, T., Reichers, G., Strain, B., Tissue, D. and Vourlitis, G.L. 1994. Transient nature of CO_2 fertilization in Arctic tundra. *Nature* **371**: 500-503.

Poorter, H. 1993. Interspecific variation in the growth response of plants to an elevated ambient CO_2 concentration. *Vegatatio* **104/105**: 77-97.

Rosenzweig, C. and Parry, M.L. 1994. Potential impact of climate change on food supply. *Nature* **367**: 133-138.

Ruttan, W. (ed.). 1994. *Agriculture, Environment, Climate and Health: Sustainable Development in the 21st Century*. University of Minnesota Press, Minneapolis.

Index

abandoned land 286
abstraction
   and land subsidence 60
   Volga system 42–3
acclimation 67
ACGE, see Applied Computable
   General Equilibrium Model 248
acid rain, and sandy soils 55
action spectra 184
adaptation
   and crop yields 209–10, 232
   technological 257, 321
   see also farming, adaptation
adaptation potential 254–5
aerosols
   cooling effect 237
   UV scattering 190, 280
   see also sulphate aerosols
afforestation, see forestation
Africa
   cereal production 2
   climate cycles 30
   climatic patterns 30
   droughts 29
   equatorial band 30
   hydrology 29
   lake levels 30
   root and tuber crops 2
aggregation 248
agricultural activities 248
agricultural demand 281–2
agricultural experiments, long-term 307
agricultural exports 247
agricultural imports, developing countries 2, 227, 232
agricultural indicators 253
agricultural land 286, 289
agricultural policies 259
agricultural production
   global 215, 224
   marginal 237
   systems 68, 209, 306
agricultural productivity 286
agricultural research 258
agricultural response 258

agriculture
   robustness 258
   thermal limits 134
   turbulence 296
agro-ecological zones 281–2, 326
agro-ecosystems 299, 306–7, 320
agro-forestry 298, 306
agronomic practices, modification 232
Agrostis capillaris 157
AIM model 277, 289
air pollution 172
air quality 172
albedo
   feedback 268, 280, 323
   transpiration response 114
alfalfa, water-use efficiency 189
allelochemicals 306
alley cropping 305
Alyssum-Pisum interaction 157
Amaranthus 157
animal husbandry 1, 303
animal waste, methane from 284
annual changes, precipitation 25
anoxic soils 58
Antarctic, temperature rise 20
anthesis 131
   heat stress at 136
anthracnose 154
anti-desertification 12
anti-transpiration effect, CO_2 8, 12, 321
   see also transpiration
Applied Computable General
   Equilibrium Model 248
aquatic systems, greenhouse gases 284
Arabidopsis, lipids 127
arable crops, transpiration 114
arable land
   and population growth 2
   trends 3, 212
Aral Sea 43
Arctic, temperature change 4, 326
ARCWHEAT model 178
aridity
   Australia 25
   mid-latitudes 247
   soil resilience 54

assimilates, partitioning of   161, 173
Aswan Dam   24, 36, 60
Atlantic conveyor current   4, 20
Atlantic Oscillation   30
atmosphere
  radiation balance   4
  water holding capacity   18, 45
Atmosphere-Ocean System   279
atmospheric forcing   16
atmospheric-vegetation fluxes, *see*
    vegetation-atmosphere fluxes
Avena fatua   157

BACER, *see* Biological and Climatic
    Effects Research
bacterial pathogens   155
Baltic Sea   40
Basic Linked System of National
    Agricultural Models   200, 210
Beta vulgaris, *see* sugar beet
biochemistry, and temperature   124–6
biodiversity, loss of   6
biofuels   280
Biological Action factor, UV-B   9
Biological and Climatic Effects
    Research   159
biomass
  and CO_2 7, 283
  crop types   281
  energy generation   287–8
  as fuel   7, 280, 284, 287
  and ozone production   9
  productivity   322, 326
  reproductive   66
  turnover   283
BIOME model   269, 281
biopores, stability of   51, 53
biosphere, terrestrial   267
Biospheric Aspects of the Hydrological
    Cycle project   298
Blue Nile   30, 36, 38
bluegrass, *see Poa*
boreal climates, rainfall changes   54
boreal forests   309
boundary layer
  leaves   106
  planetary   114, 180
  surface   158–9
brackish-water transgression   50, 61
branching   67
bud break   179
bypass flow   51, 54

C_3 and C_4 plants

  CO_2 responses   174, 322
  temperature responses   126
C_3 pathway
  and CO_2 7, 65, 67
  and temperature   124, 126
C^{13}, in herbarium leaves   104
C/N ratios
  and caterpillars   156
  changes in   67
  forage   69
  litter   12, 51, 323
calcareous soils   59
calibration, GCMs   22, 44
CAM plants
  CO_2 response   322
  temperature response   9–10
canals, Volga system   40–1
Canje, River   22
canopy
  carbon exchange   174
  closure   75, 110
  CO_2 uptake   175–6
  conductance   89
  gas exchange   110, 112
  saturation   66
  temperature at   109–10, 323
canopy level effects   108
canopy temperature, and irrigation   135
carbamate formation   130
carbohydrates
  accumulation   323
  allocation   185
  non-structural   66
  soybean   72
carbon allocation, to roots   179
carbon cycle   12, 269, 275, 277, 282–3, 320
carbon exchange, canopy   174
carbon pools   283
carbon sequestration   281, 308
carbon sinks   31, 173, 275, 286, 308
carboxylation efficiency   185
CARM model   92
Caspian Sea   40–3
cassava, GCTE   299
catchment hydrology   22, 44
caterpillars, and C/N ratios   156
cell membranes, thermal stability   127
Cercospora   154
cereal commodities, climate change   230
cereal production
  Africa   2
  global   2, 231, 249
  regional   228

INDEX

cereals, world market prices 222–3
CERES-Maize model 204
CERES-Rice model 204
CERES-Wheat model 204
CFCs, see chlorofluorocarbons
Champlain Sea 57
channel routing, in GCMs 22
cheluviation 55
chlorofluorocarbons, as greenhouse gases 5, 202
chlorosis, and ozone 156
CIAP, see Climatic Impact Assessment Program
Cladosporium cucumerinum 154
clay, surface properties 56
clay minerals 55
climate change
   cereal commodities 230
   dynamic impact 226
   and economic adjustment 220–7
   effects 43
   and evapotranspiration 50
   historical 25, 329
   rapid 93
   rate variation 4
   scenarios 68, 94, 201–2, 218–19
   and soils 49
   timing 247
climate complex 4
climate cycles, Africa 30
climate prediction 6, 297
   short-term 256
climate shock 240, 245
climatic boundaries 240
climatic factors, interactions 11
Climatic Impact Assessment Program 159
climatic zones, rainfall changes 53–4
cloud-cover 4, 323, 325
   see also albedo
clover, crimson 157
CO_2
   anti-transpiration effect 8, 12, 321
   and biomass burning 7
   and cotton 321
   and crop phenology 66, 92
   crop responses 69–86, 152, 173
   and crop yield 174
   and deforestation 7
   fertilization effect 7–8, 10, 65–100, 146, 321–2
   forest species 87
   and fossil fuel burning 7

grassland response 324
   as greenhouse gas 5, 172
   and growing season 110–13
   and growth rates 50
   and maize 88
   and nitrogen fixation 51
   and photosynthesis 321
   physiological effects 205, 207, 223
   and plant growth 65–9
   and precipitation 45
   and productivity 51
   rate of increase 51
   and respiration 130–1
   sensitivity 152
   and soil fertility 50–3
   and soil organic matter 50
   and stomatal conductance 67
   stomatal response 101
   and transpiration 101
   vegetative response 88
   and water 179–80, 192
   and water-use efficiency 8, 50, 68, 85–6, 115, 321
   and wheat 321
CO_2 fixation
   dark 130
   enhanced 66
CO_2 levels
   biomass production 7
   and C_3 metabolism 7
CO_2 and ozone, combined effects 180–4
CO_2 and temperature 174–9
   combined effects 192
   soybean 73–4
CO_2 uptake
   canopy 175–6
   and ozone 182
CO_2 and UV-B, combined effects 184–5
coal, high sulphur 237
coastal change, and sea-level rise 59
Colletotrichum laginarium 154–5
combined effects 10, 158–60, 171–97, 328–9
commodity demand 281
communication 258
cool-season crops 126
cotton
   canopy temperature 135
   CO_2 response 321
   dark respiration 130
   heat shock proteins 128

cotton (*cont.*)
   and ozone levels 146
   water stress 188
coupled GCMs 326
cover crops 256
crop damage, pollutants 324
crop distribution 269, 281
crop diversification 255
crop energy balance 68
crop failure 241
crop losses, database 311
crop mix 257
crop modelling 68, 88–94, 202–4, 239, 297
   resolution 240
Crop Networks 299, 302, 313
crop pests, and UV-B or ozone 154–6
crop phenology, and CO_2 66, 92
crop plants
   heat tolerance 124
   water use 101
crop price index 223
crop production
   patterns 93
   regional 229
   world 153, 154
crop productivity, and CO_2 increase 8, 220, 285
crop residues, use of 58, 256
crop responses
   CO_2 69–86, 323
   in models 239, 327
crop simulation models 115, 205–6
crop types, biomass production 281
crop yield
   and adaptation 209–10
   and climate change 89, 207, 218
   and CO_2 174, 242
   CO_2 and temperature 177
   latitudinal differences 231, 326
   and ozone 191
   potential 281
   rice 80–4, 136
   site-specific 202–3
   soybean 72
   and temperature 9, 136
   and UV-B 9
   vulnerability 250
   wheat 135, 208
crop-pest combinations 311, 313
crop-weed competition 156–8
cropping intensities, Asia 2

crops
   climatic boundaries 240
   growth cycles 50, 68
   photosynthetic pathways 8
crowding coefficients 157
cucumber, pests of 154–5
Cucumis sativus, see cucumber
cultivar differences
   ozone responses 144
   temperature sensitivity 129–30
   UV-B responses 184
   yield 241
cuticle thickness 143
cyclones, tropical 4

Dactylis glomerata 131
dark respiration 10
   cotton 130
   rice 84
   tomato 130
data sets 302
databases 310–11, 328
decision support systems 202
decomposition, organic matter 53, 58, 306, 322
deep water circulation 31
deforestation
   and CO_2 7
   global 284, 286
   Latin America 2
   tropical 275
deltaic aggradation 60
deltas, and sea-level rise 59–60
deterministic models 20, 44
developed countries
   CO_2 emission 329
   crop productivity 220, 223, 231
developing countries
   agricultural imports 2, 227, 232
   crop productivity 220, 224, 231, 326
   food security 329
   livestock production 2
   oil crops in 2
   planning in 328
   population growth 212
development, sustainability 1
development stages
   rice 78–80
   temperature sensitivity 131, 176
   wheat 131–2, 176
Diplocarpon rosae 154
disaster planning 257
Diversitas 290

INDEX

downregulation 67, 70, 182, 324
drought stress 109, 327
drought tolerance 269
droughts
  Africa 29
  frequency 233, 297, 327
  regional differences 324
dry matter accumulation
  and CO_2 174
  cotton 135
  and transpiration 181
  wheat cultivars 134
Dust Bowl 258, 297
dynamic impact, climate change 226

earth system models 268, 276
earthworms 51
ecological complexity 297, 306
ecological zones, and climate change 6
economic adjustment, and climate
    change 220–7, 231, 240
economic growth 212–15
economic modelling 247
economic relocation 254
economic systems 68, 93, 210
economic vulnerability, regional 251–2
economic yield 281
ecosystem function 297
ecosystem structure 297
education 258
EFEDA programme 108
effective CO_2 doubling 202
electricity generation 288
embankments 60
emission models 275, 277, 279, 284–6
emissions, global 285
employment, alternative 252
energy balance, global 280
energy consumption 279
energy demands 280
energy generation 287
Energy-Industry System 279, 281
environmental resources, pressure on 2
environmental stress 159
enzymes, thermal factors 124–6
Epilachna varivestis 156
equilibrium models 19
erosion, coastal 60
ESCAPE model 277
ethanol, from sugar cane 288
ethylene, and ozone 156
eustatic movements 59
evaporative cooling 76

evaporation
  as climate variable 4
  from soil 108
  land 18
evaporative demand 54, 110
evapotranspiration
  and climate change 50
  in GCMs 18, 22
  impact 44
  potential 4
  rice 84–5
  soybean 74
exhaust gases 172
experimental conditions, water use 102
experimental treatments 309–10
exports, cereals 2, 215
extreme temperature effects 134–6
extremes, *see* weather, extreme

faba bean 110, 111, 112–13
FACE experiments, *see* Free Air CO_2
    Enrichment
famines 258
FAO *see* Food and Agriculture
    Organization
farm performance 239
farmer vulnerability 251
farming
  adaptations 202, 206, 223–4, 240, 243,
      248, 257
  income 246
  planning 6, 239
  semi-intensive 252
  systems 297
fatty acid ratios 127
feed requirements 281
feedback processes 323
  in GCMs 101, 115
  geophysical 268
  hydrological 29, 30
ferrolysis, clay minerals 55
fertilization effect, CO_2 7–8
fertilizer regimes 257
fertilizers
  increase in use 6, 256, 323
  in modelling 212, 218, 286, 328
  and runoff 53
Festuca rubra 157
fisheries 1
  Caspian Sea 41
  and climate change 5
flavonoids, and UV-B 143
flooding, low-lying areas 233

floods 258, 327
floral number 133
flour, characteristics 243
flow data 44, 45
flux transfer, water 22
Food and Agriculture Organization 1, 319
food availability 215
food crops, major 299–300
food distribution 258
food production
 demands 254
 global 2, 215, 216, 246
 regional imbalance 7
 risk reduction 329
 trends 3
food programmes 258
food security 1, 217, 320, 325–6, 329
forbs 158
forcing variables, soils 50
forest clearance 275
forest damage, ozone 9
forest ecosystems 308–9, 328
forest growth 310
forest products 286, 307
forest species, CO_2 fertilization effect 87
forestation 286–7
forestry 1, 298
forests
 managed 298, 307
 transpiration 114
fossil fuel burning, and CO_2 7, 173
Free Air CO_2 Enrichment 106, 109, 173, 299, 315, 321
frost periods, reduction of 54
frost-hardiness, and ozone 187
fruit set 179
fuel cells 288
fuelwood 280
fungal pathogens, and ozone 155

Ganges-Brahmaputra system 59
gas exchange
 canopy 110, 112
 stomata 104
GCAM model 277
GCTE, see Global Change and Terrestrial Ecosystems
General Circulation Models 12, 15, 277
 characteristics 21
 detail in 101
 and hydrology 17, 43
 rainfall patterns 324

sensitivity 280, 300
temperature in 201
transient 19–20, 31, 44, 51, 202
and water resources 43
genetic engineering 255
genotypes
 heat tolerant 130
 stress resistant 327
geographic information systems 248, 309, 310–11
germplasm modification, adaptive 93, 130
Geum 157
GEWEX, *see* Global Energy and Water Cycle Experiment
GFDL model 88–92, 201, 207, 220, 223
GIS, *see* geographic information systems 248
GISS model 18, 88–92, 201, 207, 220, 223
Global Change and Terrestrial Ecosystems 295–8, 312, 330
Global Circulation Models, *see* General Circulation Models
Global Energy and Water Cycle Experiment 18, 114
global production 215
global trade 215
global warming
 and atmospheric CO_2 92
 autocatalysis 123
 and climate variability 4
 and CO_2 doubling 172
 hemispheric differences 20, 134–5
 and hydrological cycle 15
 latitudinal differences 123, 134, 320
 and permafrost boundary 49
 positive feedback 123
 regional variation 200
 and Sahel drought 30
 secondary processes 200
Global Warming Potential 5
Glycine max, *see* soybean
goethite 54, 55
Gossypium hirsutum, *see* cotton
grain development, and temperature 133–4
grain drying 256
grain oil content 182
grain weight, and heat stress 134
grain yield 182
grass/clover mixtures 157
grassland demand 281

grazing systems   298, 326
green manure   256
Green Revolution   199, 305
greenhouse effect, enlarged   320
greenhouse gases   4, 267
   agricultural   5–7
   anthropogenic   5
   aquatic systems   284
   emission reduction   5, 212
   in GCMs   280, 283
   global emissions   285, 320
   natural sources   5
   and plant physiology   7–10
   and rainfall   25
greenhouse plants, cuticle thickness   143
Greenland Ice-core Project   93
Gross Domestic Product   211, 213
ground cover   108
   increase in   53
groundnuts   174, 180, 299
groundwater recharge, modelling   20
groundwater storage, in GCMs   22
groundwater table   54
growing period
   extension of   50, 90
   minimum   281
   shortening   207
growing season
   changes to   255
   CO_2 effects on   110–13, 207
   and temperature   135
growth, rice   80–4
growth cycles, crops   50
growth modification factor   73, 87, 178
growth rates
   and CO_2 50, 323
   soybean   72

Hadley Centre   20
haematite   54, 55
halloysite   55
HAPEX, see Hydrologic Atmospheric Pilot Experiment
harvest index   66, 72, 80, 174
harvestable product   298
HDP, see Human Dimensions Programme
heat shock proteins   127–9
heat stress
   at anthesis   136
   and grain weight   134
heat tolerance, crop plants   124, 129
heat transfer, ocean warming   31

hemispheric differences, global warming   20, 134–5
herbarium leaves, C13 in   104
herbivores   306
host susceptibility   155
human activities, and climate change   43
Human Dimensions Programme   290, 304
humid tropics, rainfall   53
hunger
   reduction in   2, 215–17
   scenarios   227, 231
   vulnerability   252
Hurst effect   29, 30
hydrocarbon emissions   172
Hydrologic Atmospheric Pilot Experiment   113–14
hydrological cycle   4, 324
   and global warming   15, 16
hydrological feedback processes   29
hydrological models   20–23, 325
hydrology
   Africa   29
   timescales in   22
hydrolysis, clay minerals   55

IBSNAT, see International Benchmark Sites Network for Agrotechnology Transfer
ice ages
   global temperatures   16
   triggering mechanisms   17–18
ice sheets, stability of   4
IGBP, see International Geosphere-Biosphere Programme
IGCP-BAHC programme   108, 114
illite   57
illuviation   56
IMAGE model   277
IMAGE2 model   268, 277–80, 284, 289
impact assessment   239–41, 247, 275
impact estimates, United States   244
impact models   277
impact studies   23, 44, 297
import quotas   257
industrial production   279
infiltration
   in GCMs   22
   in soils   51
infrared radiation, absorption   320
insect attack, and ozone   155–6
instantaneous transpiration efficiency   102, 110

inter-tropical front   29
interaction of factors   146, 152
interactive change   303
intercropping   305, 306
interflow, in GCMs   22
interglacial period   93
Intergovernmental Panel on Climate
    Change   43, 123, 200, 247, 284, 319
International Benchmark Sites Network
    for Agrotechnology Transfer   202,
    204, 302
International Geosphere-Biosphere
    Programme   290, 330
International Global Atmospheric
    Chemistry project   298
International Union of Forestry
    Research Organisations   308
inundation
    and sea-level rise   50
    seasonal   56
investment requirements   223
IPCC, see Intergovernmental Panel on
    Climate Change
irrigation
    Asia   2
    and canopy temperature   135
    and climate change   93
    global   15
    Lake Chad   33
    in models   89–90
    Nile basin   40
    River Volga   40–1
    trends   3
    USA   93
    and water supply   256, 327
irrigation efficiency   40, 256
isostasy   60

joint effects, see combined effects

Kariba reservoir   24
Keiferia lycopersicella   156
Kootenay catchment   24

lagoon margins   60
Lake Chad   33–6
Lake Nasser   36, 37, 39
Lake Victoria   36
land classes   245
land cover change   277, 279, 280–2
land preservation   254
land resources, availability   211
land subsidence   60

land suitability, for forestry   287
Land Use Emissions model   283
Land Use and Land Cover Change
    project   298, 304
land-use change   269, 284, 295
land-use planning   257
Latin America, deforestation   2
latitudinal differences
    crop yields   231
    global warming   123, 134, 209, 320
leaching
    and climate change   59
    and rainfall   53, 54
leaf ageing   106
leaf appearance rates   78
leaf area   108
leaf area index   67, 85, 108–9, 300
leaf production, vegetative phase   108
leaf surface, temperature   106, 323
leaf thickness   108
leaf weight   108
leaves
    boundary layer   106
    nitrogen content   104
levees   59, 60
light, growth-limiting factor   87
lime, leaching   59
lipids
    oxidation   127
    phase changes   124
litter chemistry   322, 323
litter decay   268
litter production   12, 51, 322
livestock production
    developing countries   2
    in GCMs   281
    range of types   305
Lolium multiflorum   157
loss, adaptation to   255
loss function   250
Lycopersicon esculentum, see tomato

macrofauna, soil   51
maize
    CO_2 responses   88, 90, 207
    GCTE   299
    and temperature   126
malnutrition, projection   2
mangroves   60, 61
manure crops   256
marginal production   237
market integration   258
market prices   240

INDEX

mass movement 53
Medicago sativa 131, 157
meltwater 54
meristems 131
metadata 302
methane
  animal waste from 284
  as greenhouse gas 5, 202, 275
Mexican bean beetle 156
microbial activity
  and CO_2 51
  and soil temperatures 50, 322
micrometeorology 102
Milankovitch cycles 17
millet 135
Miscanthes 288
Mississippi river 60
mitigation options 286
mixed farming systems 305
mixed plantings, differential stress 157
model assessment 301–2
monocrops 298, 305
monsoon, Indian 324
monsoon climates, rainfall response 53
monsoonal circulation, and African climate 30
Montreal Treaty 284
mudflows 57
multi-species cropping 298, 304–5
mycorrhiza, and CO_2 12, 51, 323

NAPAP, *see* National Acid Precipitation Assessment Program
National Acid Precipitation Assessment Program 144
National Crop Loss Assessment Network Program 144, 186
national crop yields 209
national scenarios 68, 200
natural land, reduction in 6
navigation, Volga system 41
NCLAN, *see* National Crop Loss Assessment Network Program
Net Ecosystem Productivity 282–3, 287
net evaporation, Caspian Sea 40
net photosynthesis 75–7, 173–5, 322
Net Primary Productivity 282–3
new crops 255
Nile basin 24, 40
  irrigation efficiency 40
Nile River
  delta 60

floods 29
  and Sahel drought 36–7
Nile Waters Agreement 37
El Niño, *see* Southern Oscillation
nitrogen, fertilizer 256
nitrogen balance 327
nitrogen content
  leaves 104
  soybean 72
nitrogen fixation, and CO_2 51, 323
nitrogen oxides, as greenhouse gases 5, 202, 275
no-regrets policies 329
nucleic acids, and UV-B 161
nutrient cycles 306
nutrient release 51
nutrient supply 12
nutrient uptake 322, 327
nutrients, conservation of 58
nutrition, improvements in 1

oats, *see Avena*
ocean circulation 268
ocean warming, heat transfer 31
ocean-atmosphere interactions 267
oceans, freshwater input 18
oil crops, in developing countries 2
Open-Top Chambers, calibration 300
organic matter, decomposition 53, 58
Oryza sativa, *see* rice
ovule development 179
oxidation, lipids 127
ozone
  boundary layer 159
  combined effects 158–60
  and cotton 146
  and crop pests 154–6
  crop responses 141–6, 147–51, 185–7, 324
  cultivar differences 144
  cumulative effects 186
  and ethylene 156
  and fungal pathogens 155
  and insect attack 155–6
  and pathogens 155
  phytotoxicity 9, 144
  sensitivity 152, 186
  species differences 144, 186
  and whitefly 156
ozone and CO_2, *see* CO_2 and ozone
ozone (stratospheric), depletion 8, 158, 172, 324
ozone and temperature 187

ozone (tropospheric)
  as greenhouse gas 5
  stagnant weather systems 172
  UV-B absorption 189
ozone and UV-B 189–92
ozone and water 188–9

palaeoclimate, models 16
palisade cells 67
PAR, *see* Photosynthetically Active Radiation
partitioning, soybean 71
pasture and grazing systems 298
pastures
  improvement 303–4
  transpiration response 114
patch-scale modelling 304
pathogens
  bacterial 155
  fungal 155
  and global change 314
peat 60, 61
Penman-Monteith formula 24, 89, 114
Pennisetum glaucum, *see* millet
percolation, in GCMs 22
permafrost boundary, retreat of 49, 54, 61
peroxidation, lipids 127
pesticide use 254
pests, winter kill 10
pests and diseases, and climate change 6, 298, 311–13
phenology
  wheat 131
  see also crop phenology
phosphate fixation 54
phosphate uptake 51
photoassimilates
  allocation 66, 67, 71, 94
  partitioning 161
photochemical smog 146, 158
photorespiration 322
photosynthates, accumulation 186, 324
photosynthesis
  and CO_2 321
  and light 66, 105
  and ozone 191
  photon flux density 75
  rice 74–6
  and temperature 66, 129–30
photosynthesis ratios 205
photosynthetic acclimation, rice 76–8
photosynthetic pathways, major crops 8

photosynthetic rates, soybean 69–70
Photosynthetically Active Radiation 143
Phragmites 60, 61
phyllochron interval 78
physical processes, modelling 29
Pisum 157
planetary boundary layer 114, 180
plankton, and UV-B 9
plant communities 86, 306
Plant Functional Types 281, 283
plant growth
  and CO_2 65–9
  rates 268
plant interactions, competitive 157
plant nutrients, management 61–2
plant pathogens 154–6
plant physiology, and greenhouse gases 7–10, 321
plant response 239
plant water status 109
Plantago major 158
plantation forestry 308, 309
planting dates, as adaptive strategy 92
plasmalemma 127
Poa pratensis 158
Poa-Geum competition, UV-B effects 157
poleward shift 134, 238
policy-oriented goals 277–8
pollutants
  crop damage 324
  and stomatal openings 8
polysaccharides 51
population dynamics 68, 213
population growth
  and agricultural production 286, 320
  and arable land 2
  and food production 199, 212
  and poverty 2
  rates 2
  trends 3, 238, 320
potatoes, GCTE 299
potential evapotranspiration 23–4
potential grain number 133
potential production 239, 241–2
potential vegetation 274, 282
potential yield 281
poverty
  and population growth 2
  rural 1
  in warmer areas 254
power generation, Volga system 41

INDEX 341

precipitation
  annual changes   25, 28
  in GCMs   324–5
  global   25–7
  seasonal   25, 115
  Volga catchment   43
price changes   223–4
process models   277
production potential   252
productivity
  biomass   322
  and CO_2 51, 61
  and temperature   131
  vegetation   303
profitability, potential   239
protectionism   211
protein content   182
proteins
  heat shock   127–9
  soluble   70
pseudomycelia   57
Puccinia helianthi   155
pyrite, in soils   50

quick clays   57

radiation balance, atmosphere   4
radiation path length   190
radiation sinks   31
radiative forcing   25, 268, 277
radishes   182
rain, as climate variable   4
rainfall
  in feedback mechanisms   30
  in GCMs   324
  global   25
  intensity   22
  interception   22
  Kenya and Uganda   36
  prediction of   16, 44
  recycling   18
  tropical   50
rainfall changes, climatic zones   53–4, 320
rainfall conditions, Sahel   10
rangelands   303
reforestation, *see* forestation
regional characteristics, IMAGE2   279
regional economic vulnerability   251
regional scales   239
relocation, and economics   254
renewable energy   288
repair, ozone injury   185

replacement series   157
reproductive phase
  rice   78
  wheat   131
reservoirs   24
resource changes   248
resource degradation   258
respiration
  and CO_2 130–1
  maintenance   131
  soybean   70–1
revenue weights   245
reverse weathering, clay minerals   55
Rhine river plain   57
Rhizophora   60, 61
rice   69
  dark respiration   84
  development stages   78–80
  evapotranspiration   84–5
  GCTE   299
  growth and yield   80–4
  Indica   243
  Japonica   243
  photosynthesis   74–6
  photosynthetic acclimation   76–8
  and temperature   126
  and UV-B   146
  water-use efficiency   84–5
  yield changes   241, 285
Rice Network   302
risk analysis, water resource planning   19
risk reduction, food production   329
river basins   19
  flow regimes   45
  human modifications   24
river flow
  into Lake Chad   35, 325
  modelling   20, 325
root density   109, 322
root depths   109, 323
root exudation   51
root mass   51
root structure   322
root and tuber crops
  Africa   2
  global production   2
root/shoot ratios   315
rooting patterns   109
rotational systems   298, 304
roughage demand   281
rubisco   66, 70, 76, 323
  and temperature   124

runoff, global 33–43
runoff ratios, forest-savanna 22
runoff response 44, 325
rust fungi
  smog injury 155
  and UV-B 154
ryegrass
  Italian 157
  water-use efficiency 179

Saccharum spontaneum, see sugar cane
Sahel
  drought 29, 33, 36–7
  and global warming 30
  rainfall conditions 10, 25
  and sea surface temperature 21, 31–2
salt tolerance 54
sandy soils, and acid rain 55
savannas 289
scab 154
sea ice extent 31
sea surface temperature 4, 20
  GCMs 44
  hemispheric differences 31
  and Sahel 21, 31–2
sea-level rise
  and coastal agriculture 6
  effects on soils 59–60
  in GCMs 277
  and inundation 50
seasonal variation, precipitation 25, 115
secondary forest 275
sediment supply 59, 60
seed nitrogen content 67
selection 68, 115, 255
self-sufficiency, agricultural 215
semi-arid regions 251
senescence, and ozone 156, 185–6, 191
sensitivity analysis 300–1
single component analysis 172
sink strength 324
site selection 309
site-specific effects 241, 327
SLURP model 24
smectite 57
smog, photochemical 146
smog injury, rust fungi 155
social security 258
socio-economic trends 279, 304, 328
softwood plantations 309
soil biology 316
soil communities 306
soil degradation 57, 316, 322

soil erosion 258
soil evaporation 108
soil fertility, improvements in 49, 61
soil formation, factors 49
soil loss, erosional 12
soil moisture, in GCMs 22, 201
soil organic matter 315–16
  decomposition 12
  increases in 50, 322
  loss of 232
soil pH 58–9
soil processes 55–6
soil properties
  and CO_2 52
  improvements in 49, 61
soil reddening 54
soil reduction 58, 61
soil respiration 282
soil salinity 54
soil stabilizers 51
soil temperatures 50, 232, 322, 326
soil types, changes in 56
soil water deficits 128
soil water reserves 110
Soil-Vegetation-Atmosphere-Transfer
  models 114
soils
  and climate change 49, 298, 315
  decalcification 57
  forcing variables 50
  gas exchange 53
  human effects on 61
  internal drainage 54
  macrofauna 51
  mechanical disturbance 57
  resilience 57–8, 61
solar radiation, and climate change 25
sorghum
  and CO_2 174
  GCTE 299
  and temperature 126, 135
Sorghum bicolor, see sorghum
South Chad Irrigation Project 33
Southern Oscillation 20, 25, 30, 256
sowing dates, changing 255
soybean 67, 69–74
  carbohydrates 72
  CO_2 and ozone responses 180
  CO_2 and temperature responses
    73–4, 88
  crop yield 72, 94, 207, 241
  evapotranspiration 74
  growth rates 72

INDEX

heat shock proteins 128
nitrogen content 72
ozone and UV 191
partitioning 71
photosynthetic rates 69–70
respiration 70–1
UV-B sensitivity 143
water stress 188
SOYGRO model 204
specialization, increase in 215
species choice 255
species differences
   ozone levels 144
   stomatal response 104
spider mites 155
spikelets 131, 243
stagnant weather systems, ozone 172
starch 67, 72
   synthesis 124
static climate change impact 222, 224
static impact 220–1
Stellaria gramineae 158
steppes 289
sterility, and temperature 80
stomata
   gas exchange 104
   and pollutants 8
stomatal closure, CO_2 response 104
stomatal conductance
   acclimation 105
   and CO_2 67, 106, 173, 179, 321
   combined gases 189
   and light 105
stomatal density 103–4
stomatal functioning 104–5
stomatal response
   CO_2 101
   species differences 104
subsistence farming 303
subtropical areas, rainfall changes 53
sugar beet 154
sugar cane 126, 288
sugars 67, 72
sulphate aerosols 31, 237, 280
sulphate soils 58, 60
sulphur dioxide 324
sulphur emission 237
supply gap 224
support prices 257
surface roughness, and transpiration 114
surface runoff 18
   in GCMs 22, 44

sustainability 328
SVAT models, *see* Soil-Vegetation-Atmosphere-Transfer
swidden systems 275

tannins 323
technical progress 211, 218
technological adaptation 257
technological innovation 254
temperate climates, rainfall changes 54
temperate forests 309
temperature
   and biochemistry 124–6
   buffering 288
   and C_3 pathway 124, 126
   canopy level 109–10
   as climate variable 4
   and CO_2, *see* CO_2 and temperature
   cold-season 10
   and crop yield 9
   effects on enzymes 124–6
   extreme effects 134–6
   and grain development 133–4
   and growing season 135
   leaf surface 106
   long-term effects 136–7
   night 9
   and photosynthesis 66, 129–30
   and productivity 131
   and rice 126, 129
   sea surface 4, 20
   and sterility 80
   and water-use efficiency 86
   and wheat 126, 129
temperature change
   climatic zones 53–4
   and crop production 6–7
temperature and ozone 187
temperature response
   C_3 and C_4 plants 126
   CAM plants 9–10
temperature rise, Antarctic 20
temperature stress 124
termites 58
Terrestrial Environment System 279, 280
thermal kinetic window 124, 126
thermal stability, cell membranes 127
tillage 256
tillering 67, 131
timber, *see* forest products
timescales
   population trends 238

timescales (*cont.*)
  yield impacts 231
timing, climate change 247, 321, 328
tomato
  dark respiration 130
  heat shock proteins 129
tomato pinworm 156
training 258
transient vegetation 269
transpiration
  and CO_2 101, 180, 321-2, 323
  and dry matter 181
  regional 180
  stomatal control 84
  and surface roughness 114
transpiration response, albedo 114
transportation 258
tree crops, perennial 177
trees, dehardening 187
Trialeurodes vaporariorum 156
Trifolium incarnatum 157
T. repens 157
Triticum aestivum, *see* wheat
tropical forests 309
tropics
  food security 326
  rainfall 50
tundra 326

UKMO model 93, 203, 206, 209-10, 220, 223
UKTR 20
ultraviolet radiation, and agriculture 8-9
uncertainties 327, 328
under-nutrition, reduction of 2
undernourishment 217
United Nations Environment Programme 319
United States, impact estimates 244
Uromyces phaseoli 155
UV-B
  adverse responses 144-5
  and crop pests 154-6
  and crop yield 9
  effects on crops 141-6
  experimental variables 142
  and flavonoids 143
  flux density 158
  growth chambers 143
  joint effects 158-60
  monitoring 160, 172
  and nucleic acids 161

and plankton 9
and rice 146
and rust 154
sensitivity 152
UV-B and CO_2, *see* CO_2 and UV-B
UV-B and ozone, *see* ozone and UV-B

vapour pressure deficit 76, 85, 106, 110
  and ozone 187
variety, choice of 255, 257
vegetation cover, basins 44
vegetation zones 50, 268
vegetation-atmosphere fluxes 113-14
vegetative growth, CO_2 fertilization effect 67
vegetative phase
  leaf production 108
  rice 78
  wheat 131-3
vegetative response, CO_2 increase 88
vernalization 207
Veronic chamaedris 158
Volga, River 40
vulnerability 249-54, 258

warm-season crops 126
water
  competition for 15
  data collection 15, 45
  flux transfer 22
  lateral transfer 18, 325
  pollution 15
water abstraction 60
water availability, decrease in 207
water balance 22, 327
water exchange 110
water holding capacity, atmosphere 18
water potentials 109
water pricing 256
water quality 254
water reserves, depletion of 108
water resource planning, risk analysis 19
water resource systems 2, 15, 45
water shortage
  agricultural impact 43
  seasonal 110
water storage
  and agricultural development 45
  and increased precipitation 11
water stress 106, 109, 188
water supply, and irrigation 256
water transfer 107

INDEX

water use
  crop plants   101
  definition   102
  experimental conditions   102
water vapour deficit   104
water-use efficiency
  alfalfa   189
  and CO_2 8, 50, 68, 85–6, 110, 115, 174, 179, 268, 321
  daytime   110, 112
  definition   102
  and ozone   189
  rice   84–5
  ryegrass   179
  seasonal   110–11, 113, 179
  and temperature   86
watershed management   257
WCRP   290
weather
  extreme   258, 325
  predictability   6, 327
weather data
  baseline   88
  daily   239
weathering, and CO_2 53
weeds   156–8, 311, 314
wetland, management regimes   58
wheat
  and CO_2 321
  crop failure   241
  development   131–2
  GCTE   299
  global production   227
  Great Plains   90
  Lake Chad   33
  phenology   131
  and temperature   126
Wheat Network   308
wheat-oat competition   157
White Nile   30, 36, 38
White Sands   22
whitefly, and ozone   156
whole-plant transpiration rate   102
wind, as climate variable   4
winter chilling   179
winter hardiness, and ozone   187
wood, see forest products
world crop production   153, 199
world food system   210–17
  scenarios   218–27
world market prices, cereals   222–3
World Soil Database   315
world trade   68, 93

yield, see crop yield, economic yield
yield impact scenarios   203, 219–20, 231
yield response functions   218–19
yield stability   305

Zambezi River   24
Zea mays, *see* maize